An Overview on Toxigenic Fungi and Mycotoxins in Europe

An Overview on Toxigenic Fungi and Mycotoxins in Europe

Edited by

Antonio Logrieco
CNR,
Bari, Italy

and

Angelo Visconti
CNR,
Bari, Italy

KLUWER ACADEMIC PUBLISHERS
DORDRECHT / BOSTON / LONDON

A C.I.P. Catalogue record for this book is available from the Library of Congress.

ISBN 1-4020-2645-5 (HB)
ISBN 1-4020-2646-3 (e-book)

Published by Kluwer Academic Publishers,
P.O. Box 17, 3300 AA Dordrecht, The Netherlands.

Sold and distributed in North, Central and South America
by Kluwer Academic Publishers,
101 Philip Drive, Norwell, MA 02061, U.S.A.

In all other countries, sold and distributed
by Kluwer Academic Publishers,
P.O. Box 322, 3300 AH Dordrecht, The Netherlands.

Printed on acid-free paper

CONTENTS

PREFACE

The growing interest in health risks associated with toxigenic fungi and related mycotoxins has led to the development of a number of research projects during the past decade in several European countries and a great deal of information has been produced on the natural occurrence of mycotoxins in plants, foods and feeds. The differences in environmental conditions (temperature, light, rainfall etc.) and in exposed cultivated plants (type of cereals, cultivars etc.) in the distinct European countries have significantly influenced the distribution of specific toxigenic fungi and related mycotoxicological problems. In addition, biogeographically structured lineages within some important toxigenic fungi have been recently discovered, showing a long evolutionary history of reproductive isolation. Transglobal transposition of plant products seems to have significantly contributed to the spreading of toxigenic species and lineages worldwide. Therefore, the migration of agriculturally important toxigenic fungi generated by trade exchanges may represent a major source of inoculum for new plant diseases in Europe and for a wider genetic diversity of local populations.

It was our goal to provide the reader with an update of researches and surveys on the natural occurrence of toxigenic fungi and mycotoxins performed at a European level during the last 5-10 years.

The editors are extremely grateful to the EU-COST-835 "Agriculturally important toxigenic fungi" for the financial support that allowed a number of European scientists to meet several times over the past five years to collaborate and exchange information about research advances on toxigenic fungi and mycotoxins. This volume summarises the contribution from experts from 18 European countries presented at the closing meeting of COST-835 held in Martina Franca, Italy, on 24-25 October 2003. It provides a detailed overview on the occurrence of toxigenic fungi and mycotoxins in plants, foods and feeds that can serve as a basis for risk management based regulatory decisions.

The editors are indebted to all authors for their support and kind contribution and wish to thank Mrs Annamaria Pascali for her valuable assistance in formatting and assembling all manuscripts.

Antonio Logrieco
Angelo Visconti

R. ÖHLINGER, A. ADLER, O. KRÄUTLER & H. LEW

OCCURRENCE OF TOXIGENIC FUNGI AND RELATED MYCOTOXINS IN CEREALS, FEEDS AND FOODS IN AUSTRIA

1. INTRODUCTION

Work on the occurrence of mycotoxins and toxigenic fungi in Austria started in the mid-seventies with research on the role of the estrogenic mycotoxin zearalenone (ZON) as a feed contaminant causing non-specific sterilities in cattle in the province of Upper Austria (Lengauer, 1977). However, the main stimulus for an in-depth study of the mycotoxin complex in Austrian agriculture, over the following years, emanated from massive feeding problems in porkers, caused by *Fusarium* toxins and affecting large stretches of Southeast Austria's maize growing areas in the winter of 1977/78.

Due to the unusual weather conditions (early frost in September) there was heavy infestation of the maize ears with *Fusarium graminearum* and exceptionally high deoxynivalenol concentrations (up to 20 ppm) in harvested maize. The consequence was that in numerous pig farms the animals either refused intake of the ensiled maize kernel feed altogether or would throw up the fodder they had very reluctantly eaten (Lew *et al.*, 1979).

In view of the great importance of maize for feeding and considering the sheer extension of the areas involved (about 20 % of Austria's overall arable land), mycotoxin research in Austria is heavily concentrated on this plant, with the influence of *Fusarium* species and their toxins on selection of varieties, fertilisation, time of harvest and pest infection being the main points of interest (Lew *et al.*, 1991, Lew 1993, Hinterholzer 1993, Lew *et al.*, 1997, Lew *et al.*, 2001a).

2. IMPORTANT *FUSARIUM* SPECIES AND THEIR TOXINS IN CEREALS FROM AUSTRIA

At the end of the 1980s, the spectrum of *Fusarium* species and toxins in Austrian field crops was largely known (see updated Tables 1 and 2 from Lew, 2000). Apart from maize, in Austria mostly oats and wheat, especially durum wheat, are at risk from *Fusarium*. Although in oats and wheat sometimes other species of *Fusarium* dominate, the toxin spectrum that emerges is nearly identical. We can safely assume that deoxynivalenol and its derivatives, as well as zearalenone, moniliformin, beauvericin and nivalenol are the most common toxins contaminating local cereals. Those species of *Fusarium* that would produce more highly toxic metabolites such as T-2 toxin and HT-2-toxin, as well as the carcinogenic fumonisins play a minor role in Austria (Lew *et al.*, 2001b).

A. Logrieco and A. Visconti (eds.),
An Overview on Toxigenic Fungi and Mycotoxins in Europe, 1-10.

Table 1. Important Fusarium species *and relevant toxins in cereals from Austria*

Dominating *Fusarium* species on			Toxins produced
Maize	Wheat	Oats	
F. subglutinans			Moniliformin (MON), Beauvericin
F. graminearum	*F. graminearum*	*F. graminearum*	Zearalenone (ZON), Deoxynivalenol (DON) 15-Acetyldeoxynival-enol (15-ADON) 3-Acetyldeoxynival-enol (3-ADON)
	F. avenaceum	*F. avenaceum*	Moniliformin (MON)
	F. poae	*F. poae*	Nivalenol (NIV)

The researchers involved were able to show at a very early stage that under the prevailing climatic conditions production technology used by the Austrian farming sector, aflatoxins were not found in cereals, while potentially toxigenic strains of *Aspergillus flavus/parasiticus* did occur on nationally grown stored cereals (Lengauer 1977, Oeschlmüller 1987).

Generally speaking, infections of standing cereal crops by *Aspergillus* but also by *Penicillium* are very rare under the climatic conditions observed in Austria. Since the cereals are usually harvested dry and storage conditions meet a relatively high standard, toxins of storage fungi are unlikely to develop. Unlike neighbouring countries to the north and south, ochratoxin A contamination is of minor importance in Austria (Lew *et al.*, 2001b).

3. SHIFTING OF SPECIES AND TOXIN CONTAMINATION

Based on representative analyses of Austrian cereals, a distinct shift in the spectrum of *Fusarium* toxins and *Fusarium* species has been observed since the middle of the eighties (Lew *et al.*, 2001b, Adler *et al.*, 2002).

Although in the case of maize *F. subglutinans* – apart from *F. graminearum* – proved to be the most frequent and constant contaminant over the entire range of test series, there has been a shift in the spectrum of species, which is not to be explained simply by seasonal variations or by the varying degree of occurrence of the European corn borer, which in Austria is considered to be the main vector for infections involving *Fusarium* species of the *Liseola* section (Lew *et al.*, 1991). Compared to the results from earlier vegetation periods, the number of infections with *F. proliferatum increased significantly in the nineties*. In all likelihood, this shift in the spectrum of species is due to the changed climatic conditions now prevailing in Austria – milder and more humid winters vs. drier and warmer summers – which favour the progress of *F. proliferatum* (Table 3). As a consequence to the appearance of *F. proliferatum* the production of fumonisins increases.

Table 2. Additionally occurring Fusarium *species and their toxins on cereal kernels*

Maize	Wheat	Oats	Toxins produced
F. avenaceum			Moniliformin
F. poae			Nivalenol Fusarenon X
F. proliferatum			Moniliformin Beauvericin Fumonisins
F. tricinctum	*F. tricinctum*	*F. tricinctum* *F. pulverosum* *(F. langsethae)*	Moniliformin T-2 Toxin HT-2 Toxin
F. equiseti	*F. equiseti*	*F. equiseti*	Zearalenone, T-2- Toxin, HT-2 Toxin, Fusarochromanon
F. culmorum		*F.culmorum*	Zearalenone Deoxynivalenol, 3-Acetyl- deoxynivalenol
F. cerealis *(F. crookwellense)*	*F.cerealis*	*F.cerealis*	Zearalenone, Nivalenol
F. verticillioides *(F. moniliforme)*	*F.verticillioides*	*F.verticillioides*	Fumonisins
F. oxysporum	*F.oxysporum*	*F.oxysporum*	Moniliformin
F. sporotrichioides	*F.sporotrichioides*	*F.sporotrichioides*	T-2 Toxin, HT-2 Toxin

Table 3. Shifting of species – F. proliferatum

Year of harvest	1986-1989	1996-1998	2000-2003
Ears examined	13203	13805	16386
Ears with visible *Fusarium* infection	1409	1209	978
Fusarium strains isolated	735	934	1020
Section *Liseola*, number of isolates			
F. subglutinans	363	291	207
F. proliferatum	0	40	155
F. verticillioides	8	7	5

The principal toxin-producing fungus on cereals in Austria is *F. graminearum*. On maize, its respective populations are exclusively those which produce 15-acetyl-DON as a precursor to DON (deoxynivalenol). Whilst in the 1980s, *F. graminearum* isolates from wheat yielded both 15-acetyl-DON (15-ADON) and 3-acetyl-DON (3-ADON) types, only 15-acetyl-DON populations could be detected in the last few years (Table 4). One possible explanation for this phenomenon is the continual intensification of maize-wheat crop rotations. In the light of the above observations, the frequently used argument whereby European *F. graminearum* isolates produce mainly 3-acetyl-DON and American strains prevalently 15-acetyl-DON will have to be reviewed.

Table 4. Shifting of species – F. graminearum *(different chemotypes)*

Host	Years	Isolates	3-ADON-types (DON + 3-ADON)	15-ADON-types (DON + 15-ADON)
Wheat	1981 – 1990	13	6	7
	1991 – 2002	17	0	17
Oats	1981 – 1990	5	5	0
	1991 – 2001	19	19	0
Maize	1981 – 1990	5	0	5
	1991 - 2001	26	0	26

It is true, on the other hand, that oats continue to yield exclusively 3-acetyl-DON producers. But although oats do not immediately follow on maize in standard crop rotation, other causes (not yet known) should be considered to explain the different contamination of oats as compared to wheat.

4. AUSTRIAN MYCOTOXIN REGULATIONS

Due to serious problems with mycotoxin contamination of cereals, different limits are recommended for feed and food (Tables 5 and 6). The limits in feed were established from a working group of the Austrian Association of Agricultural Research Institutions (published in Razzazi-Fazeli *et al.*, 2003). Obeying these limits for feed and providing good conditions in feeding and livestock husbandry no harm is expected for animals.

Table 5. Austrian recommended limits in feeds (mg/kg)

Animal	DON	ZON
Prepubertarian breeding sows	0,5	0,05
Fattening pigs, breeding sows	0,5	0,15
Breeding poultry, layers, fattening turkey	1	-
Broilers	2	-
Fattening cattle	1	-

Austrian recommended limits in cereals for human consumption were published in *Mitteilungen der österreichischen Sanitätsverwaltung (1993).* The recommendations are in a similar magnitude as the discussed maximum levels for EU regulation (2003). The limits for ochratoxin A (OTA) should not be in force anymore since the Commission regulation No 472/2002 regulates OTA concentrations in cereals.

Table 6. Austrian recommended limits in cereals for human consumption (mg/kg)

Cereals	DON	ZON	OTA
Wheat	0,5	0,06	0,005
Durum wheat	0,75	0,06	0,005
Rye	0,5	0,06	0,005

5. MYCOTOXIN DATA

In the following tables the results of mycotoxin analysis are mainly describing the situations of the last two years. According to no normal distribution of mycotoxin concentrations, median and range are used for data description. Tables mostly inform only about positive mycotoxin samples. Mycotoxins such as T-2, HT-2, monoacetoxyscirpenol, diacetoxyscirpenol or fusarenon X, which were also determined in most samples, were mostly below their detection limit.

5.1. Wheat

Wheat samples mainly originated from Upper Austria (Table 7), where wheat is mostly used for feeding stuff production. DON data showed significant increased concentrations in 2003 (median of about 1,3 mg/kg) in comparison to 2002 (median 0,14 mg/kg). A similar increase is indicated in the southern parts of Austria (Styria). Although in 2003 the climate was very dry and warm, *Fusarium* infestation was partly abundant. The main reason of these strong infestations was a short period of warm and humid conditions in may 2003 during flowering-time. The maximum DON-concentration of 20 mg/kg was a unique value rising from typical promoting agricultural conditions like no tillage, lodged grain and unfavourable crop rotation (maize before wheat). Some observations from Lower Austria (Pannonic area) demonstrate no heavy contamination with DON (median about 0,5 mg/kg).

Zearalenone or nivalenol concentrations did not show a similar increase. In contrast to 2002, values above 0,02 mg/kg were very seldom detected in 2003. ZON production begins after DON production and it seems that the specific climate conditions of 2003 prevented ZON production until harvest.

As mentioned above the shifting of species is observed in the 15-ADON formation without any 3-ADON production on wheat.

According to the Austrian recommended limits for feed most of DON concentrations of 2003 might cause negative effects on health, especially for pigs if such wheat is directly used as feed.

Table 7. Fusarium *toxins in wheat mainly for feed from Upper Austria in mg/kg*

	Year	DON	ZON	15-ADON	NIV
Median	2002	0,14	<0,02		
	2003	1,25	<0,02	0,1	<0,05
Mean	2002	0,58	0,03		
	2003	2,6	<0,02	0,3	<0,1
Range	2002	<0,05-2,3	<0,02-0,24		
	2003	<0,05-20	<0,02-0,04	<0,05-0,25	<0,05-0,12
n	2002	144	94		
	2003	124	48	52	52

5.2. Oats

Oats (also barley or rye) would not cause any problems as feeding stuff according to our investigations on mycotoxins in 2002 and 2003 (Table 8).

Medians (oats) from 0,06 to 0,12 mg/kg DON showed a slight increase in 2003. In contrast to wheat, more frequent contamination with nivalenol (about 50 % of samples analysed) and HT-2 toxin (about 30 % samples analysed) were characteristic for oats due to *F. poae* and respectively *F. sporotrichioides* and *F. pulverosum* infestations (Petterson, 1991 and Lew *et al.*, 1997b, Adler *et al.*, 2003).

Table 8. Fusaria *toxins in oats (mg/kg)*

	Year	DON	NIV	HT-2
Median	2002	0,06	0,10	
	2003	0,12	0,18	0,07
Mean	2002	0,07	0,25	
	2003	0,15	0,33	0,09
Range	2002	<0,05-0,21	<0,02-0,8	
	2003	<0,05-0,64	<0,02-1,3	<0,05-0,3
n	2002	60	60	
	2003	32	32	32

5.3. Maize (kernels)

Depending on weather variations during the period of vegetation mean contaminations with mycotoxins (especially DON) differ from year to year. Humid periods caused an in increased formation of DON, e.g. in 1996 (mean: 0,65 mg/kg) and in 2002 (mean: 0,95 mg/kg). On the other hand, warm and dry summers diminished DON production (Lew *et al.*, 2001b). So the mean concentration of DON in 1997 was 0,14 mg/kg and 0,2 mg/kg in 2003 (Table 9). As mentioned above we observed a shifting of species and toxin contamination also on maize. In the nineties the number of infections with *F. proliferatum increased significantly*. The appearance of fumonisins B1 + B2 (FUM) confirms these observations. Especially

F. subglutinans or *F. proliferatum* are responsible for moniliformin (MON) and beauvericin contaminations (Lew *et al.*, 2001b). In 2002 MON concentrations ranged from <0,1 to 0,6 mg/kg (1998: <0,1 to 0,8 mg/kg). Analyses on beauvericin are not very frequent.

Table 9. Fusarium *toxins in maize kernels (mg/kg)*

	Year	DON	ZON	FUM	MON
Median	2002	0,68	0,04	<0,1	0,2
	2003	0,17	<0,02	0,13	
Mean	2002	0,95	0,10	0,3	0,3
	2003	0,19	<0,02	0,22	
Range	2002	<0,05-6,3	<0,02-1,4	<0,1-1,1	<0,1-0,6
	2003	<0,05-0,5	<0,02-0,28	<0,1-1,1	
n	2002	156	156	25	25
	2003	30	30	29	

In contrast to wheat, maize mycotoxin contamination of 2003 would not cause problems for animal health according to our analyses.

5.4. Feeds

Following data (Table 10) originate from feed samples drawn from officials during their controlling activities. The feeding stuffs were manufactured of various cereals mainly originating from Austria. As shown in Table 10 recent medians ranged from 0,16 to 0,23 mg/kg, so that the rate of objections was far below 10 %. Perhaps this rate will increase the next time when high DON concentrations of 2003 would become obvious in feeding stuffs.

Table 10. Fusarium *toxins in feeds (mg/kg)*

	Year	DON	ZON
Median	2002	0,23	<0,02
	2003	0,16	<0,02
Mean	2002	0,38	0,12
	2003	0,37	0,04
Range	2002	<0,05-4,5	<0,02-1,1
	2003	<0,05-2	<0,02-0,25
n	2002	176	102
	2003	64	40

Aflatoxin B1 or ochratoxin A were very seldom found in Austrian cereals (median <0,0005 respectively <0,0002 mg/kg).

5.5. Food

Recent DON and ZON analyses of food made of cereals mostly showed medians below their detection limits (table 11). Therefore objections are rare.

Table 11. Fusarium *toxins in cereals for human consumption in 2002-2003 (mg/kg)*

	DON	ZON
Median	<0,05	<0,02
Range	<0,05-0,7	<0,02
n	41	44

On the other hand, aflatoxin contamination in special kinds of imported food sometimes exceeds EU maximum levels. The rate ranges from 12 to 47 % depending on the product (Table 12).

Because of its toxicity Ochratoxin A (OTA) is also taken into consideration with different maximum levels or recommended limits. In Austria OTA is still an "imported" problem. There its appearance is very seldom in cereals because cereals are usually harvested dry and storage conditions meet a relatively high standard. In addition, Austrian wine and juice were scarcely contaminated with OTA (Eder *et al.*, 2002). Meat samples (kidney, liver) derived from pig, horse, cow, sheep, poultry or fish showed no OTA contamination (table 13).

Table 12. Aflatoxins (sum of aflatoxin B1, B2, G1, G2) in food (2002 – 2003). (LQ = limit of quantification; TL = EU threshold limit for direct consumption: 4 µg/kg for nut and shell products; 10 µg/kg for spices)

Product	Range in µg/kg	> TL	n
Peanut, peanut products, hazelnut, hazelnut products	< LQ - 128	14 %	75
Fig, fig products	< LQ - 135	(17 %)	42
Pistachio	< LQ - 110	47 %	5
Spices	< LQ - 18	12 %	26

Table 13. Ochratoxin A in food in 2002-2003 (µg/kg)

	Cereals	Juice	Coffee (import)	Austrian Wine	Meat
Median	<0,3	<0,05	<0,3	<0,01	<0,5
Range	<0,3-0,7	<0,05-0,2	<0,3-1,6	<0,01-0,02	<0,5
n	34	24	36	117	75

Comparing the DON-medians (2002-2003) of unprocessed cereals, feed and food, we can estimate mycotoxin outcome in the food chain (fig.1). Beginning with a median of about 0,4 mg/kg DON contamination decreases to 0,2 mg/kg in feed followed by a median below the limit of quantification in food. Main reasons for this distribution are diluting effects and cleaning and selecting processes, respectively during food and feed production.

Figure 1. Comparison of DON medians (2002-2003)

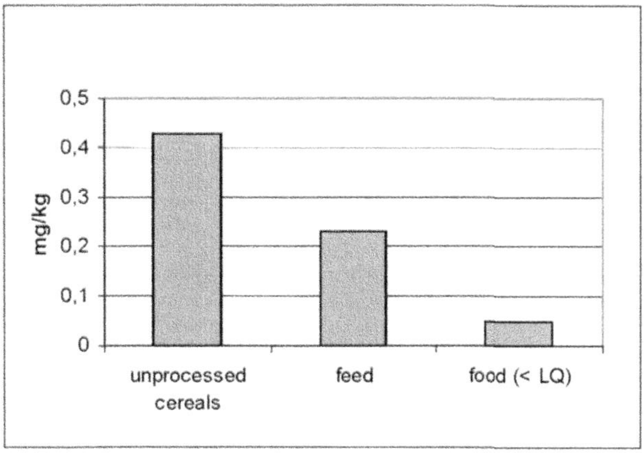

6. ACKNOWLEDGEMENTS

We wish to thank our laboratory colleagues Elisabeth Kiendler, Wolfgang Brodacz, Walter Edinger, Manuela Höfler, Alexandra Kiroje and Marianne Bachl.

7. AFFILIATIONS

Richard Öhlinger
Austrian Agency for Health and Food Safety GmbH, CC Cluster Chemie
Wieningerstaße 8, A-4020 Linz, Austria
Tel: +43 (0)732 381261 269; Fax: +43 (0)732 385482
richard.oehlinger@ages.at

Andreas Adler
Austrian Agency for Health and Food Safety GmbH,
Wieningerstaße 8, A-4020 Linz, Austria
Tel: +43 (0)732 381261 257; Fax: +43 (0)732 385482
andreas.adler@ages.at

Oswald Kräutler
Austrian Agency for Health and Food Safety GmbH,
Kinderspitalgasse 15, A-1090 Wien, Austria
Tel: +43 (0)1 40490; Fax: +43 (0)1 40490-9278

oswald.kraeutler@ages.at

Hans Lew
Freistädterstr. 98
A-4040 Linz, Austria

8. REFERENCES

Adler, A., Lew, H., Edinger, W. (1990): Incidence and toxigenicity of *Fusarium* species in cereals from Austria. Die Bodenkultur 41, 145 - 152.

Adler, A., Lew, H., Brunner, S., Oberforster, J., Hinterholzer, J., Kullnig-Gradinger, C., Mach, R., Kubicek, C. (2002): *Fusaria* in Austrian cereals – change in species and toxins spectrum. J. Appl. Genet. 43 A, 11-16.

Adler, A., Lew, H., Moudry, J., Sterba, Z., Vratilova, K., Edinger, W., Kiendler, E. (2003): Microbiological and mycotoxicological quality parameters of naked and covered oats with regard to the production of bran and flakes. Die Bodenkultur 51, 41 – 48.

Eder, R., Paar, E., Edinger, W., Lew, H. (2002): Untersuchungen über den Gehalt an Ochratoxin A (OTA) in Weinen, insbesondere Prädikatsweinen aus Österreich. Mitt. Klosterneuburg 52, 125-131.

Hinterholzer, J. (1993): Pflanzenbauliche Aspekte zum Toxinproblem bei Mais. Veröffentlichungen der Bundesanstalt für Agrarbiologie Linz 21, 69-80.

Lengauer, E. (1977): Zur Stellung und Aktualität von Mykotoxinen in der heimischen Landwirtschaft. Veröffentlichungen der landwirtschaftlich-chemischen Bundesversuchsanstalt Linz 11, 33-70.

Lew, H., Müllner, E., Hager, R., Gregor, M. (1979): Fütterungsprobleme bei Mastschweinen verursacht durch fusarientoxinhältigen Mais. Die Bodenkultur 30, 309-316.

Lew, H., Adler, A., Edinger, W. (1991): Moniliformin and the European corn borer (Ostrinia nubilalis). Mycotoxin Research 7A, 71 – 76.

Lew, H. (1993): Die aktuelle Mykotoxinsituation in der heimischen Landwirtschaft. Veröffentlichungen der Bundesanstalt für Agrarbiologie Linz 21, 5-26.

Lew, H., Adler, A., Edinger, W. (1997a): Dynamics of the *Fusarium* toxin distribution in maize plants affected by stalk rot. Cereal Research Communications 25, No. 3/1, 467-470.

Lew, H., Adler, A., Brodacz, W., Edinger, W. (1997b): Zum Vorkommen von Nivalenol in Getreide und Mais, Proceedings, 19. Mykotoxin-Workshop, München, 6-9.

Lew, H. (2000): Umweltschonende Strategien zur Bekämpfung des Mykotoxinproblems. Veröffentlichungen des Bundesamtes für Agrarbiologie Linz 22, 103-114.

Lew, H., Adler, A., Edinger, W., Brodacz, W., Kiendler, E., Hinterholzer, J. (2001a): Fusarien und ihre Toxine bei Mais in Österreich. Die Bodenkultur 52(3), 199-207.

Lew, H., Adler, A., Thimm, N., Krska, R., Wiedner, G., Schuh, M. (2001b): Occurrence of toxigenic fungi and related mycotoxins in plants, food and feed in Austria. In: Occurrence of toxigenic fungi and mycotoxins in plants, food and feed in Europe. A. Logrieco, Ed., European commission, Cost Action 835, p 3.

Oeschlmüller, R. (1987): Toxinogene *Aspergillus flavus*- und *Aspergillus parasiticus*- Stämme auf Mais und Getreide. Dissertation an der Technischen Universität Wien.

Petterson, H. (1991): Nivalenol production by *Fusarium poe*. Mycotoxin Research 7A, 26-30.

Razzazi-Fazeli, E., Böhm, J., Adler, A., Zentek, J. (2003): Fusarientoxine und ihre Bedeutung in der Nutztierfütterung: eine Übersicht. Vet. Med. Austria 90, 202-210.

A. CHANDELIER, J.Y. MICHELET, E.K. TANGNI, K. BAERT, E. MOONS & C. VINKX

MYCOTOXINS SURVEY IN BELGIUM AND TOXIGENIC *FUSARIUM* IN BELGIAN WHEAT

1. INTRODUCTION

Mycotoxins are naturally occurring secondary metabolites produced by fungi, which develop on agricultural crops. While fungi are destroyed during processing, mycotoxins mostly remain in the final product (Creppy, 2002). Such mycotoxin-contaminated commodities are implicated in toxic effects in animals and humans (Riley, 1998). The nature and extent of the health problems depend on the amount and the spectrum of mycotoxins contaminating the product. One toxigenic fungus may be capable of producing different mycotoxins, while a specific toxin can be produced by more than one fungal species.

Toxigenic fungi from three genera (*Fusarium*, *Penicillium*, and *Aspergillus*) are widespread in various agricultural products and considered to be economically important worldwide. In temperate regions, *Fusarium* species are the main cause of cereal contamination in the fields, while species from the genus *Penicillium* are probably the most commonly occurring fungi on agricultural products under storage conditions. In warm humid climates, numerous agricultural products are prone to invasion by *Aspergillus* and, to a lesser extent, by *Penicillium* species.

A wide range of mycotoxins produced by these fungi has been identified. However, only a few appear to be of importance due to their high toxicity and/or incidence in agricultural products. Among these, the trichothecenes (notably deoxynivalenol and its derivatives, nivalenol, T2 and HT2 toxins), zearalenone and the fumonosin B1 produced by *Fusarium* species have been commonly detected in cereals (Bottalico, 1998). The main *Penicillium* mycotoxins, including ochratoxin A and patulin as well as the aflatoxins produced by some *Aspergillus* species, have been found on a large number of agricultural products (Smith *et al.*, 1994).

To minimise human and animal exposure to mycotoxins, most European countries carry out regular analyses on food and feed supplies. This paper presents an overview of mycotoxin analyses carried out in the food and feed chain in Belgium, as well as a study of the toxigenic *Fusarium* species encountered in Belgian winter wheat.

A. Logrieco and A. Visconti (eds.),
An Overview on Toxigenic Fungi and Mycotoxins in Europe, 11-32.

2. ORGANISATION OF THE OFFICIAL CONTROL

Official control of mycotoxin is coordinated by the Belgian Federal Agency for the Safety of the Food Chain (AFSCA/FAVV). This agency is the responsible authority for the control of all feed and food products and their components throughout the food chain. It is also responsible for official control of imports at the points of entry into Belgium for products originating from non-EU member states. The CONSUM (Contaminant Survey System) data bank holds all data concerning the official control of contaminants, notably mycotoxins. Each year, the Federal Agency establishes a monitoring programme, which specifies the number of samples, food and feed commodities, schedule, sampling method and the contaminant to analyse. From a practical point of view, samples are collected from all over Belgium throughout the year. Analyses are carried out in laboratories that are approved and accredited according to ISO17025 for the purposes of the official control. Maximum limits for the analyses described in this paper (data collected from 2001 to 2002) were those set by EC legislation (regulation EC 466/2001 as amended for foodstuffs and Council directive 1999/29/EC for feed products, Official Journal of the European Communities). Sampling methods for food commodities can be summarised as follows. In case a European directive was of application, it was applied (directive 98/53/EC for aflatoxins). For other mycotoxins in 2001-2002, the following sample sizes were used, composed by several incremental samples: at least 2 packages of prepacked foodstuffs (raisins, coffee, wine, grape juice, cacao products, apple juice, several cereal products, baby food) resulting in a sample of 0.5 to 3 kg, or for bulk foodstuffs (malt, cereal grains) 1-3 kg made from 10-30 incremental samples. The entire sample was homogenised in the laboratory to ensure representative results.

3. MYCOTOXIN OCCURRENCE

3.1. Aflatoxins

3.1.1. Aflatoxins B1, B2, G1 and G2

There are four naturally occurring aflatoxins (AFB1, AFB2, AFG1 and AFG2) produced by some strains of *Aspergillus* in tropical environments, at both pre-harvest and post-harvest on a wide range of agricultural commodities (Smith *et al.*, 1994). These molecules are carcinogenic to most animal species (Sabbioni and Sepai, 1998) and the International Agency for Research on Cancer (IARC) has placed naturally occurring aflatoxins on their list of human carcinogens (IARC, 1993). The Scientific Committee for Food of the European Commission (SCF) evaluated aflatoxins in 1996 and concluded that there is no safe dose for these genotoxic carcinogens.

Within the framework of the CONSUM programme, monitoring of aflatoxin B1 (AFB1) has been carried out on food samples, such as different kinds of nuts, dried fruits and spices (Table 1). Each sample was extracted with a mixture of methanol and water, and the extract purified by cleaning through an immunoaffinity column (Easy-Extract Aflatoxin, R-Biopharm Rhone Ltd). A high performance liquid chromatography method (HPLC) with fluorescence detection was then used to detect and quantify the mycotoxin. This methodology follows the recommendations of the European Committee for Standardisation (procedure CEN/TC 275/WG5 N253).

For a vast majority of samples, the level of AFB1 was below the limit of quantification. However, eight contaminated samples exceeded the EC maximum limit of 2 μg/kg (Official Journal of the European Communities, 2001): one sample of Brazil nuts (in 2001), two samples of hazelnuts (one in 2001 and the other in 2002) and five samples of pistachios (in 2001). The highest level of contamination was found in pistachios (one sample analysed in 2001 contained 52.5 μg/kg). Analysis of the four aflatoxins (B1, B2, G1 and G2 as a sum) on the same samples led, in this case, to the same conclusions (data not shown).

Table 1. Official control of aflatoxin B1 in food products (CONSUM data bank, 2001 - 2002) N = number of samples analysed; LOQ = limit of quantification

Product	N	LOQ[a,c]	<LOQ	Mean ± SD[a,b]	Median[a,b]	Max[a]
Almonds	19	0.2 - 0.5	13	0.3 ± 0.4	< 0.2	1.5
Brazilnuts	11	0.2 - 0.5	10	1.8 ± 5.4	0.3	18.1
Dried fruits	11	0.5 - 0.7	8	< 0.5	< 0.5	0.9
Groundnuts	9	0.5 - 0.7	9	< 0.5	< 0.5	< 0.5
Hazelnuts	15	0.1 - 0.4	9	0.9 ± 2.1	< 0.1	8.4
Coconuts	3	0.3	3	< 0.3	< 0.3	< 0.3
Peanuts	7	0.2 - 0.5	6	< 0.2	< 0.2	0.5
Pistachios	24	0.2 -1	15	3.9 ± 11.2	0.4	52.5
Spices	5	0.5	1	1.0 ± 0.7	0.9	2
Walnuts	4	0.2 - 0.3	4	< 0.2	< 0.2	< 0.2

[a]: LOQ, mean, median and maximum value expressed in μg/kg
[b]: Arithmetic mean and median values of all samples. Values less than LOQ considered as LOQ/2
[c]: Different LOQ according to the yield of analysis

Foodstuffs entering the EU from non-EU countries and suspected of contamination by aflatoxins were also analysed. They were sampled at random, except for those products that have to comply with specific EC decisions. In 2001, 20 consignments from Argentina (peanuts), China (peanuts), India (walnuts), South Africa (peanuts) and Turkey (dried figs and fig paste) were analysed. None of these samples were contaminated above the maximum limits established for aflatoxins B1, B2, G1 and G2. In 2002, 54 consignments were submitted for analysis, of which 3 samples of hazelnuts from Turkey were non-compliant (with concentrations ranging from 10.2 to 13.8 μg/kg). The corresponding lots were refused for importation.

Feed samples were investigated for the occurrence of AFB1. Analyses were carried out on raw material and compound feeding stuffs for cattle, pigs, poultry and sheep using an official method (Official Journal of European Communities, 1992) (Table 2).

Table 2. Official control of aflatoxin B1 in feed products (CONSUM data bank, 2001-2002). N = number of sample; LOQ = limit of quantification

Product	N	LOQ[a]	<LOQ	Mean ± SD[a,b]	Median[a,b]	Max[a]
Comp. feed cattle	646	1	483	< 1	< 1	46
Comp. feed pig	7	1	7	< 1	< 1	< 1
Comp. feed poultry	10	1	10	< 1	< 1	< 1
Comp. feed sheep	12	1	12	< 1	< 1	< 1
Raw material	252	1	93	8.8 ± 19.9	3	270

[a]: LOQ, mean, median and maximum value expressed in µg/kg
[b]: Arithmetic mean and median values of all samples. Values less than LOQ considered as LOQ/2

AFB1 content was below the limit of quantification in 65% of feed products analysed. Regarding compound feeding stuffs for pig, poultry and sheep, none of the samples exceeded the limit of quantification. In contrast, AFB1 was quantifiable in 25% of compound feeding stuffs for cattle (157 samples contained between 1 and 5 µg/kg, and 6 samples were more contaminated). About 60% of the positive raw material products (92 samples) had AFB1 content between 1 and 10 µg/kg. The highest level of contamination (270 µg/kg) was found in a groundnut-based sample. Based on the Council directive 1999/29/EC (Official Journal of the European Communities, 1999), 13 samples were non-compliant.

3.1.2. *Aflatoxin M1*

In ruminants, aflatoxin B1 is metabolised by various enzymes and can be excreted in the milk as aflatoxin M1 (AFM1) (Sabbioni and Sepai, 1998). AFM1 is less toxic than AFB1 but, because of high level of consumption of milk, notably by children, AFM1 is of great concern. Although fermentation can decrease the level of AFM1, yoghurt can still be contaminated (Govaris *et al.*, 2002).

Within the framework of the CONSUM programme, analyses have been carried out on both raw and consumption milk, as well as on dairy products such as cheese and yoghurt. The analytical procedure to quantify AFM1 included an immunoaffinity column extraction (AflaprepM, R-Biopharm Rhone Ltd) and HPLC coupled with fluorometric detection. For milk, the procedure described by the European Committee on Standardisation (CEN/TC 275/WG5 N203) was used, while for dairy products the procedure followed the recommendation of the International Dairy Federation (procedure IDF 171 (1995), corresponding to procedure ISO/CD 14501). More than 1300 samples were analysed over two years (2001–2002). The results are shown in Table 3.

As far the maximum limit of 50 ng/kg was concerned (Official Journal of the European Communities, 2001), one raw milk sample (0.1%) was non-compliant. Some samples of powdered milk (20%) also exceeded 50 ng kg^{-1} (concentration ranging from 50 to 115 ng/kg). However, based on the water content in those products (about 4% compared to 80-90% in milk, Nubel, 2003), none of these samples were rejected, as the maximum limit applicable has to be modified taking into account the concentration process linked to the preparation of powdered milk.

Table 3. Official control of aflatoxin M1 in milk and dairy products (CONSUM databank, 2001 –2002); N = number of samples; LOQ = limit of quantification

Product	N	LOQ[a,c]	<LOQ	Mean ± SD[ab]	Median[a,b]	Max[a]
Consumption milk	189	1.5 - 3.0	35	3.5 ± 1.7	3	9
Raw milk	787	1.5 - 5.0	490	6.0 ± 6.5	2.5	94
Powdered milk	110	15 - 30	12	38 ± 17	39	115
Cheese	175	6 - 10	33	15 ± 9	14	46
Yoghurt	66	5 - 10	59	< 5	< 5	< 5

[a]: LOQ, mean, median and maximum value expressed in ng/l or ng/kg
[b]: Arithmetic mean and median values of all samples. Values less than LOQ considered as LOQ/2
[c]: Different LOQ according to the yield of analysis

3.2. Ochratoxin A

Ochratoxin A (OTA) can be produced by some *Penicillium* and *Aspergillus* species in a vast majority of agricultural products (Abarca *et al.*, 2001; Pitt, 1987). Contamination can occur in the field (as for grapes, Battilani & Pietri, 2002) or during storage of inadequately dried products. In experimental trials on animals, OTA is nephrotoxic and induces tumours in the kidneys of rats and mice. Regarding humans, OTA has been linked with Balkan endemic nephropathy, a renal disease confined to some rural regions of the Balkan Peninsula (Plestina *et al.*, 1990). However, recent studies indicate that other environmental factors might contribute to the disease (Tatu *et al.*, 1998). OTA has been classified as "possibly carcinogenic to humans" (group B) by the International Agency for Research on Cancer (IARC, 1993). The Scientific Committee for Food (SCF) evaluated ochratoxin A in 1998 and recommended to reduce the intake below 5 ng/kg bw/day. SCF identified carcinogenic, nephrotoxic, teratogenic, immunotoxic and possibly neurotoxic properties.

3.2.1. Official control

Different commodities including raw foodstuffs and processed food were analysed during 2001-2002, using HPLC method with fluorescence detection as described by the European Committee on Standardisation (procedure CEN/TC 275/WG5 N203). Out of the 184 samples investigated, OTA contamination was above the limit of quantification in 73 samples (40%) (Table 4).

Two results (cereal grain samples contaminated at 12 and 25.4 µg/kg respectively) exceeded the current maximum level of 5 µg/kg. However, these analyses were

carried out in 2001 before the entry into force of the EC regulation setting maximum limits for OTA in cereals and cereal-based products (Official Journal of the European Communities, 2002).
The presence of OTA was also investigated in feed products. Extracts were purified by immunoaffinity column (Ochratest, Vicam) and OTA determination was performed by HPLC with fluorescence detection. Table 5 shows results of a survey conducted on 210 samples including raw material and compound feeding stuffs for cattle, pigs and poultry. Concerning compound feeding stuffs, one sample for cattle and two samples for pigs exceeded 10 μg/kg. With regard to raw material, thirty-one samples had OTA content ranging from 1 to 10 μg/kg and one sample exceeded this value.

Table 4. Official control of ochratoxin A in food products (CONSUM data bank, 2001 - 2002); N = number of samples analysed ; LOQ = limit of quantification

Product	N	LOQ[a,c]	<LOQ	Mean ± SD[a,b]	Median[a,b]	Max[a]
Fruit juice	25	0.05	7	0.16 ± 0.19	0.10	0.90
Dried fruit	6	0.40	5	<0.40	< 0.40	0.80
Cereal-based prod.	30	0.10	8	0.21 ± 0.16	0.20	0.60
Cereal grains	57	0.10/0.30	43	0.93 ± 3.67	0.15	25.40
Coffee	27	0.10/0.30	19	0.24 ± 0.27	0.15	1.20
Malt	10	0.10	10	< 0.10	< 0.10	< 0.10
Wine	29	0.05	19	0.18 ± 0.28	< 0.05	0.90

[a]: LOQ, mean, median and maximum value expressed in μg/kg
[b]: Arithmetic mean and median values of all samples. Values less than LOQ considered as LOQ/2
[c]: Different LOQ according to the yield of analysis

Table 5. Official control of ochratoxin A in feed products (CONSUM data bank, 2001 - 2002) N = number of samples ; LOQ = limit of quantification

Product	N	LOQ[a,c]	<LOQ	Mean ± SD[a,b]	Median[a,b]	Max[a]
Comp. feed cattle	19	1.0	12	1.1 ± 1.3	< 1.0	6.1
Comp. feed pigs	17	1.0	8	2.3 ± 4.4	< 1.0	19
Comp. feed poultry	51	1.0	33	2.1 ± 5.5	< 1.0	38
Raw material	123	1.0 / 5.0	91	1.4 ± 2.3	< 1.0	16

[a]: LOQ, mean, median and maximum value expressed in μg/kg
[b]: Arithmetic mean and median values of all samples. Values less than LOQ considered as LOQ/2
[c]: Different LOQ according to the yield of analysis

Animal studies have shown that OTA can accumulate in meats and in various organs such as the kidney and liver (Jorgensen and Petersen, 2002). Between 2001 and 2002, a survey was carried out on 89 animal kidneys using the analytical procedure described above for food commodities. The samples included kidneys from poultry (n=32), cattle (n=27) and pigs (n=30). The percentage of samples with

OTA above the limit of quantification was 12%, 11% and 27% respectively (Table 6). The levels found were very low, as the maximum concentration was 0.4 µg/kg.

Table 6. Official control of ochratoxin A in animal kidneys (CONSUM data bank, 2001 – 2002). N = number of samples ; LOQ = limit of quantification

Product	N	LOQ[a]	<LOQ	Mean ± SD[a,b]	Median[a,b]	Max[a]
Poultry kidney	32	0.1	28	< 0.1	< 0.1	0.2
Cattle kidney	27	0.1	24	< 0.1	< 0.1	0.1
Pig kidney	30	0.1	22	< 0.1	< 0.1	0.4

[a]: LOQ, mean, median and maximum value expressed in µg/kg
[b]: Arithmetic mean and median values of all samples. Values less than LOQ considered as LOQ/2

3.2.2. Occurrence in Belgian and imported beers

A Belgian survey of ochratoxin A (OTA) in commercial beers was carried out on 62 domestic and 20 imported beers from Denmark, France, Germany, Ireland, Mexico, The Netherlands and Scotland. These samples were marketed in Belgium and sampled at the retail market during 1998-2002 (Tangni *et al.,* 2002a,b). Determination of OTA concentration was performed using OchraTest immunoaffinity column and liquid chromatography with fluorescence detection. For the complete analytical procedure, the limit of detection (LOD) was 3 ng/l, whereas the limit of quantification (LOQ) was 10 ng/l.

The overall incidence of OTA in domestic beers was 97% and 100% in all the imported samples . However, none of these samples exceeded a level of 200 ng/l, making it highly unlikely that the malt and cereals used to make the beer would have exceeded the maximum level of 3 µg/kg. The highest level was 185 ng/l, with the average levels of contamination being 33 and 32 ng/l, respectively (Table 7).

It is noted that these levels are comparable with those observed in other European countries, showing the widespread detection of low levels of OTA in beers in Europe. On the basis of the established tolerable daily intake (TDI) of 5 ng/kg body weight, accepted by the European Commission's Scientific Committee for Food, the survey indicates that beer consumption in Belgium (average of 0.3 l/day per capita) is not likely to contribute to more than a few percent of the TDI. This level of exposure to OTA from beer does not represent a serious health risk for the consumer, but should be included in calculations of total exposure to OTA. Despite the moderate mean levels of OTA contamination found in beers in the present survey, the risk of highly contaminated batches occurring from time to time should not be underestimated.

3.3. Patulin

Patulin (PAT) is a mycotoxin produced by *Penicillium*, *Aspergillus* and *Byssochlamys* species. In our temperate climate, patulin is produced mainly by *Penicillium expansum*, a fungus causing rot on fruits (Beretta *et al.*, 2000;

McCallum *et al.*, 2002). As such products (particularly apples) are often used in the preparation of food for children, patulin has been well documented. It is reported to be acutely toxic to mice (Broom *et al.*, 1944), teratogenic to chicken embryos (Ciegler *et al.*, 1976), immunosuppressive in mice and rabbits (Escoula *et al.*, 1988) and carcinogenic in rats (Dickens & Jones, 1961). This mycotoxin is classified in category 3 (not classifiable regarding its carcinogenicity to humans) by the International Agency for Research on Cancer (IARC, 1993). Based on a temporary tolerable daily intake of 0.4 μg/kg body weight/day (Scientific Committee for Food, 2000), patulin is since 1 November 2003 regulated in the European Union at levels of 50, 25 and 10 μg/kg respectively in fruit juices and fruit nectar, solid apple products and apple-based products for infants and young children (Official Journal of the European Communities, 2003a).

Table 7. Levels of OTA in Belgian and imported beers (Catholic University of Louvain, data collected in 2001). LOD = limit of detection; LOQ = limit of quantification

Occurrence	Belgian beer	Imported beer	Total
Number of samples	62	20	82
< LOD	2 (3%)	0	2 (2%)
LOD-LOQ	9 (15%)	3 (15%)	12 (15%)
10 - 100 ng l^{-1}	46 (74%)	17 (85%)	63 (77%)
101 - 200 ng l^{-1}	5 (8%)	0	5 (6%)
> 200 ng l^{-1}	0	0	0
Overall incidence	60 (97%)	20 (100%)	80 (98%)
Level of contamination			
Mean ± SD[a,b]	33 ± 37	32 ± 26	
Median[a]	19	21	
Range[a,c]	10 - 185	12 - 87	

[a]: Mean, median and range values expressed in ng/l
[b]: Arithmetic mean of all samples. Values less than LOD considered as LOD/2 and values between LOD and LOQ considered as (LOD + LOQ)/2
[c]: Range of samples with ochratoxin A concentration ≥ 10 ng/l

3.3.1. Official control

A survey was carried out in Belgium on 86 fruit derivatives including baby food products (n=21), apple juice (n=45), cider (n=5) and grape juice (n=15). PAT level was determined by HPLC coupled with ultra-violet detection (CEN/TC 275/WG5 N205). PAT was quantified in one baby food product (15 μg/l), nine of the apple juice samples (ranging from 5 to 21 μg/l) and one grape juice sample (36 μg/l) (Table 8).

3.3.2. Incidence in organic and conventional apple juices marketed in Belgium

Due to food crises in the recent past, organic food products have become more popular in Belgium. However, the influence of organic cultivation methods on the presence of natural toxins, such as patulin, is not yet clear. Therefore Baert *et al.* (2004) compared the patulin contamination of organic and conventional apple juices consumed by the Belgian population. A reversed-phase high performance liquid chromatography (HPLC) method with UV-detection at 276 nm was optimised to quantify patulin. A limit of detection (LOD) of 5.2 and 8.1 μg/l respectively, was obtained for clear and cloudy apple juices. The limit of quantification (LOQ) was 8.6 and 16.0 μg/l respectively, for clear and cloudy apple juice.

Baert *et al.* (2004) analysed 58 apple juice samples, of which 22 samples were organic and 36 conventional. Patulin was detected in seven samples (12%) and in four of these (7%) the level was higher than the LOQ (Table 9). In all tested samples, the patulin content was lower than the new legal limit of 50 μg/l. The study showed that although the incidence of patulin contamination in organic and conventional apple juice was not statistically significant ($p = 0.578$), the level of contamination was higher in organic apple juices than in conventional juices ($p = 0.035$).

Table 8. Official control of patulin in fruit products (CONSUM data bank, 2001 - 2002). N =number of samples ; LOQ = limit of quantification

Product	N	LOQ[a,c]	<LOQ	Mean ± SD[a,b]	Median[a,b]	Max[a]
Babyfood	21	5 - 7	20	< 5	< 5	15
Apple juice	45	4 - 8	36	5 ± 4	4	21
Cider	5	7	5	< 7	< 7	< 7
Grape juice	15	6 - 8	14	< 6	< 6	36

[a]: LOQ, mean, median and maximum value expressed in μg/kg or μg/l
[b]: Arithmetic mean and median values of all samples. Values less than LOQ considered as LOQ/2
[c]: Different LOQ according to the yield of analysis

3.3.3. Occurrence in domestic and imported apple-based drinks

Data was gathered on the occurrence and levels of patulin (PAT) in 43 commercial apple juices (Belgian (29) and imported (14)) and seven ciders (fermented products) sampled at random from supermarkets and from handicraft producers in Belgium during 2001 (Tangni *et al.*, 2003). These samples represent all types and brands of domestic and imported (from France, Germany and Switzerland) products sold in Belgium. PAT detection and quantification was performed by HPLC with UV detection. The analytical procedure of PAT determination had limits of detection (LOD) of 0.67 and 1.33 μg/l for clear and cloudy apple juices respectively, whereas the respective limits of quantification (LOQ) were 2.21 and 4.42 μg/l. For ciders, LOD and LOQ were similar to those for clear apple juices. The survey indicates that both domestic and imported apple juices and ciders consumed in Belgium are PAT-contaminated. Indeed, PAT was detected in 79 %, 86 % and

43% of these tested samples, respectively. However, none of the contaminated samples exceeded the safe level of 50 μg PAT/l. Levels of PAT contamination were comparable in Belgian and imported juice samples (Table 10).

The overall mean PAT concentrations were 9.0 μg/l and 3.4 μg/l for contaminated apple juices and ciders, respectively. The survey also indicates that there is no statistically significant difference in the mean PAT contamination between handicraft-made (14.6 μg/l) and industrial (7.0 μg/l) apple juices. These low levels of PAT in apple-based drinks are a clear parameter for judging the quality of the raw materials as well as the manufacturing processes used. On the basis of the mean results of the survey and the established TDI of 0.4 μg/kg bw, consumer exposure assessment indicates that a daily intake of 0.2 l of apple juice could contribute up to 45% of the TDI, the worst case being the consumption of apple juice by children of 10 kg bw. As a result, the relevant EC regulation has established the maximum level of PAT at 10 μg/l in apple juice and solid apple products (compote, apple puree) intended for infants and young children (Official Journal of the European Communities, 2003a) and recommended some precautionary strategies to deal with this challenge (Official Journal of the European Communities, 2003b).

3.4. Fumonisin B1

Fumonisins are a group of mycotoxins divided into four series named A, B, C and D, the most important being fumonisin B1 (FB1) produced mainly by *F. verticillioides* (formerly *F. moniliforme*) and *F. proliferatum*, two fungi that develop on maize (Chelkowski, 1998). In horses, FB1 can cause a disease called leukoencephalomalacia (Ross *et al.*, 1993). Regarding humans, it has been suggested that in some regions of the world where there is a potential for regular fumonisin intake (notably southern Africa and China), there is a high incidence of oesophageal cancers (Rheeder *et al.*, 1992). In 1993, IARC classified FB1 in group 2B, i.e. "possibly carcinogenic to humans". Although fumonisins are mainly found in maize (Logrieco *et al.*, 2002), other agricultural products can potentially be contaminated (Martins *et al.*, 2001). FB1 is stable during most types of processing (Bullerman *et al.*, 2002). In 2000, Scientific Committee for Food (SCF) established a TDI of 2 μg/kg bw/day.

A survey by AFSCA was carried out on 45 maize-based products, including popcorn, cornflakes, tinned maize, tortilla chips, beer and corn starch products using a HPLC method with fluorescence detection as recommended by the European Committee on Standardization (procedure CEN/TC 275/WG5 N238). Two samples (tortilla chips) were contaminated at 400 and 800 μg/kg respectively.

These results have been published in the "Collection of occurrence data of *Fusarium* toxins in food and assessment of dietary intake by the population of EU Member States" (SCOOP task 3.2.10, 2003).

Table 9. Patulin levels in conventional and organic apple juices (University of Ghent, data collected in 2003). LOD = limit of detection; LOQ = limit of quantification

Occurrence	Convention.	Organic	Total
Number of samples	36	22	58
< LOD	32 (89%)	19 (86%)	51 (88%)
LOD-LOQ	2 (5%)	1 (5 %)	3 (5%)
LOQ-25 $\mu g\ l^{-1}$	2 (5%)	0	2 (3%)
25-50 $\mu g\ l^{-1}$	0	2 (9%)	2 (3%)
> 50 $\mu g\ l^{-1}$	0	0	0
Overall incidence	4 (11%)	3 (14%)	7 (12%)
Level of patulin			
Concentration of samples > LOQ[a]	8.5; 10.7	39.9; 42.0	
Mean ± standard deviation[a,b]	8.1 ± 1.9	31.4 ± 16.6	

[a]: Mean and range values expressed in µg/l

[b]: The mean value of the samples with a patulin content > LOD. For samples with a level lower that the LOQ, but above the LOD, it was assumed that the concentration was half the LOD+LOQ

Table 10. Patulin contamination in domestic and imported apple juices and ciders available in Belgium (Catholic University of Louvain, data collected in 2001). LOD = limit of detection; LOQ = limit of quantification

	Apple juice			Cider		
Occurrence	Belgian	Imported	Total	Belgian	Imported	Total
Nber of samples	29	14	43	5	2	7
<LOD	6	2	8 (19%)	3	1	4 (57%)
LOD-LOQ	7	1	8 (19%)	1	0	1 (14%)
LOQ-25 $\mu g\ kg^{-1}$	14	11	25 (58%)	1	1	2 (29%)
25 - 50 $\mu g\ kg^{-1}$	2	0	2 (4%)	0	0	0
> 50 $\mu g\ kg^{-1}$	0	0	0	0	0	0
Overall incidence	23 (79%)	12 (86%)	35 (81%)	2 (40%)	1 (50%)	3 (43%)
Level of contamination						
Mean ± SD[a,b]	10.3 ± 9.8	6.4 ± 2.8	9.0 ± 8.2	2.1 ± 0.9	6.1	3.4 ± 2.4
Median[a]	5.8	6.2	6.0	2.1	-	2.8
Range [a]	2.9-38.8	2.5-10.6	2.5-38.8	2.8	6.1	2.8-6.1

[a]: Mean, median and range values expressed in µg/l

[b]: Levels represent the mean or median values of contaminated samples. For this purpose, trace level was assumed to indicate that the sample contains half the LOD + LOQ. Concentration range takes only the samples with PAT concentration > LOQ into account.

3.5. Deoxynivalenol

Deoxynivalenol (DON), also known as vomitoxin, is probably the *Fusarium* toxin that is detected most frequently and in highest concentration. It is mainly produced by *Fusarium graminearum* and *F. culmorum*, two pathogenic fungi causing head blight in small grain cereals and ear rot in maize (Bottalico, 1998). As DON is a very stable compound during processing, final products such as bread, pasta and biscuits can potentially be contaminated. DON is concentrated mainly in the outer parts of cereal grains. Consequently, milling leads to fractionation of DON content, with increased levels in later middling and bran fractions and decreased levels in white flour (Schollenberger *et al.*, 2002). In animals, acute effects include vomiting, diarrhoea and reduced feed intake. DON, as other trichothecenes, is a potent inhibitor of protein synthesis in eukaryotic cells. In 1999, the Scientific Committee for Food (SCF) expressed an opinion on DON and established a temporary tolerable daily intake (TDI) of 1 µg/kg body weight. The EC recommended an action level of 500 µg/kg for cereal foodstuffs as consumed and other cereal products at retail stage, and an action level of 750 µg/kg for flour used as raw material in food products. Belgium used a guideline level of 5000 µg/kg for cereals and cereal-based products used as raw material for animal feeding stuffs (FAO/WHO, 2002).

3.5.1. Official control

Cereal-based foodstuffs were sampled and analysed for DON as part of the CONSUM monitoring programme. Samples included products such as wheat grains (n=47), baby food (n=5), bread (n=28) and pasta (n=19). Each sample was extracted with a mixture of water and poly-ethylene-glycol and the extract purified by cleaning through an immunoaffinity column (DONprep, R-Biopharm Rhone Ltd). A HPLC method was the used to quantify the mycotoxin. DON was found in five wheat samples (maximum value: 504 µg/kg), in one bread sample (350 µg/kg) and in four pasta samples (maximum value: 716 µg/kg). The results have been published in the "Collection of occurrence data of *Fusarium* toxins in food and assessment of dietary intake by the population of EU Member States" (SCOOP task 3.2.10, 2003).

Compound feeding stuffs for cattle, pigs and poultry, and raw material were also submitted for analysis of DON levels within the framework of the CONSUM monitoring programme. DON was extracted in a mixture of methanol and water, and the extracts were purified by immunoaffinity column (DONPrep, R-Biopharm Rhone Ltd). HPLC with UV detection was then used to determine DON content in the different samples. Table 11 summarises the results. Based on a guideline level of 5000 µg/kg, none of the samples were rejected.

3.5.2. Survey linked to the milling industry

A large proportion of the cereal-based products processed by the milling industry are composed of wheat grains of different origins. These products are used for either the food or feed industries. A study was undertaken at the Agricultural Research Centre of Gembloux to investigate whether DON was present in Belgian (n=31) and imported (n=86) wheat grain samples, with the imported samples being from

Sweden, Germany, USA, United Kingdom, Germany and France. The grain samples, harvested in 2002 from 14 mills were considered representative of the milling industry in Belgium (Table 12). An enzyme-linked immunosorbent assay (ELISA) was carried out on the crude flour using the Veratox for DON 5/5 kit (Neogen Corporation, USA). The ELISA method was compared to an AOAC HPLC method (Trucksess *et al.*, 1996) in an intra-laboratory validation procedure (AFNOR standard NF V 03-110, 1998). The limit of detection (LOD) was 120 μg/kg and the limit of quantification (LOQ) was 400 μg/kg.

Table 11. Official control of deoxynivalenol in feed products (CONSUM data bank, 2001–2002). N = number of samples ; LOQ = limit of quantification

Product	N	LOQ[a,c]	<LOQ	Mean ± SD[a,b]	Median[a,b]	Max[a]
Comp. feed cattle	39	200-250	21	321 ± 416	< 200	2570
Comp. feed pigs	55	200-250	18	397 ± 317	330	1800
Comp. feed poultry	62	200-250	16	434 ± 329	365	1600
Raw material	129	200-250	63	399 ± 413	250	2500

[a]: LOQ, mean, median and maximum value expressed in μg/kg
[b]: Arithmetic mean and median values of all samples. Values less than LOQ considered as LOQ/2
[c]: Different LOQ according to the yield of analysis

Table 12. DON levels in wheat grains for feed and food from the milling industry (Agricultural Research Centre of Gembloux, data collected in 2002). Wheat B = Belgian wheat; Wheat I = Imported wheat. LOD = limit of detection ; LOQ = limit of quantification

Occurrence	Wheat B	Wheat I	Total
Number of samples	31	86	117
<LOD	5 (16%)	28 (32%)	33 (28%)
LOD-LOQ	6 (19 %)	21 (24%)	27 (23%)
LOQ-1000 $\mu g\ kg^{-1}$	15 (48%)	29 (34%)	44 (38%)
1000-2000 $\mu g\ kg^{-1}$	5 (16 %)	6 (7%)	0
> 2000 $\mu g\ kg^{-1}$	0	1 (1%)	0
Overall incidence	26 (39%)	57 (66 %)	83 (71%)
Level of DON ($\mu g\ kg^{-1}$)			
Mean ± standard deviation [a,b]	573 ± 419	408 ± 402	
Median[a,b]	510	253	
Maximum[a]	1690	2100	

[a]: Mean, median and maximum values expressed in μg/kg
[b]: Arithmetic mean and median values of all samples. Samples less than LOD considered as LOD/2

On-going control of DON analyses was monitored using reference materials from the Food Analysis Performance and Assessment Scheme (FAPAS, United Kingdom) and by participating in inter-comparison trials. DON was detected respectively in 39% (Belgian) and 66% (imported) of the tested wheat samples. The mean DON content was comparable in both types of products. DON could be quantified in 64% of Belgian wheat, and in 41 % of imported wheat.

4. TOXIGENIC *FUSARIUM* IN WINTER WHEAT

Fusarium species are widespread on agricultural products and most of them are capable of generating toxins of variable chemical composition and toxicity (Bottalico, 1998). In Europe, three toxigenic *Fusarium* species are predominantly found to be associated with head blight in wheat: *Fusarium graminearum*, *F. culmorum* and *F. avenaceum* (Bottalico and Perrone, 2002). *Microdochium nivale* (formerly *F. nivale*) is also a causal agent of head blight in winter wheat. However, grains contaminated by this fungus do not contain the typical mycotoxins produced by *Fusarium* species (Logrieco *et al.*, 1991).

Fusarium graminearum and *F. culmorum* isolates mainly produce type B-trichothecenes (deoxynivalenol and its derivatives, and nivalenol) and zearalenone. *Fusarium graminearum* (sexual stage *Gibberella zeae*) predominates under warmer conditions, while *F. culmorum* is more frequent in cooler wheat-growing areas (Parry *et al.*, 1995). *F. avenaceum* (sexual stage *G. avenacea*) is also common on wheat and notably produces moniliformin (Abramson *et al.*, 2002). *F. poae* is generally less frequent in wheat but capable of producing toxic compounds such as nivalenol and beauvericin (Bottalico and Perrone, 2002).

Weather conditions at the flowering stage of the cereal, local temperature, frequency of rain and wetness duration, play a major role in the development of the disease (Parry *et al.*, 1995). As a result, wide variations in the incidence and population of *Fusarium* species on grains have been observed at both temporal (from year to year) and spatial (from one region to another) scales. An agrometeorological model integrating leaf wetness duration has been developed to predict the risk of head blight infection in Belgian winter wheat on a regional scale. Standard meteorological data were obtained from electronic weather stations, while precipitation data were collected from a weather radar covering the complete Belgian territory (Detrixhe *et al.*, 2003).

4.1. Fusarium *population: survey over three years (2001–2003)*

Winter wheat (*Triticum aestivum*) grain samples were randomly collected over three years (2001–2003) in production fields located in the central part of Belgium, i.e. the winter wheat growing area. A total of 355 production fields were sampled, consisting of 67 fields in 2001, 91 fields in 2002 and 197 fields in 2003 and representing the different situations found in Belgium in terms of agricultural practices based on conventional farming systems.

The mycoflora present on those samples was determined by plating 208 grains per sample onto potato dextrose agar (PDA) after surface sterilisation. *Fusarium* species were first identified according to the density and colour of aerial mycelium. Isolates that could not be identified on PDA were transferred on synthetic nutrient agar (SNA). The *Fusarium* species were identified microscopically according to Burgess *et al.* (1994). Presumed *Fusarium* that did not sporulate was classified as "undetermined *Fusarium*". The collected samples were also analysed for the major toxin, deoxynivalenol, using competitive direct enzyme-linked immunosorbent assay (CD-ELISA) after in-house validation as described previously (Table 13).

In 2001, the weather conditions were not conducive for the development of the disease and fungal species from the head blight complex represented only 3.4% of the total mycoflora found on the grains. The percentage of *Fusarium* infected seeds varied from 0 to 15%. The DON content did not exceed 400 µg/kg and 8.4% of samples were contaminated above 120 µg/kg, the limit of detection of ELISA after in-house validation. In contrast, in 2002, heavy rainfall at the flowering stage caused abundant head blight infection in winter wheat and a higher mean DON content (660 µg/kg). The percentage of *Fusarium* infected seeds in the analysed samples varied from 0 to 39%, and *F. graminearum* was the prevalent species. In 2003, the proportion of *Fusarium* species in the total mycoflora reached 20%, and the percentage of *Fusarium* infected seeds ranged from 0 to 92%. *F. graminearum* and *F. avenaceum* were the prevalent species. In spite of an increased level of *F. graminearum*, the mean DON content (270 µg/kg) was lower than in 2002. *F. poae* and *F. culmorum* were infrequently identified, while *Microdochium nivale* represented less than 4% of the total mycoflora in 2002, and less than 2% in 2003. The high incidence of *F. graminearum and F. avenaceum* in years conducive to head blight is in agreement with surveys carried out in other European countries (Birzele *et al.*, 2002; Eskola *et al.*, 2001; Waalwijk *et al.*, 2003).

As several cultivation factors have been shown to influence *Fusarium* head blight severity and contamination of grains by *Fusarium* mycotoxins (Parry *et al.*, 1995), information regarding the previous crop and the wheat variety used were collected for fields sampled in 2003. Of the 197 wheat fields, 30% were cultivated after maize. Analysis of variance revealed that the DON content in wheat grains collected from these fields was significantly higher ($P<0.05$) than that of samples collected from fields cultivated after sugar beet, small grain cereals or other crops. These results are in agreement with other studies which demonstrate that maize can provide a higher risk of DON contamination than sugar beet, potato, small-grains cereals or soybean as previous crop (Dill-Macky and Jones, 2000).

Of the fields sampled, 43% were concerned by seven varieties [Corvus (n=28); Dekan (n=18); Drifter (n=15); Baltimore (n=11); Claire (n=11); Centenaire (n=11); Meunier (n=10)], the other fields being concerned by 29 varieties. The distribution of the *Fusarium* species over the seven varieties most frequently used did not reveal any clear interaction (fig. 1). The mean DON content in the corresponding grain samples was not significantly different.

4.2. Chemotype identification and distribution

Strains of *F. graminearum* and *F. culmorum* have been classified according to their toxigenic potential, some isolates producing DON and acetyl-DON (DON-chemotype) while others produce NIV and fusarenon-X (NIV-chemotype) (Ichinoe *et al.*, 1983; Muthomi *et al.*, 2000). Due to the high toxicity of nivalenol (Ryu, 1988), the NIV-chemotype is of great concern in countries where it has been identified.

In 2002 at the Agricultural Research Centre of Gembloux, an inoculation test was conducted on a susceptible wheat variety (Cactus, Belgian variety) whereby five

F. culmorum and five *F. graminearum* collected in Belgium in 2001 were compared in terms of mycotoxin production capacity. The main type B-trichothecenes deoxynivalenol (DON), nivalenol (NIV), 3-acetyl-DON and 15-acetyl-DON were separated on a reversed phase C18 column and determined by high performance liquid chromatography-diode array detection (HPLC-DAD) using a multi-linear gradient. The sample preparation was performed by extraction with acetonitrile-water (84:16, v/v) and clean-up was done by applying a Mycosep column. Of the 10 evaluated, three *F. graminearum* isolates and one *F. culmorum* isolate mainly produced mainly deoxynivalenol, while the others produced nivalenol (Chandelier *et al.*, 2003).

Table 13. Proportion of the different head blight agents in the total mycoflora found on winter wheat grain samples (mean in percent) and DON content over three years (Agricultural Research Centre of Gembloux, data from 2001 to 2003). LOD = limit of detection ; n = number of samples analysed

Fungal species	**2001**	**2002**	**2003**
	n=67	n=91	n=197
Microdochium nivale	0.15	3.30	1.79
Fusarium avenaceum	0.67	2.05	7.65
Fusarium culmorum	0.54	0.48	0.83
Fusarium graminearum	0.38	5.72	7.70
Fusarium poae	0.47	0.80	1.64
Undetermined Fusarium	1.36	3.03	2.19
Total *Fusarium* (%)	3.4	12.1	20.0
Range *Fus*. infected seeds (%)	0 - 15	0 - 39	0 - 92
DON content			
Mean ± standard deviation[a,b]	< LOD	660 ± 750	270 ± 450
Median [a,b]	<LOD	400	120
Maximum[a]	400	3300	2720
Overall incidence (%)[c]	8.4	74.7	51

[a] : Mean, median and maximum value expressed in µg/kg
[b]: Arithmetic mean and median values of all samples
[c]: Percentage of samples with DON concentration ≥ 120 µg/kg

The classification based on the mycotoxin potential was confirmed by PCR methods.

Using the primers Gztri7F/R developed by Lee *et al.* (2001), *F. graminearum* NIV-chemotypes generated a ~160 bp fragment on 2% agarose gels stained with ethidium bromide, and DON-chemotypes generated fragments of 173 and 195 bp respectively, depending on the number of 11 bp repeats within each sequence.

The primers N1-2/N1-2R developed by Bakan *et al.* (2002) were used to differentiate high-producing and low-producing *F. culmorum*. All isolates corresponding to the NIV-chemotype based on chemical analysis did not produce any signal, while the isolate classified as DON-chemotype generated a 200 bp amplification fragment, as expected.

As studies have demonstrated ecological differences in chemotypes distribution (Jennings *et al.*, 2003), the frequency and distribution of *F. graminearum* chemotypes was evaluated on isolates sampled in 2003. Forty isolates taken from fields evenly distributed around Belgium were selected. A pure culture of each isolate was obtained through single spore isolation. The isolates were then screened by PCR to determine their toxigenic potential using the procedure described by Lee *et al.* (2001). Based on this analysis, all but one *F. graminearum* isolate generated an amplification product. Three *F. graminearum* isolates were classified as NIV-chemotypes and 36 as DON-chemotypes (Table 14). The distribution of the two chemotypes was not confined to a specific region of the country.

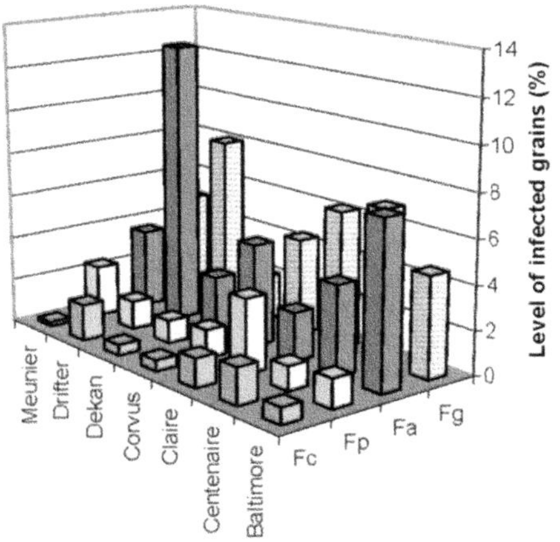

Figure 1. Percent recovery of Fusarium graminearum *(Fg),* F. avenaceum *(Fa),* F. poae *(Fp) and* F. culmorum *(Fc) on grains from seven varieties (Agricultural Research Centre of Gembloux, data collected in 2003)*

Due to their toxigenic potential, NIV-chemotype isolates are thought to be of greater concern than DON-chemotype isolates. The survey carried out in 2003 revealed that they did not represent a threat in terms of risk of nivalenol contamination. However, as *Fusarium* populations are regulated by environmental factors, a shifting of DON-chemotype towards NIV-chemotype can not be underestimated, therefore reinforcing the need to carry out regular surveys at the species and chemotype levels.

Table 14. Frequency of NIV- and DON-chemotypes of F. graminearum *(Agricultural Research Centre, data collected in 2003)*

Species	strains	NIV (%)	DON (%)
F. graminearum	40	3 (7.5)	36 (90)

5. CONCLUSION

Since 2001, the results of the official control of mycotoxins in food and feed products have been registered in a databank (CONSUM).

During 2001-2002, seven mycotoxins (aflatoxin B1, aflatoxins B1, B2, G1 and G2 (as a sum), aflatoxin M1, patulin, ochratoxin A, deoxynivalenol and fumonisin B1), were analysed in food commodities and three mycotoxins (aflatoxin B1, ochratoxin A and deoxynivalenol) were investigated in feeding stuffs.

Regarding food products, the analyses reveal some non-compliant results in both imported and local food products, thereby reinforcing the need to improve control of these mycotoxins in the future, and highlighting the need to implement precautionary measures during all stages of food production, - "from the farm to the fork"- within a philosophy of an HACCP-based integrated mycotoxin management programme (i.e. good agricultural practices and good manufacturing practices).

Regarding legal limits, there is a need to continue the work of setting maximum levels for more mycotoxins in more foodstuffs.

The surveys undertaken in Belgium can help to form a scientific basis for the setting of maximum limits.

The overall incidence of DON in cereal and cereal-based products is also a matter of concern due to the staple food status of cereals in the human diet.

The correct application of adapted agricultural practices (crop rotation, fungicide application, tillage) and the use of more resistant cereal varieties should significantly reduce DON content in cereals.

Moreover, the development of a forecasting model aiming at identifying the risk level at a field scale, and consequently to avoid the mixing of contaminated and sound lots, should also decrease the overall incidence of DON in food commodities.

Regarding *Fusarium* populations found on Belgian winter wheat, the study carried out over three years (2001–2003) demonstrates the predominance of *Fusarium graminearum* in infected kernels as well as the occurrence of *Fusarium* species or isolates that do not produce DON, but generate other mycotoxins such as nivalenol (produced by *F. graminearum* and *F. culmorum*, NIV-chemotype) or moniliformin (produced by *F. avenaceum,* a species frequently found on infected kernels in 2003).

Due to their toxicity, moniliformin and nivalenol surveys should be carried out on cereals cultivated in Belgium. In this respect, it is important to note that there is an urgent need for routine tests applicable to these mycotoxins.

6. ACKNOWLEDGEMENTS

We wish to thank the Belgian Federal Agency for the Safety of the Food Chain (AFSCA/FAVV) for kindly supplying the data on official control of mycotoxins in food and feed supplies. We are also grateful to Pierre Detrixhe for kindly supplying data on *Fusarium* populations in winter wheat.

7. AFFILIATIONS

Chandelier, Anne, Agricultural Research Centre, Department Biocontrol and Plant Genetic Resources (Gembloux); e-mail: chandelier@cra.wallonie.be

Michelet, Jean-Yves, Scientific Institute of Public Health; e-mail: Jean-Yves.Michelet@iph.fgov.be

Tangni, Emmanuel K., Veterinary and Agro-Chemical Research Centre, Department of Quality and Security (Tervuren); e-mail: emtan@var.fgov.be

Baert Katleen, Ghent University, Department of Food Technology and Nutrition; e-mail: Katleen.Baert@UGent.be

Moons, Emmanuelle, Belgian Federal Agency for the Safety of the Food Chain; e-mail: emmanuelle.moons@afsca.be

Vinkx, Christine, Federal Public Service, Public Health, Food Chain Safety and Environment; e-mail: Christine.Vinkx@health.fgov.be

8. REFERENCES

Abarca, M.L., Accensi, F., Bragulat, M.R., & Cabanes, F.J. (2001). Current importance of ochratoxin A-producing Aspergillus. Journal of Food Protection, 64, 903-906

Abramson, D., McCallum, B., Smith, D.M., & Tekauz, A. (2002). Moniliformin in barley inoculated with *Fusarium avenaceum*. Food Additives and Contaminants, 19, 765-769

AFNOR (1998). Analysis of agri-foodstuffs – Intralaboratory validation procedure for an alternative method compared to a reference method. NF V 03-110, December 1998, pp 1-40

Baert, K., Kasase, C., De Meulenaer, B., Devlieghere, F., & Huyghebaert, A. (2004). Incidence of Patulin in Organic and Conventional Apple Juices Marketed in Belgium. (in press)

Bakan, B., Giraud-Delville, C., Pinson, L., Richard-Molard, D., Fournier, E., & Brygoo, Y. (2002). Identification by PCR of *Fusarium* culmorum strains producing large and small amounts of deoxynivalenol. Applied and Environmental Microbiology, 68, 5472-5479

Battilani, P., & Pietri, A. (2002). Ochratoxin A in grapes and wine. European Journal of Plant Pathology, 108, 639-643

Beretta, B., Gaiaschi, A., Galli, C.L., & Restani, P. (2000). Patulin in apple-based foods: occurrence and safety evaluation. Food Additives and Contaminants, 17, 399-406

Birzele, B., Meier, A., Hindorf, H., Krämer, J., & Dehne, H.W. (2002). Epidemiology of *Fusarium* infection and deoxynivalenol content in winter wheat in the Rhineland, Germany. European Journal of Plant Pathology,108, 667-673

Bottalico, A. (1998). *Fusarium* diseases of cereals: species complex and related mycotoxin profiles in Europe. Journal of Plant Pathology, 80, 85-103

Bottalico, A., & Perrone, G. (2002). Toxigenic *Fusarium* species and mycotoxins associated with head blight in small-grain cereals in Europe. European Journal of Plant Pathology, 108, 611-624

Broom, W.A., Bulbring, E., Chapman, C.J., Hampton, J.W.F., Thomson, A.M., Ungar, J., Wien, R., & Woolfe, G. (1944). The Pharmacology of Patulin. The British Journal of Experimental Pathology, 25, 195-207

Bullerman, L.B., Ryu, D., & Jackson L.S. (2002). Stability of fumonisins in food processing. Advances in Experimental Medicine and Biology, 504, 195-204

Burgess, L.W., Summerell, B.A., Bulock, S., Gott, K.P., & Backhouse, D. (1994). Laboratory Manual for *Fusarium* Research, 3rd ed. University of Sidney, Sidney, Australia

Chandelier, A., Kestemont, M.H., Detrixhe, P., Pierard, J.Y. & Cavelier, M. (2003). Fungal pathogens associated with head blight in wheat: a three-year analysis in Belgium. Aspects of Applied Biology, 68, 187-191

Chelkowski, J. (1998). Distribution of *Fusarium* species and their mycotoxins in cereal grains. In: Mycotoxins in agriculture and food safety, Sinha, K.K. and Bhatnagar, D. (eds), Marcel Dekker Inc., pp 45 - 64

Ciegler, A., Beckwith, A.C., & Jackson, L.K. (1976). Teratogenicity of Patulin and Patulin Adducts Formed with Cysteine. Applied and Environmental Microbiology, 31, 664-667

Creppy, E.E. (2002). Update of survey, regulation and toxic effects of mycotoxins in Europe. Toxicology Letters, 127, 19-28

Detrixhe, P., Chandelier, A., Buffet, D., Oger, R., & Cavelier, M. (2003). Development of an agrometeorological model integrating leaf wetness duration estimation to assess the risk of head blight infection in wheat. Aspects of Applied Biology, 68, 199-204

Dickens, F., & Jones, H.E.H. (1961). Carcinogenic Activity of a Series of Reactive Lactones and Related Substances. British Journal of Cancer, 15, 85-100

Dill-Macky, R., & Jones, R.K. (2000). The effect of previous crop residues and tillage on *Fusarium* head blight of wheat. Plant Disease, 84, 71-76

Doohan, F.M., Parry, D.W., Jenkinson, P., & Nicholson, P. (1998). The use of species-specific PCR-based assays to analyse *Fusarium* ear blight of wheat. Plant Pathology, 47, 197-205

Escoula, L., Thomsen, M., Bourdiol, D., Pipy, B., Peuriere, S., & Roubinet, F. (1988). Patulin Immunotoxicology: Effect on Phagocyte Activation and the Cellular and Humoral Immune System of Mice and Rabbits. International Journal of Immunopharmacology, 10, 983-989

Eskola, M., Parikka, P., & Rizzo, A. (2001). Trichothecenes, ochratoxin A and zearalenone contamination and *Fusarium* infection in Finnish cereal in 1998. Food Additives and Contaminants, 18, 707-718

Food and Agriculture Organisation / World Health Organisation (FAO / WHO) (2003). Discussion paper on deoxynivalenol CX/FAC 03/35, November 2002. Joint FAO/WHO Codex Committee on Food Additives and Contaminants, Thirty-fifth Session, 17-21 March 2003, pp 1-12, Retrieved from: http://europa.eu.int/comm/food/fs/sc/scf/out44_en.htm

Govaris, A., Roussi, V., Koidis, P.A., & Botsoglou, N.A. (2002). Distribution and stability of aflatoxin M1 during production and storage of yoghurt. Food Additives and Contaminants, 19, 1043-1050

Ichinoe, M., Kurata, H., Sugiura, Y., & Ueno, Y. (1983). Chemotaxonomy of *Gibberella zeae* with special reference to production of trichothecenes and zearalenone. Applied and Environmental Microbiology, 46, 1364-1369

International Agency for Research on Cancer (IARC) (1993). Some naturally occurring substances; food items and constituents, heterocyclic aromatic amines and mycotoxins. IARC Monographs on the evaluation of carcinogenic risks to humans (Lyon:WHO/IARC), vol 56, pp 489-521

Jennings, P., Coates, M., Turner, J.A., & Nicholson, P. (2003). Distribution of deoxynivalenol/nivalenol producing isolates of *Fusarium culmorum* and *F. graminearum* in England and Wales. Aspects of Applied Biology, 68, 43-48

Jorgensen, K., & Petersen, A. (2002). Content of ochratoxin A in paired kidney and meat samples from healthy Danish slaughter pigs. Food Additives and Contaminants, 19, 562-567

Lee, T., Oh, D.W., Kim, H.S., Lee, J., Kim, Y.H., Yun, S.H., & Lee, Y.W. (2001). Identification of deoxynivalenol and nivalenol producing chemotypes of *Gibberella zeae* by using PCR. Applied and Environmental Microbiology, 67, 2966-2972

Logrieco, A., Vesonder, R.F., Peterson, S.W., & Bottalico, A. (1991). Re-examination of the taxonomic disposition of and deoxynivalenol production by *Fusarium nivale* NRRL 3289. Mycologia, 83, 367-370

Logrieco, A., Mulè, G., Moretti, A., & Bottalico, A. (2002). Toxigenic *Fusarium* species and mycotoxins associated with maize ear rot in Europe. European Journal of Plant Pathology, 108, 597-609

Martins, M.L., Martins, H.L., & Bernardo, F. (2001). Fumonisins B1 and B2 in black tea and medicinal plants. Journal of Food Protection, 64, 1268-1270

McCallum, J.L., Tsao, R., & Zhou, T. (2002). Factors affecting patulin production by *Penicillium expansum*. Journal of Food Protection, 65, 1937-1942

Muthomi, J.W., Schütze, H., Dehne, H.W., Mutitu, E.W., & Oerke, E.C. (2000). Characterisation of *Fusarium culmorum* isolates by mycotoxin production and aggressiveness to winter wheat. Zeitschrift für Pflanzenkrankheiten und Pflanzenschutz, 107, 113-123

Nubel, asbl (2003). Table belge de composition des aliments. Third edition, pp 25

Official Journal of the European Communities (1992). Commission Directive (EC) No 92/95/EC of 9 November 1992 amending the Annex to the Seventh Directive 76/372/EC establishing Community methods of analysis for the official control of feedingstuffs, L327, pp 54-62.

Official Journal of the European Communities (1999). Council directive 1999/29/EC of 22 April 1999 on the undesirable substances and products in animal nutrition, L115/32, pp 32-46.

Official Journal of the European Communities (2001). Commission Regulation (EC) No 466/2001 of 8 March 2001 setting maximum levels for certain contaminants in foodstuffs, L77, pp 1-13.

Official Journal of the European Communities (2002). Commission Regulation (EC) No 472/2002 of 12 March 2002 amending Regulation (EC) No 466/2001 as regards ochratoxin A, L75, pp 18-20.

Official Journal of the European Communities (2003a). Commission Regulation (EC) No 1425/2003 of 11 August 2003 amending Regulation (EC) No 466/2001 as regards patulin, L203, pp 1-3

Official Journal of the European Communities (2003b). Commission Recommendation 2003/598/EC notified under document number C(2003) 2866 of 11 August 2003 on the prevention and reduction of patulin contamination in apple juice and apple juice ingredients in other beverages, L203, pp 54-59

Parry, D.W., Jenkinson, P., & McLeod, L. (1995). *Fusarium* ear blight (scab) in small grain cereals – a review. Plant Pathology, 44, 207-238

Pitt, J.I. (1987). *Penicillium viridicatum, Penicillium verrucosum*, and production of ochratoxin A. Applied and Environmental Microbiology, 53, 266-269

Plestina, R., Ceovic, S., Gatenbeck, S., Habazin-Novak, V., Hult, K., Kokby, E., Krogh, P., & Radic, B. (1990). Human exposure to ochratoxin A in areas of Yugoslavia with endemic nephropathy. Journal of Environmental Pathology, Toxicology and Oncology, 10, 145-148

Rheeder, J.P., Marasas, W.F.O., Thiel, P.G., Sydenham, E.W., Shepard, G.S., & van Schalkwyk, D.J. (1992). *Fusarium moniliforme* and fumonisins in corn in relation to human oesophageal cancer in Transkei. Phytopathology, 82, 353-357

Riley, R.T. (1998). Mechanistic interactions of mycotoxins: theoretical consideration. In: Mycotoxins in agriculture and food safety, Sinha, K.K. and Bhatnagar, D. (eds), Marcel Dekker Inc., pp 227-253

Ross, P.F., Ledet, A.E., Owens, D.L., Rice, L.G., Nelson, H.A., Osweiler, G.D., & Wilson, T.M. (1993). Experimental equine leukoencephalopathy, toxic hepatitis, and encephalopathy caused by corn naturally contaminated with fumonisins. Journal of Veterinary Diagnostic Investigation, 5, 69-74

Ryu, J.C., Ohtsubo, K., Izumiyama, N., Nakamura, K., Tanaka, T., Yamamura, H., & Ueno, Y. (1988). The acute and chronic toxicities of nivalenol in mice. Fundamental and Applied Toxicology, 11, 38-47

Sabioni, G., & Sepai, O. (1998). Determination of human exposure to aflatoxins. In: Mycotoxins in agriculture and food safety, Sinha, K.K. and Bhatnagar, D. (eds), Marcel Dekker Inc., pp 183 – 226

SCOOP Task 3.2.10. (2003). Collection of occurrence data of *Fusarium* toxins in food and assessment of dietary intake by the population of EU Member States. European Communities, Directorate-General Health and Consumer Protection (April 2003), p 609, Retrieved from: http://europa.eu.int/comm/food/fs/scoop/task3210.pdf

Schollenberger, M., Jara, H.T., Suchy, S., Drochner, W., & Muller, H.M. (2002). *Fusarium* toxins in wheat collected in an area in southwest Germany. International Journal of Food Microbiology, 72, 85 - 89

Smith, J.E., Lewis, C.W., Anderson, J.G., & Solomos, G.L. (1994). Mycotoxins, occurrence and toxicity. In: Mycotoxins in human nutrition and health. European Commission, Directorate-General XII, Science, Research and Development, Agro-Industrial Research Division (eds), pp 1-55

Tangni, E.K., Ponchaut, S., Maudoux, M., Rozenberg, R., & Larondelle, Y. (2002a). Ochratoxin A in domestic and imported beers in Belgium : occurrence and exposure assessment. Food Additives and Contaminants, 19, 1169-1179

Tangni, E. K., Ponchaut, S., Maudoux M., Rozenberg R., & Larondellle Y. (2002b). Exposure assessment to ochratoxin A through beer consumption in Belgium. Toxicology Letters, 135 (Suppl. 1), 106-107

Tangni, E.K., Theys, R., Mignolet, E., Michelet, J.Y., & Larondelle, Y. (2003). Patulin in domestic and imported apple-based drinks in Belgium: occurrence and exposure assessment. Food Additives and Contaminants, 20, 482-489

Tatu, C.A., Orem, W.H., Finkelman, R.B., & Feder, G.L. (1998). The etiology of Balkan endemic nephropathy: more questions than answers. Environmental Health Perspectives, 106, 689-700

Trucksess, M., Ready, D.E., Pender, M.K., Ligmond, C.A., Wood, G.E., & Page, S.W. (1996). Determination and survey of deoxynivalenol in white flour, whole wheat flour, and bran. Journal of AOAC International, 79, 883-887

Waalwijk, C., Kastelein, P., de Vries, I., Kerenyyi, Z., van der Lee, T., Hesselink, T., Köhl, J., & Kema, G. (2003). Major changes in *Fusarium* spp. in wheat in the Netherlands. European Journal of Plant Pathology, 109, 743-754

S. PEPELJNJAK & M. ŠEGVIĆ

AN OVERVIEW OF MYCOTOXINS AND TOXIGENIC FUNGI IN CROATIA

1. INTRODUCTION

Mycotoxins are secondary fungal metabolites that contaminate food and feed and can affect human and animal health. Contamination occurs through the fungal growth before harvest depending on consistent environmental conditions, or after harvest as consequence of improper storage. Mycotoxins differ in their chemical characteristics and mode of action in biological systems. Between few hundred known mycotoxins aflatoxins, ochratoxins, zearalenone, trichotecens (T-2, HT-2, DON, DAS) and fumonisins have a great toxicological and economical importance.

The "Golden Age" of mycotoxin research started with the discovery of aflatoxins, which caused fatal turkey outbreak in Great Britain known as "turkey X disease". These toxins are produced by storage moulds *Aspergillus flavus* and *A. parasiticus*, frequently detected in peanuts, cereals, cotton, oilseed and dried meat products from hot and humid countries. Aflatoxins (B_1, B_2, G_1, G_2) are strong hepatotoxins and carcinogens for humans and animals. In the ochratoxin group, ochratoxin A (OTA) is most frequently found in cereals from tropical and temperate countries. It is mainly produced by some *Penicillium* and *Aspergillus* species. Ochratoxin A showed nephrotoxic, immunosuppressive, neurotoxic, carcinogenic and teratogenic potential in all experimental animals. Due to a similarity of kidney damage in OTA-induced porcine nephropathy, this toxin has been proposed as one of the causative agents of endemic nephropathy. This kidney disease occurs in some regions of Bulgaria, Romania, Yugoslavia, Bosnia and Herzegovina and in Croatia (Middle Posavina).

Zearalenone, trichotecens (T-2, HT-2, DON, DAS) and fumonisins are produced by *Fusarium* species, which contaminate cereals during harvest and storage. Chronic exposure of domestic animals to zearalenone causes oestrogenism, reproductive organ disorders, abortion and sterility. Trichotecenes cause weight loss, vomiting, skin irritation, immunosuppression internal bleeding and animal death. Trichotecenotoxicosis known as "alimentary toxic aleukia" with fatal outcome in humans has been observed in the USSR (1932-1947) and was the result of *Fusarium sporotrichoides* contamination of grain.

Fumonisins are a "new" group of fusariotoxins, described in 1988. They are involved in several animal toxicosis including equine leukoencephalomalacia, porcine pulmonary oedema, hepatotoxicity, nephrotoxicity, immunosuppression and have carcinogenic, genotoxic and apoptotic (Figure 1) properties. Most of these particular mycotoxins have been studied in Croatia over a thirty years period,

A. Logrieco and A. Visconti (eds.),
An Overview on Toxigenic Fungi and Mycotoxins in Europe, 33-50.

especially ochratoxin A (OTA) as possible aetiological factor of endemic nephropathy. Occurrence of fungi and mycotoxins in different substrates (field and stored crops, dried meat products, animal feed) and air, mycotoxin residues in humans and animals, and toxigenic potential of fungal strains isolated over a past three decades in Croatia will be discussed in this review.

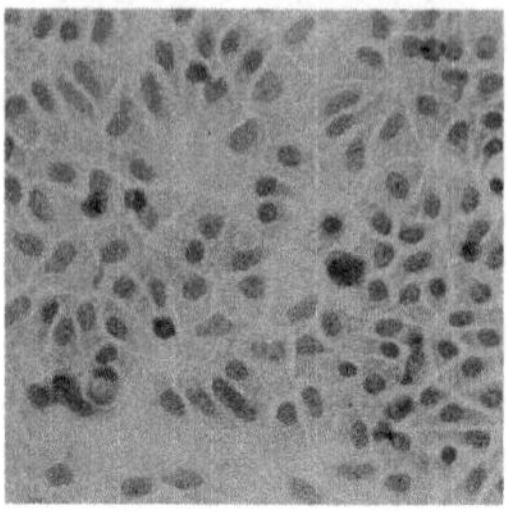

Figure 1. Apoptosis of RK13 cell line exposed to 0,2μM of fumonisin B_1 during 48 hours (Rumora et al.*, 2002)*

2. FUNGAL CONTAMINATION OF HUMAN FOODS AND ANIMAL FEEDS

2.1. Field and Stored Crops for Human Nutrition

The fungal contamination of crops and biosynthesis of mycotoxins depend on environmental conditions, crops health before harvesting (contamination with field fungi), meteorological conditions (humidity and temperature oscillations), harvesting techniques, grain damage and storage conditions (Balzer *et al.,* 1977; Pepeljnjak *et al.,* 1979).

In past three decades over 3000 samples of field and 4000 of stored crops (mostly maize, cereals and beans) from endemic (Middle Posavina) (EN) and non-endemic (non-EN) areas in continental and Mediterranean Croatia (Istra and Dalmacija) have been mycologicaly analysed. Qualitatively and quantitatively field crops mycoflora from endemic (EN) and non-endemic (non-EN) continental areas did not reveal differences. The most dominant fungal species were field fungi *Fusarium* (60-100%), *Alternaria* (70-100%) and *Cladosporium* (33-96%) (Čuturić *et al.,* 1979; Pepeljnjak & Balzer, 1982; Pepeljnjak, 1984; Pepeljnjak & Cvetnić, 1985a, 1985b). Extremely high levels of *Fusarium* were found on maize samples in 1978 when the relative humidity was high and rain was frequent. Maize samples contained up to 45% of water and over 80% of maize was visually contaminated with moulds. These particular fungi were detected in much lower frequencies on plants in vegetation in Mediterranean parts: *Fusarium* (up to 25%), *Alternaria* (16-60%) and *Cladosporium* (10-20%) (Pepeljnjak & Cvetnić, 1985b, Pepeljnjak & Šegvić, 2003). Pepeljnjak *et al* (1999) reviewed mycological contamination of stored cereals and maize in the EN and non-EN region. In stored maize samples the most frequently found were *Penicillium* (40-60%), *Rhizopus* (32-50%) and *Fusarium* (40-47%). *Penicillium* and *Rhizopus* species were more frequent in samples from endemic area (Figure 2). Contamination with *Absidia*, *Aspergillus*, *Alternaria* and *Cladosporium* species were between 15 and 26% in both areas. Other species including *Trichoderma*, *Trichothecium*, *Mucor*, *Botrytis*, *Chaetomium* and *Epicoccum* were present in lower frequencies (1-14%). In other cereals including wheat, oat, barley and rye (Figure 3) *Alternaria* (58-67%), *Fusarium* (47-63%) and *Penicillium* (44-45%) were dominant fungal species, followed by *Aspergillus* (39%) and *Rhizopus* (33%). Levels of *Fusarium* contamination were considerably higher in samples from endemic region only.

During the wartime in Croatia (1991-1995) fungal contamination of maize and other cereals were elevated for 14,5% particularly with *Fusarium* (8,9%) and *Penicillium*, *Alternaria*, *Cladosporium* and *Trichoderma* species (4,2%). The increase of mycological contamination was probably due to the war destruction, deserted storages and crop fields, and organic decay (Pepeljnjak *et al.*, 1999, 2002a). In the last few years, *Fusarium* was the dominant fungus in both field and stored maize (86% and 54% respectively). Among fusaria most frequently found was *F. verticilliodes* (= *moniliforme*) (32,4 and 16,2 respectively). Other isolated fungal genera including *Absidia, Alternaria, Aspergillus, Chaetomium, Cladosporium, Mucor, Nigrospora, Penicillium, Rhizopus, Trichoderma* and *Trichothecium* were present in a smaller percent of occurrence (1.5-35.3%), in both field and stored maize (Šegvić & Pepeljnjak, 2003).

On the other side, fungal flora of stored herbal drugs revealed high contamination with storage fungi *Aspergillus* sp. (62,3%), *Rhizopus* sp. (42,3-50%) and *Penicillium* sp. (41,1%) while *Fusarium* was found in a low percentage (8,2%) (Pepeljnjak & Cvetnić, 1998; Cvetnić & Pepeljnjak, 1999).

2.2. Dried Meat Products

From 1980 until now more than 5000 samples of dried meat products were screened on mycological contamination.

These samples (country-cured ham, dried bacon, pork sausage, smoke-dry ham) were taken from individual households in Middle Posavina (EN), Northern Posavina, Bilogora region and Istria.

These screenings revealed domination of *Penicillium* (66-86%) and *Aspergillus* species (59-70%) over other identified fungi: *Alternaria* (50%), *Absidia* and *Rhizopus* (25%), *Trichoderma* (20,5%) and *Fusarium* (0-8%).

Among *aspergilla,* the most frequently found were *A. glaucus* (24-40%), *A. flavus* and *A. parasiticus* (10-37%), *A. niger* (16-32%) and *A. ochraceus* (10%) (Cvetnić & Pepeljnjak, 1986, 1990, 1995, 1998). These species are potential producers of aflatoxins and ochratoxins and they are present in high frequency in home-made dried meat products, which are consumed on daily basis by the rural population of Croatia. Toxigenic potential of these species will be discussed later.

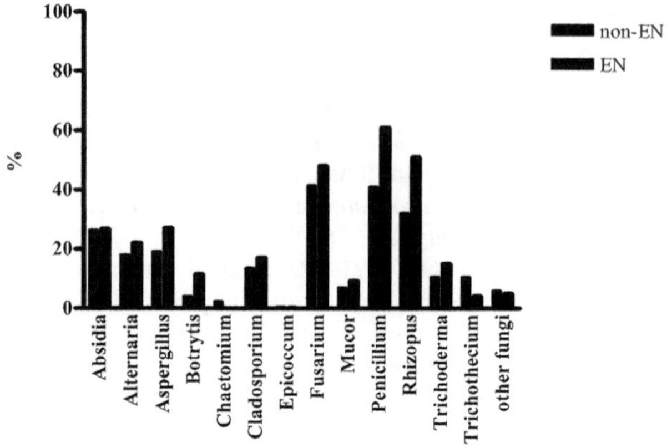

Figure 2. Mycological contamination of stored maize collected in the non-endemic (non-EN) and endemic (EN) regions over the period 1975-2001.

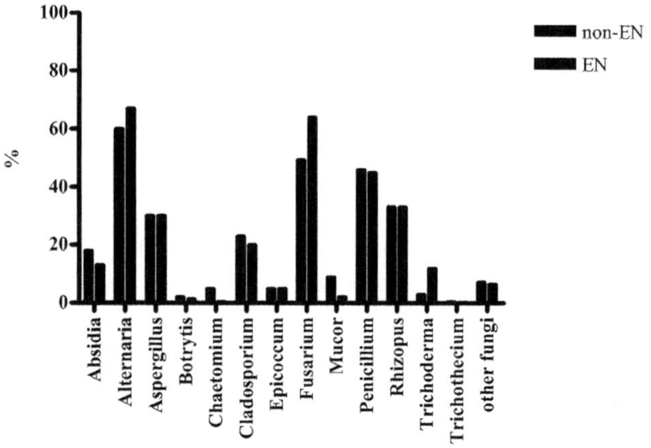

Figure 3. Mycological contamination of stored wheat, oat, barley and rye collected in the non-endemic (non-EN) and endemic (EN) regions over the period 1975-2001.

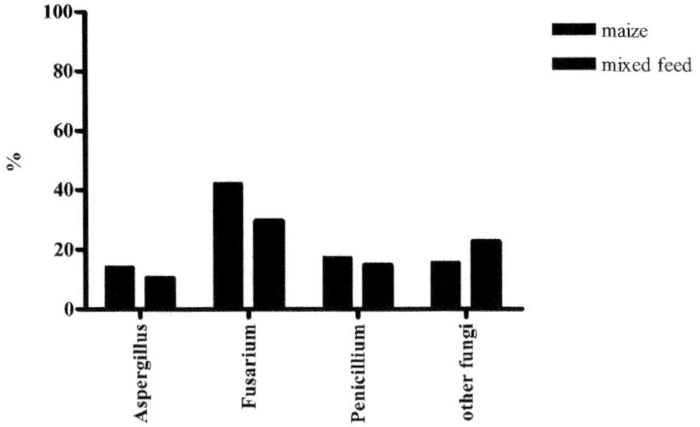

Figure 4. Mycological contamination of animal feed collected in the storage-houses over the period 1988-1999

2.3. Animal Feed

In the last decade samples of maize for animal feed and mixed feed were also mycologicaly analysed.

In these samples (Figure 4) the prevalence of fusaria was found as well (42,3 and 29,6% respectively). *Fusarium* species as field fungi are most frequent contaminants of maize and other cereals in the vegetation phase with continual development in storage conditions. Except fusaria, typical storage fungi were detected: *Penicillium* sp. (17,4% and 14,9% respectively) and *Aspergillus* sp. (14,2% and 10,5% respectively) (Pepeljnjak & Cvetnić 1986; Pepeljnjak *et al.,* 1999, 2002a).

Increase of feed fungal contamination (for 8%) was observed in 1994 and 1995, which was probably influenced by war destruction and deserted storage houses.

3. AIRBORNE FUNGI IN CONTINENTAL AND MEDITERRANEAN CROATIA

Fungal spores have been found in soil, plants in vegetation, stored crops, animal remains and other organic material and from there they are airborne. Exposure to airborne fungi is known to induce numerous human respiratory diseases including rhinitis, bronchitis, hypersensitivity pneumonitis, asthma and aspergillosis (Horner *et al.,* 1995). Besides time of day, vegetation type, air pollution, industrial and other human activities, airspora concentrations mostly depend on seasonal climatic factors of a particular region.

Screenings of airborne fungi were conducted in period from 1989 to 2002 in both continental and Mediterranean Croatia (Cvetnić & Pepeljnjak, 1991, 1997; Pepeljnjak & Šegvić, 2003). These studies indicated that mycoflora in the outdoor air of both regions includes 4 constant genera: *Cladosporium* (44-47%), *Penicillium* (22-34%), *Alternaria* (2-58%), *Aspergillus* (2-21%). Increase of airspora concentrations in spring and summer (up to 40 CFU/plate) and decrease in autumn and winter (up to 10 CFU/plate) were observed in both areas, but with significantly lower levels in the Mediterranean part of Croatia.

Airspora contributing to such a difference belong to allergologically important species; *Cladosporium, Alternaria* and *Penicillium*. Since the presence and concentrations of airspora in the outdoor environment depend on seasonal climatic factors, reduced rainfall and dry Mediterranean climate can be one of the reasons for the reduction of fungal spora in this area. On the other hand, the presence of plants, known to produce essential oils and aerosol, which showed sporistatic, fungistatic, and fungicide activity (Pepeljnjak *et al.,* 2003), possibly reduce airspora concentrations.

4. MYCOTOXINS IN HUMAN FOODS AND ANIMAL FEEDS

4.1. Mycotoxins in human foods from endemic and non-endemic regions in Croatia

Various samples of human food (cereals, maize, beans, dried meat products, apples) were obtained from individual households in EN (Middle Posavina) and non-EN regions in Croatia and analysed on the presence of mycotoxins, including ochratoxin A (OTA), zearalenone (ZEN), aflatoxin B_1 (AFB_1) skin irritating factors (SIF, T-2 toxin) and citrinin (CTN) (Table 1). These mycotoxins were detected by previously described TLC methods (Eppley, 1968; Bogdanić *et al.,* 1976; Balzer *et al.,*1978; Pepeljnjak *et al.,* 2002b). Deoxynivalenol (DON) and nivalenol (NIV) in haricot beans were detected by the HPLC method (Tanaka *et al.,* 1988)

Significant findings of OTA in maize and cereals occurred in 1979 and 1980/81 (56 and 81%) with highest concentration observed in 1979 in the nephropatic area (up to 68,9 mg/kg). In spite of the high incidence of OTA in 1980/81, the concentration range was lower (0,01-4,7 mg/kg) and similar to levels found in 1976/77 (0,01-2,1 mg/kg). In the period from 1976-1981, 45% of OTA positive samples contained more than 2mg/kg and 7,5% of these samples contained over 20 mg/kg of this toxin.

In the same period 15% of OTA positive samples taken from non-endemic region contained a maximum concentration of 2,1 mg/kg. A relatively high number of OTA positive samples of dried meat products (12-29%) was found in endemic area in a range from 0,01 to 0,92 mg/kg. In the period 1992-1995, which includes the wartime in Croatia, levels of OTA in maize were higher (up to 3,6 mg/kg) compared to those observed from 1981 to 1985 in non-endemic regions. In the last six years including reports by Jurjević *et al* (1999, 2002), concentrations of OTA in maize were lower in both endemic and non-endemic area.

Similar results were observed for ZEN and SIF (T-2 toxin). From 1975 to 1979 occurrence of ZEN and SIF varied from 2,6 to 46% and 5,5 to 9,5% of samples with the highest amount detected in 1978/79 (275,8 mg/kg and 20,52 mg/kg respectively). Findings of high levels of both ZEN and SIF correspond to increased contamination of crops with *Fusarium* species observed in 1978/79. Higher amounts of ZEN (up to 19,9 mg/kg) and DON (12,4 mg/kg) were observed in cereals and beans in 1991 and from 1992 to 1995 (wartime in Croatia). Levels of ZEN (up to 10,7 mg/kg) and SIF (up to 25,5 mg/kg) were also higher in maize samples.

Aflatoxin B_1 was found in 4 maize samples only in the non-endemic region during 1978/79 (0,005-0,05 mg/kg), but a relatively high number of *Aspergillus* aflatoxin-producer strains was isolated from various substrates, which will be discussed later.

Apples obtained from households in different parts of continental Croatia were screened on the presence of CTN and this toxin was detected in 19% of samples in a range from 0,05 to 0,24 mg/kg.

In addition, Jurjević *et al* (1999,2002) reported high frequency of fumonisins (FB_1+FB_2) in maize from both endemic and non-endemic (95%) areas in a range of 0,012-9,903 mg/kg. Beauvericin (BEA) was also found in these samples with a frequency of 7,5% in non-endemic (0,034-1,346 mg/kg) and 11% in endemic areas (0,025-1,864 mg/kg). Beauvericin co-occurred with FB_1+FB_2 and OTA in 4 samples, while FB_1+FB_2 and OTA co-occurred in 21%.

Co-occurrence of nephrotoxins (OTA and FB-s) and their synergism may contribute to the development of endemic nephropathy. In addition, simultaneous occurrence of fusariotoxins (trichotecens, ZEN and FB-s) and OTA, which are all immunosuppressive compounds, can contribute to the development of secondary infections, apoptosis and carcinogenesis, depending on dose and time of exposure.

4.2. Mycotoxins in animal feeds

Pepeljnjak *et al* (1999) reviewed fusariotoxin contamination of stored maize and mixed animal feed for the period from 1988 to 1997.

The mean frequency of T-2 was 15,3% in maize and 27,8% in mixed feed with the same mean concentration range from 0,1-0,5 mg/kg. Similar findings were observed in the period from 1978 to 1987 for mixed feed when 26,6% of samples contained T-2 in concentrations from 0,18 to 0,96 mg/kg (Pepeljnjak, 1989). In the same period (1988-1997) diacetoxyscirpenol (DAS) was found with a similar mean percent of occurrence in maize (29%) and mixed feed (30,85%) with a mean range of 0,1-0,4 mg/kg and 0,1-0,31 mg/kg respectively.

Increased T-2 and DAS contamination frequency was observed in maize (up to 65%) and mixed feed (up to 75%) during the wartime in Croatia. The mean frequency of ZEN was lower in maize (10%) than in mixed feed (19,4%) ranging from 0,15 to 1,2 mg/kg. Contrary to T-2 and DAS, increased occurrence of ZEN was observed in the 1988/89 and after war in the 1996/97 in mixed feed (67%).

Before 1988, Nemanič *et al* (1986) reported on occurrence of ZEN (29%) and OTA (10,5%) in mixed feed with concentrations 0,1-6,88 mg/kg and 0,02-0,18 mg/kg respectively, while aflatoxin B_1 was qualitatively detected in 8,5% of samples. Pepeljnjak & Cvetnić (1986) also founded OTA (0,05-1,62 mg/kg) and AFB_1+B_2 (0,02-1,12 mg/kg) in 65,6% and 9,3 % of mixed feed collected from individual households.

Table 1. Mycotoxins in human foods detected from 1975 to 2001 in endemic (EN) and non-endemic (non-EN) regions of Croatia

Period	*Region*	*Samples (N)*	*Mycotoxin/concentration*	*Reference*
1975	non-EN EN	maize (157+15)	25% OTA 0,01-5,1 mg/kg 13% OTA 0,650 mg/kg	Mužic *et al.*, 1976
1975/76	non-EN EN	maize (191)	26% OTA 0,045-5,21 mg/kg 2,6% ZEN 0,043-10 mg/kg 8,3% SIF 0,5-5,0 mg/kg	Balzer *et al.*, 1977
1976/77	non-EN EN	cereals (161)	6,2% ZEN 0,36-4,20 mg/kg 5,5% SIF 5,5-15 mg/kg	Pepeljnjak & Čuturić, 1978
1976/77	non-EN EN	cereals (111)	16,1% OTA 0,04-2,1 mg/kg 5,8% ZEN 0,051-2,5 mg/kg 5,9% SIF 0,5-7,0 mg/kg	Pepeljnjak & Čuturić, 1978
1977	EN	maize (267)	16% ZEN 0,1-4,0 mg/kg	Pepeljnjak *et al.*, 1979
1978/79	EN non-EN	cereals (116)	56% OTA 0,01-68,9 mg/kg 46,5% ZEN 0,01-275,8 mg/kg 9,5% SIF 0,55-20,52 mg/kg 3,4% AFB_1 0,005-0,05 mg/kg	Pepeljnjak & Balzer, 1982; Pepeljnjak, 1983; Pepeljnjak & Cvetnić, 1985a
1978-80	EN	dried meat products (206)	12-29% OTA 0,01-0,92 mg/kg	Pepeljnjak & Blažević, 1982
1980/81	EN	cereals (126)	82% OTA 0,01-4,7 mg/kg	Pepeljnjak & Cvetnić, 1985a
1980-83	EN	haricot beans (1300)	8% OTA 0,017-0,053 mg/kg 2% ZEN 0,16 mg/kg 14% SIF 0,0005-0,01 mg/kg	Pepeljnjak, 1984
1984/85	non-EN	cereals (46)	23,9 % OTA 0,02-0,32 mg/kg 15,2 % ZEN 0,56-3,0 mg/kg 13% SIF 0,16-0,48 mg/kg	Pepeljnjak & Cvetnić, 1986
1991	EN	cereals and beans (111)	70,2 % DON 0,002-12,4 mg/kg 45 % NIV 0,006-0,4 mg/kg 43,2% ZEN 0,001-19,9 mg/kg	Pepeljnjak *et al.*, 1992
1992/95	non-EN	cereals, maize, beans (234)	10 % OTA 0,001-3,2 mg/kg 10 % ZEA 0,023-10,7 mg/kg 10 % SIF 0,028-25,5 mg/kg	Pepeljnjak *et al.*, 1999
1996/97	non-EN	maize (119)	16,8 % OTA 0,00026-0,2236 mg/kg 95 % FB_1+FB_2 0,012-7,956 mg/kg 7,5 % BEA 0,034-1,346 mg/kg	Jurjević *et al.*, 1999, 2002
1996/97	EN	maize (82)	29 % OTA 0,00029-0,6137 mg/kg 95 % FB_1+FB_2 0,012-9,903 mg/kg 11 % BEA 0,025-1,864 mg/kg	Jurjević *et al.*, 1999, 2002
1998-01	non-EN	maize (105)	8,5 % OTA 0,01-1,4 mg/kg 6,6 % ZEN 0,5-1,7 mg/kg 1,9 % SIF 0,4-1,2 mg/kg	Pepeljnjak *et al.*, 2002a
1998-01	EN	maize (31)	16,1% OTA 0,03-1,2 mg/kg 12,9 % ZEN 0,2-2,7 mg/kg 6,4 % SIF 0,6-1,3 mg/kg	Pepeljnjak *et al.*, 2002a
1999	non-EN	apples (100)	19 % CTN 0,05-0,24 mg/kg	Pepeljnjak *et al.*, 2002b

In the period from 1990 until 1999 ZEN was detected in 21,19% of various feedstuffs and 14,7% in mixed feed, in levels ranging from 0,05 to 1,2 mg/kg (Mitak *et al.*, 2001). Higher frequency of ZEN was observed in 1992 (32,6%) and in 1997 (41,3%), while in other years the frequency was lower than 20%. These results indicated a frequent contamination of animal feed with ZEN and trichotecens at lower rates, which could cause feeding and reproductive disorders in animals.

However, higher incidence of mycotoxins in stored crops and mixed feed during the wartime in Croatia is evident. In the last few years mycological and mycotoxin contamination of such substrates is decreasing probably due to dry environmental conditions and improvement of silage.

Table 2. Mycotoxins in maize and mixed animal feeds

Period	*Samples (N)*	*Mycotoxins/concentrations*	*Reference*
1978-87	mixed feed (90)	26,6 % T-2 0,18-0,96 mg/kg	Pepeljnjak, 1989
1984/85	mixed feed (32)	65,6% OTA 0,05-1,62 mg/kg 25% ZEN 0,1-1,2 mg/kg 9,3 T-2 0,08-0,2 mg/kg 9,3% AFB_1+AFB_2 0,02-1,12 mg/kg	Pepeljnjak & Cvetnić, 1986
1985	mixed feed (180+136+47)	10,5% OTA 0,12-0,18 mg/kg 29% ZEN 0,1-6,88 mg/kg 8,5% AFB_1	Nemanič *et al.*, 1986
1988-97	mixed feed (156)	27,8% T-2 $\bar{x}$ 0,1-0,5 mg/kg 30,8% DAS $\bar{x}$ 0,1-0,31 mg/kg 19,4% ZEN $\bar{x}$ 0,15-1,2 mg/kg	Pepeljnjak *et al.*, 1999
1988-97	maize (378)	15,3% T-2 $\bar{x}$ 0,1-0,5 mg/kg 29% DAS $\bar{x}$ 0,1-0,4 mg/kg 10% ZEN $\bar{x}$ 0,15-1,2 mg/kg	Pepeljnjak *et al.*, 1999
1990-99	mixed feed (459)	21,19% ZEN 0,05-1,2 mg/kg	Mitak *et al.*, 2001

5. TOXIGENIC FUNGI IN CROATIA

Over the last two decades toxigenic potential of fungi isolated from various substrates was examined under laboratory conditions on liquid (YES-media) and solid (wheat, 40% humidity) media. Isolated fungal strains were tested on production of OTA, AF-s, cyclopiazonic acid (CPA), sterigmatocystin (ST), citrinin (CTN), ZEN, DAS, T-2 and FB_1 (Table 3).

Since *Aspergillus* species were the most frequent contaminants of dried meat products and were also detected in cereals and beans in endemic and non-endemic areas, their ability to produce OTA and AF-s was tested on few occasions. Toxigenic potential of 130 *A. ochraceus* and 227 *A. flavus* and *A. parasiticus* strains, isolated mostly from dried meat products, was examined. OTA-producing ability was observed in 41,5% of *A. ochraceus* strains ranging from 0,07 to 240 mg/kg and/or mg/L. Maximum OTA-concentration produced strains isolated from dried meat products from endemic regions. Aflatoxin production was observed in 9,6% of strains from *flavi*-section (*A. flavus, A. parasiticus*), also isolated from dried meat products. *Aspergillus flavus* strains produced mainly AFB_1 with various levels of toxigenicity (0,09-240 mg/kg and/or mg/L). Among these isolates, some strains were able to produce both AFB_1 and AFG_1 (120-220 mg/L). Besides AF-s, CPA-production was observed in 5,2% of *A. flavus* strains (0,5-30 mg/kg) but none of them produced both CPA and AF-s. High potent AF-producers were *A. parasiticus* strains, which produced all four aflatoxins (2 strains), or AFB_1+AFG_1 (4 strains) in the range from 0,1 to 240 mg/kg (AFB_1+AFB_2) and from 0,8 to 450 mg/kg (AFG_1+AFG_2). Besides *A. flavus*, aflatoxigenic potency was observed in strains of *Rhizopus nigricans*, which highly contaminated stored herbal drugs. Among 64 strains 40% produced both AFB_1 and AFB_2 (0,1-1,33 mg/L). Among airborne aspergilla 2 out of 38 *A. flavus* strains tested were aflatoxigenic (0,3 mg/kg AFB_1) and 9 out of 33 *A. versicolor* strains produced ST (3,8-120 mg/kg). The occurrence of airborne *Fusarium* (7,5%) and *Trichoderma* (2,5%) was not very frequent but some isolates were found to be toxigenic. Three out of 14 *F. graminearum* strains produced ZEN (0,4-2,4 mg/kg) and T-2 (16 mg/kg), while one strain of *T. viridae* produced DAS (26,6 mg/kg).

Citrinin production ability of *Penicillium expansum* was also studied, because this species is the most common contaminant of apples. In our investigation we detected 38% of CTN-producer strains (0,07-9,00 mg/kg) and 19% CTN-positive apple samples.

Among fusaria *Fusarium vreticillioides* (=*moniliforme*), a principal producer of FB-s, is the most frequent contaminant of maize. All strains isolated from such samples were able to produce high amounts of FB_1 (181-1320 mg/L), which is in agreement with findings of high frequency of FB_1 in maize samples from both endemic and non-endemic areas. However, toxin production under laboratory conditions does not necessarily indicate that the fungal strain produced toxin in substrate from which it was isolated. Such investigations can contribute in relation to certain problems with fungal occurrences in specific regions.

6. MYCOTOXIN RESIDUES IN HUMAN AND ANIMAL TISSUE

6.1. Ochratoxin A in humans from endemic and non-endemic regions in Croatia

Endemic nephropathy (EN) is the kidney disease mainly characterised by tubular degeneration, interstitial fibrosis and hyalinisation of the glomeruli. This nephropathy is endemic in the Middle Posavina region, a rural area of the Republic Croatia located along the Sava river. Ochratoxin A is suspected to be one of the aetiological agents responsible for EN because the initial stage of the disease is characterised by regressive changes in the tubular epithelium morphologically similar to the changes seen in OTA-induced nephropathy in swine.

Surveys of foodstuffs (cereals, maize, beans, dried meat products) demonstrated that 8-30% of the samples were contaminated with OTA. Hult *et al* (1982) first reported that about 7% of human blood samples collected in the endemic region were positive for OTA, with a maximum level of 40 ng/ml. In the period from 1981 until 1989, OTA was detected in 1,4% of 13797 human blood samples collected in the EN in a range from 5 to 50 ng/ml, and OTA positive samples were found every year of this investigation period. At the same time blood samples were collected from non-EN villages and OTA was found in 2,5% of 1824 samples with the same range (5-50 ng/ml) but only in the period 1981-1985, while in the last 4 years of the study none of the 1554 samples contained OTA (Fuchs *et al.*, 1991).

A 10-year study (1985-1994) also conducted in a EN village (Kaniža) in the Middle Posavina, and non-EN areas revealed similar observations (Radić *et al.*, 1997). This village is the hyperendemic with highest incidence of EN (Figure 5). The occurrence of OTA in human sera from the EN village was 0,2 to 4,5% of 4343 samples in a range from 2 to 50 ng/ml, and 0,4 to 2,4% of 2566 samples from non-EN contained 2-10 ng OTA/ml sera.

Afterwards, exposure of the general population in Croatia to OTA was investigated in five cities: Split, Rijeka (Mediterranean cities), Varaždin, Osijek and Zagreb (continental cities) (Peraica *et al.*, 1999). OTA-positive blood samples were observed in all tested Croatian cities. Mean concentrations of OTA in plasma were 0,18 ng/ml (Rijeka), 0,25 ng/ml (Split), 0,26 ng/ml (Zagreb), 0,59 ng/ml (Varaždin) and 0,68 ng/ml (Osijek). The highest frequency of OTA-positive samples and highest number of samples with the concentration exceeding 1 ng/ml were found in Osijek. This difference is probably due to the different dietary habits of the people living at the coast (consuming seafood, beef, pasta and olive oil) and inlanders who consume more fresh and smoked pork and cereals. Peraica *et al* (1999) also calculated daily intake of OTA, estimated from the mean OTA concentration of all samples in each town, from 0,24 ng/kg b.w. (Rijeka) to 0,91 ng/kg b.w. (Osijek). Repeated investigation of OTA occurrence in human blood of healthy population in these cities revealed similar findings (Peraica *et al.*, 2001). The estimated daily intake of healthy population in Croatia is lower than the tolerable daily intake (16 ng/kg b.w.) proposed by World Health Organization. Considering higher exposure of population to OTA in EN region and their dietary habits, the incidence of EN could be related to OTA contamination of foodstuffs.

6.2. Mycotoxin residues in domestic animals

Survey of OTA in the organs of swine taken from EN and non-EN villages was conducted in 1979 and 1981 in connection with findings of OTA in maize and wheat samples. A total of 38 kidney, liver and blood samples from both regions were examined in 1979, and a total of 44 samples were examined in 1981. In the year 1979 OTA was found in 16,6% of the blood samples (36-77 μg/L), in 6,6% of kidney samples (26-76 μg/kg), and it was not detected in liver of animals from EN.

Region in Croatia	*Samples*	*Species*	*Toxinogenic strains/total*	*Mycotoxin/ range of concentration*	*Reference*
EN	cereals, beans, dried meat products	*A. ochraceus*	18/24	0,1-11,5 mg/kg OTA	Pepeljnjak & Cvetnić, 1981
EN and non-EN	dried meat products	*A. flavus*	5/56	0,09-240 mg/L AFB_1	Cvetnić & Pepeljnjak, 1986
			2/56	120-220 mg/L AFG_1	
		A. ochraceus	11/36	0,3-240 mg/L OTA	
EN and non-EN	cereals, beans, dried meat products	*A. ochraceus*	10/27	0,07-240 mg/L OTA	Cvetnić & Pepeljnjak, 1990
			15/43	0,22-27 mg/L OTA	
Continental parts of Croatia	airborne	*A. flavus*	1/2	0,32 mg/kg AFB_1	Cvetnić & Pepeljnjak 1991
		A. versicolor	3/15	4,0-56 mg/kg ST	
Different parts of Croatia	smoke-dried meat products	*A. parasiticus*	8/75	0,1-240 mg/kg AFB_1	Cvetnić & Pepeljnjak, 1995
			2/75	20-34,5 mg/kg AFB_2	
			6/75	08-450 mg/kg AFG_1	
			2/75	10,4-60 mg/kg AFG_2	
Different parts of Croatia	airborne	*A. flavus*	1/36	0,3 mg/kg AFB_1	Cvetnić & Pepeljnjak,1997
		A. versicolor	6/18	3,8-120 mg/kg ST	
		F. graminearum	2/14	0,4-2,4 mg/kg ZEN	
			1/14	16 mg/kg T-2	
		T. viridae	1/6	26,6 mg/kg DAS	
Zagreb	drug plants	*R. nigricans*	6/15	0,1-1,33 mg/L AFB_1+B_2	Pepeljnjak & Cvetnić, 1998
Differnet parts of Croatia	cereals, dried meat products	*A. flavus*	5/96	0,5-30 mg/kg CPA	Cvetnić & Pepeljnjak, 1998
			9/96	0,1-14,8 mg/kg AFB_1	
Continental Croatia	apples	*P. expansum*	11/29	0,07-9 mg/L CTN	Pepeljnjak *et al.*, 2002b
EN and non-EN	maize	*F. verticillioides*	66/66	181-1320 mg/L FB_1	Šegvić & Pepeljnjak, 2003

Table 3. Toxinogenic potential of fungi isolated from various substrates

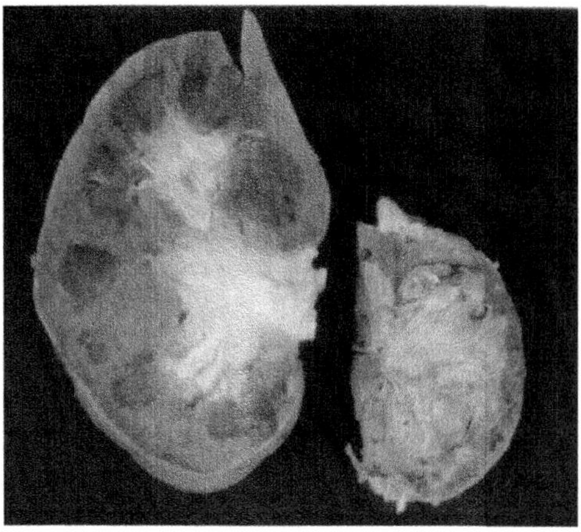

Figure 5. Human kidney of the EN patient (right) and normal kidney (left) (Ožegović & Pepeljnjak, 1995)

The finding of OTA in the kidneys of swine in EN was similar to that in 1979 (somewhat lower concentrations of 16 µg/kg), and it was also detected in 10,5% of liver samples (average of 21 µg/kg). OTA was not detected in any of the blood or organ samples taken from animals in the non-EN. Histopathological analysis of the organs observed disseminated mononuclear interstitial nephritis, degeneration of tubular epithelial cells and chronic interstitial hepatitis (Pepeljnjak *et al.*, 1982; Pepeljnjak & Cvetnić, 1985). Contrary to these findings, OTA was detected in organs and blood of swine (70 animals) taken from non-EN regions in 1984 (Pepeljnjak & Čuljak, 1986). The occurrence of OTA was detected in blood samples (28,5%) in a range of 5,3-53 µg/kg, in 12,8% of kidneys (40-114 µg/kg) and in 10% of liver samples (32-160 µg/kg). Histopathological examination of liver and kidney samples revealed observations similar to the previous study. In addition to these findings Pepeljnjak (1983) reported intestinal disorders in swine from EN regions in the period from 1976 to 1979. These disorders were associated with feed refusal, diarrhoea, apathy, skin erythema and bleeding crusts, and probably caused by trichotecens and ZEN (Figure 6).

Ramljak *et al* (1986) described intoxication of swine by zearalenol, which has been proved in the quantity of 4,25 mg/kg in the feedstuff. The animals have been refusing feed, vomited and showed signs of hyperoestogenism. Other investigators (Mitak *et al.*, 1995; Mitak & Cvetnić, 1997) reported sub-toxic concentrations of ZEN (100-250 ng/kg) in the blood samples of swine with significant reproductive disorders. Pseudoestrogenism in swine, as resulting from intoxication by ZEN in a concentration of 0,34 mg/kg was also observed.

Frequent contamination of feedstuffs with lower concentrations of mycotoxins mentioned in this review could lead to the development of secondary infections, reproductive and feeding disorders and liver and kidney damage with accumulation of mycotoxin residues in the animal body.

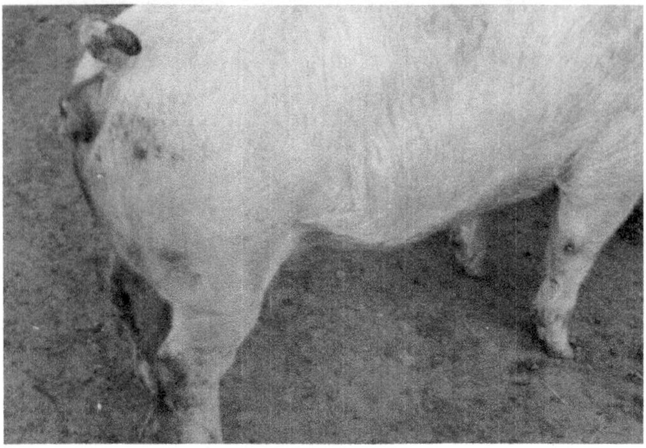

Figure 6. Zearalenone toxicosis in female swine - individual household in the EN-region (Pepeljnjak, 1983)

7. CONCLUSIONS

Over the past three decades field and stored fungal flora in EN and non-EN areas were qualitatively and quantitatively similar: dominant fungal species were *Fusarium*, *Alternaria* and *Cladosporium*. In stored maize the most frequent were *Penicillium*, *Rhizopus* and *Fusarium*, while in wheat, oat, barley and rye *Fusarium* and *Penicillium* prevalence was observed. These particular species were more frequent in samples collected in the EN region. *Fusarium* species were also major contaminants of animal feed.

Increased fungal contamination of maize, other cereals and animal feed was observed in the year 1978/79 (frequent rainfall and extremely high humidity) and during the wartime in Croatia (probably due to the war destruction, deserted storage-houses, crop fields and organic decay). Although, T-2 and DAS frequency was significantly higher in the animal feed samples during the wartime.

Screenings of dried meat products collected from individual households in the EN area revealed domination of *Penicillium* and *Aspergillus* species. *Aspergillus ochraceus* strains were found to produce significant amounts of OTA, especially strains isolated from samples collected in the EN area.

Airborne fungal flora is characterised by domination of allergologically important species of *Cladosporium*, *Alternaria* and *Penicillium*. Concentrations of these fungal entities were significantly higher in continental than in Mediterranean areas, which can be due to the environmental conditions and vegetation type of the particular region.

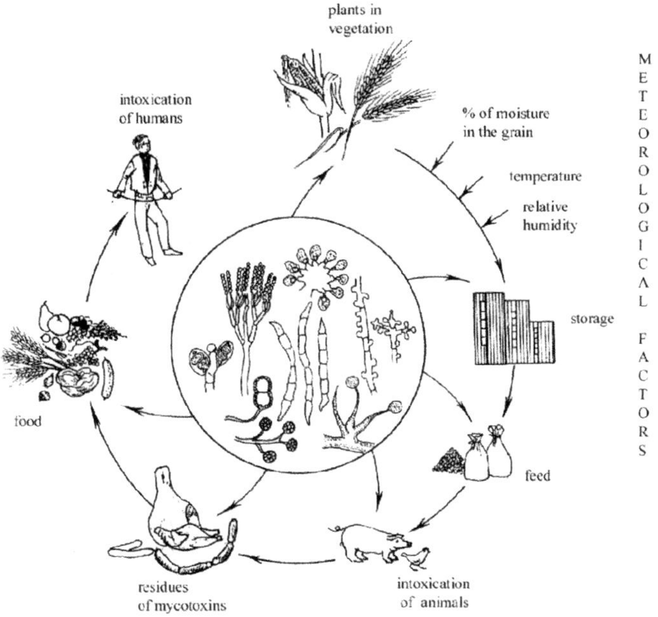

Figure 7. The chain of mycotoxicological contamination of humans and animals (Pepeljnjak & Cvetnić, 1985a)

Increased contamination of human foodstuffs with OTA, ZEN and SIF was observed in 1978 as a consequence of unfavourable environmental conditions (frequent rainfall and high humidity), and in the period from 1992 to 1995 during the wartime in Croatia. Co-occurrence of mycotoxins (OTA, FB-s, trichotecens, ZEN) and their possible synergism may contribute to the development of various diseases in humans including nephropathy and carcinogenesis (urinary tract tumours). In addition, frequent contamination of animal feed with mycotoxins at lower rates could cause feeding and reproduction disorders, and liver and kidney damage. Accumulation of mycotoxin residues in the animal body contributes to higher micotoxicological risk for humans (Figure 7) trough the possible development of chronic diseases and immunosuppression.

Considering higher frequencies of OTA positive human blood samples in EN region, higher exposure to this toxin and dietary habits of the population in this area, the incidence of endemic nephropathy could be related to OTA contamination of food. Possible synergistic action of OTA, FB-s, trichotecens and other mycotoxins can result in cell death (apoptosis) and consecutive organ dysfunction.

8. AFFILIATIONS

Stjepan Pepeljnjak, Prof. Dr. Sc.-Department of Microbiology, Faculty of Pharmacy and Biochemistry, University of Zagreb, Schrottova 39, HR-10000 Zagreb, Croatia; spep33@yahoo.com

Maja Šegvić, M. Sc. .- Department of Microbiology, Faculty of Pharmacy and Biochemistry, University of Zagreb, Schrottova 39, HR-10000 Zagreb, Croatia; msegvic@net.hr

9. REFERENCES

Balzer, I., Bogdanić, Č., Mužic, S., & Pepeljnjak, S. (1977). Factors of mycological contamination of corn, Krmiva XIX-5, 101-104.

Balzer, I., Bogdanić, Č., & Pepeljnjak, S. (1978). Rapid thin layer chromatographic method for determining aflatoxin B_1, ochratoxin A, and zearalenone in corn, Journal of the Association of Analytical Chemists, 3, 584-585.

Bogdanić, Č., Mužic, S., Pepeljnjak, S., & Balzer, I. (1976) Rat and carp skin-irritatin factors in contaminated corn, Veterinaria, 25, 91-95.

Cvetnić, Z., & Pepeljnjak, S. (1986). Distribution and finding of toxigenic kinds of mould in smoked meat products, Veterinarski Arhiv, 56, 75-82.

Cvetnić, Z., & Pepeljnjak, S. (1990). Ochratoxinogenicity of *Aspergillus ochraceus* strains from nephropathic and non-nephropatic areas in Yugoslavia, Mycopathologia, 110, 93-99.

Cvetnić, Z., & Pepeljnjak, S. (1991). Distribution and toxinogenicity of *Aspergillus* species isolated from the air, Prehrambeno-Tehnološka i Biotehnološka Revija, 29, 57-60.

Cvetnić, Z., & Pepeljnjak, S. (1995). Aflatoxin producing potential of *Aspergillus flavus* and *Aspergillus parasiticus* isolated from samples of smoked driet-meat, Die Nahrung, 39, 302-307.

Cvetnić, Z, & Pepeljnjak, S. (1997). Distribution and mycotoxin producing ability of some fungal isolates from the air, Atmospheric Environment, 31, 491-495.

Cvetnić, Z, & Pepeljnjak, S. (1998). Production of cyclopiazonic acid aflatoxinogenic and non aflatoxinogenic strains of Aspergilus flavus, Die Nahrung, 42, 321-323.

Cvetnić, Z, & Pepeljnjak, S. (1999). Mycological contamination of stored herbal drugs, Acta Pharmaceutica, 49, 201-209.

Čuturić, S., Pepeljnjak, S., & Cvetnić, Z. (1979). The prevalence of Mycromyces on cereals and other crops in the vegetation growing in a region of endemic nephropathy in Middle Posavina, Liječnički Vjesnik, 101, 525-530.

Eppley, R. (1968). Screening method for zearalenone, aflatoxin and ochratoxin, Journal of the Association of Analytical Chemists, 51, 74-78.

Fuchs, R., Radić, B., Čeović, S., Šoštarić, B., & Hult, K. (1991). Human exposure to ochratoxin A In: Castegnaro, M., Pleština, R., Dirheimer, G., Chernozemsky, I., N., & Bartsch, H. (Editors). Mycotoxins: endemic nephropathy and urinary tract tumours. No. 115 (Lyon: IARC), 131-135.

Horner, W., E., Helbling, A., Salvaggio, J.,E., & Lehrer, S.,B. (1995). Fungal allergens, Clinical Microbiology Reviews, 8, 161-178.

Hult, K., Pleština, R., Habazin-Novak, V., Radić, B., & Čeović, S. (1982). Ochratoxin A in human blood and Balcan endemic nephropathy, Archives of Toxicology, 51, 313-321.

Jurjević, Ž., Solfrizzo, M., Cvjetković, B., Avantaggiato, G., & Visconti, A. (1999). Mycotoxin Research, 15, 67-80.

Jurjević, Ž., Solfrizzo, M., Cvjetković, B., De Girolamo, A., & Visconti, A. (2002). Occurrence of beauvericin in corn from Croatia, Food Technology and Biotechnology, 40, 91-94.

Mitak, M., & Cvetnić, Ž. (1997). Estrogenism in swine as a consequence of intoxication with zearalenon, Veterinarska Stanica, 28, 213-217.
Mitak, M., Topolko, S., & Cvetnić, Ž. (1995). Determining zearalenone in guilt sera by Immunoenzymic test (ELISA), Veterinarska Stanica, 26, 11-16.
Mitak, M., Gojmerac, T., Pavičić, P., & Topolko, S. (2001). Contamination of feedstuffs and animal feed mixtures for swine with zearalenon in the period from 1990 till 1999, Veterinarska Stanica, 32, 205-209.
Mužic., S., Pepeljnjak, S., Bogdanić, Č., & Balzer, I. (1976). Occurrence of ochratoxin A in corn, Veterinarski Arhiv, 46, 69-73.
Nemanič, A., Brlek, V., Ramljak, D., & Matešić, D. (1986). Incidence of mycotoxins in feeds and feed mashes for poultry and other domestic animals. In: Ožegović, L. (Editor) Special publications: second symposium on mycotoxins. Book LXXX, Academy of sciences and arts of Bosnia and Hercegovina, Sarajevo, 51-57.
Ožegović, L., & Pepeljnjak, S. (1995). Mycotoxicosis. Školska knjiga, Zagreb.
Pepeljnjak, S., (1983). Trichothecene Problems in Yugoslavia. In: Ueno, Y., (Editor) Trichothecenes-Chemical, Biological and Toxicological Aspects. Tokyo, Elsevier Amsterdam-Oxford-New York, 265-272.
Pepeljnjak, S., (1984). Mycotoxic Contamination of haricot beans in nephropathic areas in Yugoslavia. Microbiologie Aliments Nutrition, 2, 331-336.
Pepeljnjak, S., (1989) Frequency and resistence of T-2 under natural conditions. In: Ožegović, L., (Editor) Special publications: third symposium on mycotoxins. Book LXXXIX, Academy of sciences and arts of Bosnia and Hercegovina, Sarajevo, 47-54.
Pepeljnjak, S., & Balzer, I. (1982). A survey of mycological and mycotoxicological research on seeds in the nephropathic and anephropathic areas of Yugoslavia (Croatia). In: Ožegović, L., (Editor) Special publications: symposium on mycotoxins. Book LX, Academy of sciences and arts of Bosnia and Hercegovina, Sarajevo 75-80.
Pepeljnjak, S., & Blažević, N. (1982). Contamination with moulds and occurance of ochratoxin A in smoked meat products from endemic nephropathy region of Yugoslavia. In: Pfannhauser, W., & Czedick-Eysenberg, P., B. (Editors) Proceedings of Vth International IUPAC Symposium on Mycotoxins and Phycotoxins, 1-3 September, Vienna, 102-105.
Pepeljnjak, S., & Cvetnić, Z. (1985a). The mycotoxicological chain and contamination of food by ochratoxin A in the nephropathic and non-nephropathic areas in Yugoslavia, Mycopathologia, 90, 147-152.
Pepeljnjak, S., & Cvetnić, Z. (1985b). Moulds of plants in vegetation in nephropathic and non-nephropathic area in SR Croatia, Mikrobiologija, 22, 51-57.
Pepeljnjak, S., & Cvetnić, Z. (1986). Mycological and mycotoxicological contamination of grains in a wide anephropathic area of SR Croatia. In: Ožegović, L. (Editor) Special publications: second symposium on mycotoxins. Book LXXX, Academy of sciences and arts of Bosnia and Hercegovina, Sarajevo, 29-41.
Pepeljnjak, S., & Cvetnić, Z. (1998). Aflatoxigenicity of Rhizopus nigricans strains isolated from drug plants, Acta Pharmaceutica, 48, 139-144.
Pepeljnjak, S., & Čuturić, S. (1978) Review on the investigation of mycotoxins, Veterinarski Arhiv, 48, 33-35.
Pepeljnjak, S., & Čuljak, K., (1986). Residues of ochratoxin A in organs of pigs in wider anephropathic area of SR Croatia. In: Ožegović, L. (Editor) Special publications: second symposium on mycotoxins. Book LXXX, Academy of sciences and arts of Bosnia and Hercegovina, Sarajevo, 71-76.
Pepeljnjak, S., & Šegvić, M. (2003). Occurrence of fungi in air and on plants in vegetation of different climatic regions in Croatia, Aerobiologia, 19, 11-19.
Pepeljnjak, S., Čuturić, S., Topolko, S., & Munk, M. (1979). Factors of mycotoxin conatmination of corn, Krmiva XX-5, 109-112.
Pepeljnjak, S., Blažević, N., & Čuljak, K. (1982). Histopathologic changes and findings of ochratoxins a in organs of pigs in the area of endemic nephropathy in Yugoslavia. In: Pfannhauser, W., & Czedick-Eysenberg, P., B. (Editors) Proceedings of Vth International IUPAC Symposium on Mycotoxins and Phycotoxins, 1-3 September, Vienna, 346-348.
Pepeljnjak, S., Ueno, Y., & Tanaka, T. (1992). Mycological situation and occurrence of fusariotoxins in cereals from endemic nephropathy region in Croatia, Microbiologie Aliments Nutrition, 10, 191-197.

Pepeljnjak, S., Cvetnić, Z., & Brlek, V. (1999). Storage fungi and mycotoxins in Croatian storages. In: Korunić, Z. (Editor) Protection of stored agricultural products. Korunić Press, Zagreb, 51-64.

Pepeljnjak, S., Šegvić, M., & Cvetnić, Z. (2002a). Mycotoxicological contamination of stored agricultural products in Croatia (30-year review). In: Korunić, Z. (Editor) Disinfection, disinfestation, deratization and protection of stored agricultural products, Korunić Press, Zagreb, 97-110.

Pepeljnjak, S., Šegvić, M., & Ožegović, L. (2002b). Citrininotoxinogenicity of *Penicillium* spp. isolated from decaying apples, Brazilian Journal of Microbiology, 33, 134-137.

Pepeljnjak, S., Kosalec, I., Kalođera, Z., & Kuštrak D. (2003). Natural antimycotics from Croatian plants In: Rai, M.,K., & Mares, D., (Editors) Plant Derived Antimycotics, Binghamton, The Haworth Press, 49-73.

Peraica, M., Domjan, A., M., Fuchs, R., Lucić, A., & Radić, B. (1999). The occurrence of ochratoxin A in blood in general population of Croatia, Toxicology Letters, 110, 105-112.

Peraica, M., Domjan, A., M., Matašin, M., Lucić, A., Radić, B., Delaš, F., Horvat, M., Bosanac, I., Balija, M., & Grgičević, D. (2001). Variations of ochratoxin A concentration in the blood of healthy populations in some Croatian cities, Archives of Toxicology, 75, 410-414.

Radić, B., Fuchs, R., Peraica, M., & Lucić, A. (1997). Ochratoxin A in human sera in the area with endemic nephropathy in Croatia, Toxicology Letters, 91, 105-109.

Ramljak, D., Matešić, D., & Uremović, M. (1986). Intoxication of swine by zearalenol in Yugoslavia. Residues of ochratoxin A in organs of pigs in wider anephropathic area of SR Croatia. In: Ožegović, L. (Editor) Special publications: second symposium on mycotoxins. Book LXXX, Academy of sciences and arts of Bosnia and Hercegovina, Sarajevo, 195.

Rumora, L., Kovačić, S., Rozgaj, R., Čepelak, I., Pepeljnjak, S., & Žanić-Grubišić, T. (2002). Cytotoxic and genotoxic effects of fumonisin B_1 in rabbit kidney RK13 cell line, Archives of Toxicology, 76, 55-61.

Šegvić, M., & Pepeljnjak, S. (2003). Distribution and fumonisin B_1 production capacity of *Fusarium moniliforme* isolated from corn in Croatia, Periodicum Biologorum, 105, 275-279.

Tanaka, T., Hasegawa, A., Yamamoto, S., Lee, U., S., Sugiura, Y., & Ueno, Y. (1988). Worldwide contamination of cereals by *Fusarium* mycotoxins, nivalenol, deoxynivalenol and zearalenone-Survey of 19 countries, Journal of Agriculture and Food Chemistry, 5, 979-982.

E. IOANNOU – KAKOURI, M. ALETRARI, E. CHRISTOU,
A. RALLI, A. KOLIOU & M. CHRISTOFIDOU

OCCURRENCE AND CONTROL OF MYCOTOXINS IN FOODSTUFFS IN CYPRUS

1. INTRODUCTION

The Competent authority in Cyprus for the official control of foodstuffs is the Ministry of Health with its two Departments, the State General Laboratory (SGL) and the Medical and Public Health Services (MPHs). The SGL is the official control Laboratory, for surveillance, monitoring and applied research in the field of food, water, environment, drugs, consumer products and forensic chemistry, and is accredited since 2002. The first draft of Cyprus Legislation for mycotoxins in foodstuffs was enacted in 1986 and now almost all the national food legislation has been amended in order to comply with the relevant EU legislation. The full harmonization to the EU legislation for mycotoxins will be finalized by the end of 2003.

2. SURVEILLANCE AND CONTROL

The occurrence (incidence and level) of Aflatoxins (AFs) B_1, B_2, G_1, G_2, M_1, Ochratoxin A (OTA) and other mycotoxins in locally produced and imported foodstuffs (nuts, cereals, pulses, dried fruit, seeds, etc) were monitored and controlled systematically and effectively, since 1990 in Cyprus, especially for aflatoxins. Results of AFs and OTA for the years 1992–1996 have been published (1,2) or presented (3) previously. In this report, we present the results of both local and imported foodstuffs for the years 1997-2003 for AFs, 1997-2000 for OTA and preliminary results for the years 1999 for Fumonisins B_1 and B_2 (FB_1 and FB_2) and 2002 for Zearolenone and Deoxynivelenone (ZON and DON).

The samples were collected according to a specific sampling plan (based on the EU sampling plan for Aflatoxins) at critical control points (Hazard Analysis Critical Control Point Approach, HACCP approach (1)), such as import, primary storage, processing and market.

The objectives of the study and control were: (i) To investigate the occurrence of AFs and OTA in foodstuffs (ii) to prevent unfit products (above MLs) entering the Cyprus market and to limit, as far as possible, the population exposure, (iii) to protect public health and fair trade and (iv) to validate the immunoaffinity column (IAC) approach (Aflatest, IAC/HPTLC (1) and HPLC (4)) for AFs B1, B2, G1, G2

A. Logrieco and A. Visconti (eds.),
An Overview on Toxigenic Fungi and Mycotoxins in Europe, 51-65.

and the IAC/HPLC for AFM1 (2), OTA (5,6), FB1 and FB2 (7), ZON and DON (8) as appropriate methodology for official and regulatory purposes (9).

The evaluation of the results was carried out according to the Cyprus and EU maximum level (MLs) for AFs (10), to the relevant MLs for OTA adopted by EU (11) (e.g. ML: 5 μg/kg for cereals) and to proposed EU limits for AFs and OTA for other products (12) such as food for infants and young children, coffee, wine, grape juice, dried vinefruits and other Mycotoxins such as *Fusarium* toxins: Trichothecenes, [Deoxynivelenone (DON), T-2 and HT-2 toxins], FB1, FB2 and ZON for cereals and their products (13) .

3. RESULTS AND DISCUSSION

3.1 Aflatoxins (AFs) B_1, B_2, G_1, G_2

Measurable levels of AFs, sometimes above the MLs, were found for the years 1992-1996[(1)] and for the years 1997-2003 in nuts, cereals and other products local and imported (Tables 1 – 4). In Table 1, results for peanuts, pistachios and other crops are presented, indicating that the peanuts and pistachios are the most sensitive to aflatoxin contamination, compared to other nuts, cereals, coffee and cocoa products, dried fruit, pulses, spices, tea, seeds and their products. The highest incidence of aflatoxin contamination observed, was in imported pistachios (10%) for the last years 2002 – 2003 (Table 2) and the highest level was 785 μg/kg total AFs in peanuts for the year 1997 (Table 1).

The lots of contaminated nuts (imported and local) with levels of AFs above MLs were destructed by the Medical and Public Health Services (sampling Authority of Cyprus) or re-exported to the country of origin. These lots were detected before entering the market, due to HACCP approach preventive control applied at critical control points (1) (Figures 1 – 2).

In contrast to the peanuts and pistachios and most recently Brazil nuts, other nuts (cashew nuts, walnuts, hazelnuts, coconut) cereals dried fruit, spices and other products were almost negative to AFs (Figure 3). Some samples of almond and figs were also found to be positive, but their levels were within the relevant MLs (Tables 1 - 4).

Table 1. Incidence and levels of aflatoxins in foodstuffs in Cyprus

Sample	*No. of samples*	*No. of[a] positives*	*Total Aflatoxins*	*Range (avg of positives)[a], µg/kg* B_1	B_2	G_1	G_2
			1997				
Almond	102	-	-	-	-	-	-
Brazil nuts	10	-	-	-	-	-	-
Cashew nuts	59	-	-	-	-	-	-
Chestnut	64	-	-	-	-	-	-
Peanuts	717	31	0.4–785 (86.7)	0.2–769 (67.6)	0.02–20.6 (3.9)	0.15–130 (16.5)	0.24–26.0 (3.8)
Pistachio	349	91	0.9-695 (107)	0.7-691 (104.1)	0.02-15.0 (4.34)	0.2-2.0 (1.9)	2.0-2.0 (2.0)
Walnut	120	-	-	-	-	-	-
Hazelnut	85	-	-	-	-	-	-
Corn	55	-	-	-	-	-	-
Wheat	27	-	-	-	-	-	-
Raisins	6	-	-	-	-	-	-
Dried figs	10	-	-	-	-	-	-
Beans	5	-	-	-	-	-	-
Total	1609	122					
			1998				
Almond	109	-	-	-	-	-	-
Cashew nuts	81	-	-	-	-	-	-
Chestnut	46	-	-	-	-	-	-
Peanuts	598	36	2.1–530 (93.5)	1.53–334 (27.7)	0.31–9.0 (3.3)	0.52–136 (27.5)	0.4–8.0 (3.5)
Pistachio	273	57	9.5-196 (132)	8.0-184 (126)	0.85-16 (6.8)	2.0-64.0 (51.6)	-
Walnut	165	2	1.0-19.0 (3.57)	-	-	-	-
Hazelnut	72	2	1.0-4.0 (1.2)	-	-	-	-
Corn	65	-	-	-	-	-	-
Wheat	3	-	-	-	-	-	-
Raisins	9	-	-	-	-	-	-
Dried figs	12	1	5.3–5.3 (5.3)	1.0–1.0 (1.0)	1.4–1.4 (1.4)	2.9–2.9 (2.9)	-
Beans	5	-	-	-	-	-	-
Herbs (oregano, mint)	7	-	-	-	-	-	-
Total	1455	98					

[a] Positive samples were those with detectable total aflatoxines > 0.1µg/kg by IAC/HPTLC method (1).

Table 2: Incidence and levels of aflatoxins in nuts and nut products in Cyprus, 1999 - 2003

Samples	*No of Samples*	*No of Positives*[a]	*Range (aver. of positives) μg/kg*			
			B1	*B2*	*G1*	*G2*
Brazil nuts	7	2	0.5-1.4 (69.2)	1.4-1.4 (1.4)	126-126 (126)	0.8-1.3 (1.1)
Cashew nut	107	---	---	---	---	---
Almond	226	9	0.6-27.9 (8.7)	0.3-4.1 (1.52)	0.9-21.6 (6.36)	0.3-1.7 (1.03)
Coconut	6	---	---	---	---	---
Walnut	109	---	---	---	---	---
Chestnut	11	1	(2.0-2.0) 2.0	0.6-0.6	---	---
Hazelnut	89	2	1.9-8.8 (5.3)	---	---	3.2-3.2 (3.2)
Peanut	1363	43	0.7-288.3 (42.9)	0.05-40 (5.1)	0.4-390.4 (50.7)	0.2-5 (6.4)
Peanut butter	22	7	5.0-5.0 (5.0)	0.8-0.8 (0.8)	1.4-1.4 (1.4)	1.3-1.3 (1.3)
Pistachio	373	45	0.6-170.6 (117.31)	0.15-360 (12.7)	0.3 – 29.0 (5.4)	0.7-5.5 (2.4)
Nut products	33	---	---	---	---	---
Total	**2346**	**115**				

[a]: Positive samples were those with detectable aflatoxins > 0.2 μg/kg with the IAC/AOAC 991.31 (2000) method (4)

Table 3: Aflatoxins monitoring programme 1997-2003 in Cyprus samples above limits at critical control points market-warehouse-import

	Total No. of analysed samples	Total No. of samples above limit (%)	Market Samples		Warehouse Samples		Imported Samples	
			Analysed Samples	Samples above limit (%)	Analysed Samples	Samples above limit (%)	Analysed Samples	Samples above limit (%)
2003	196	17 (8.7)	93	2 (1.1)	19	---	84	15 (17.9)
2002	272	7 (2.6)	68	---	51	---	153	7 (4.6)
2001*	523	23 (4.4)	251	4 (1.6)	65	9 (13.8)	207	10 (4.8)
2000	1538	31 (2.0)	101	3 (0.20)	378	4 (0.26)	1059	24 (1.56)
1999	1935	42 (2.1)	313	3 (0.16)	532	10 (0.52)	1074	29 (1.50)
1998	1763	86 (4.9)	225	3 (0.17)	430	12 (0.68)	1038	71 (4.03)
1997	2277	189 (8.3)	166	25 (1.10)	710	48 (2.11)	1360	116 (5.09)

* During the year 2001 the method of numbering of the samples has changed towards harmonization to EU sampling method requirements (9), so the total number of analysed samples for the years 2001 – 2003 seems to be less than the previous years.

Table 4: Incidence and level of Aflatoxins in non-dairy foodstuffs, other than nut and nut products in Cyprus, 1999-2003

Samples	No of samples	No of positives[a]	Range (av. of positives) μg/kg			
			B_1	B_2	G_1	G_2
Cereals & Cereal products						
Corn	52	-	-	-	-	-
Barley	14	-	-	-	-	-
Wheat	30	-	-	-	-	-
Malt	4	-	-	-	-	-
Breakfast cereals & others	12	-	-	-	-	-
Coffee cocoa & products						
Coffee beans	47	15	-	-	-	-
Cocoa products	21	2	-	-	-	-
Dried fruit & products						
Dried fruits	57	5	-	-	-	-
Raisins	16	6	-	-	-	-
Figs & figpie	5	1	-	-	-	-
Dates	7	-	-	-	-	-
Pulses						
Bean	1	-	-	-	-	-
Broad bean	3	-	-	-	-	-
Chick pea	6	-	-	-	-	-
Seeds & products						
Sun flower & pine seeds	1	-	-	-	-	-
Pumpkin seed	11	-	-	-	-	-
Sesame products	71	1	-	-	-	-
Sesame	15	2	-	-	-	-
Pinolia	11	-	-	-	-	-
Roasted chicken	14	-	-	-	-	-
Spices & tea						
Oregano	1	-	-	-	-	-
Pepper & others	1	-	-	-	-	-
Curry	4	2	3.4-5 (4.20)	0.76-1 (0.88)	-	-
Total	407	34				

[a] Positive samples were considered those which had detectable Aflatoxins>0.1 μg/kg (1,4).

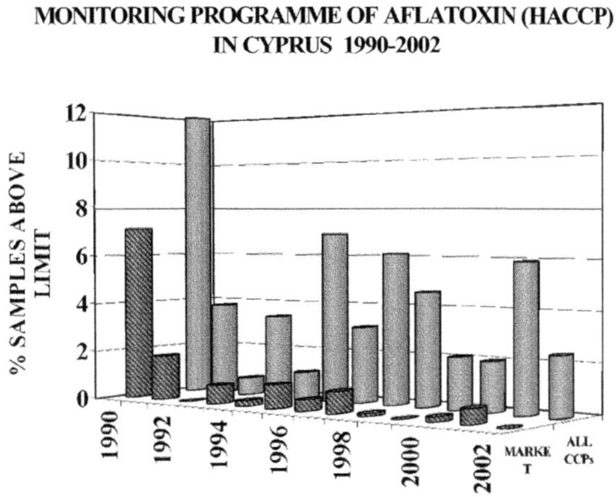

Figure 1: Aflatoxins control for the years 1990-2002 and percentage of violative samples taken at several critical control points (CCPs) such as import, primary storage and market. The samples include several nuts, dried fruit cereals and their products and pulses.

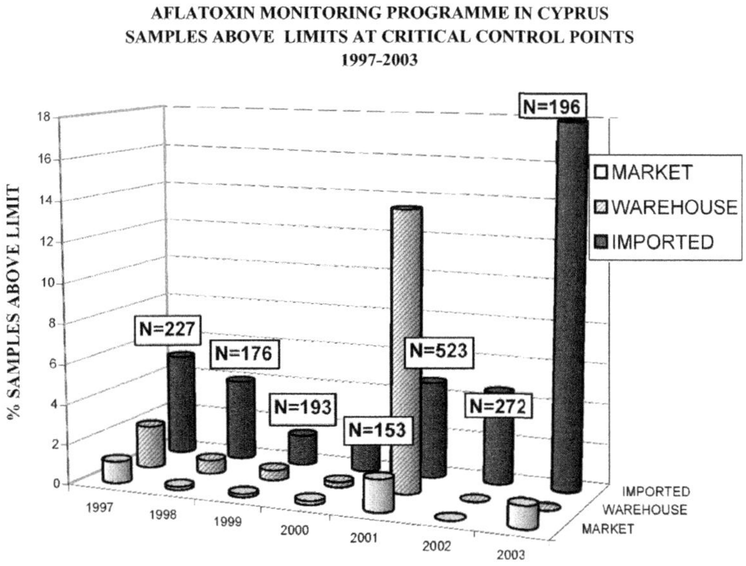

Figure 2: Aflatoxins control for the years 1997-2003. Samples are taken at several critical control points: import, primary storage in warehouses and market. In the market almost no contaminated samples were found.

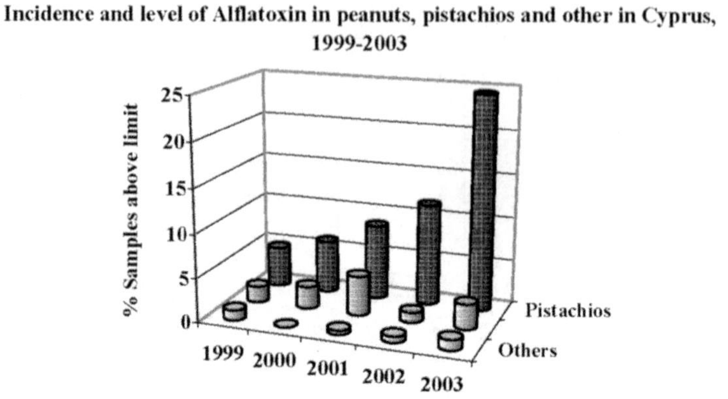

Figure 3: Incidence and control of aflatoxins B_1, B_2, G_1, G_2, for the years 1999-2003 for peanuts, other nuts and nut products

3.2 Aflatoxin M_1

The AFM_1 levels in local raw milk (cow, goat, sheep) and pasteurized milk (cow) were not only below the Cyprus ML (0.5 μg/L for M_1) but also well below the lower ML (0.05 μg/L) adopted by the EU legislation (10) (Table 5) for all of the years 1995-2003. This is a justification of the safety status of the used feeding stuffs in Cyprus at least from the concept of their Aflatoxins status.

A higher number of positive milk samples was observed in raw compared to pasteurized milk. Not one positive sample was observed in imported baby and evaporated milk and from this concept, they conform to the new proposed EU limits (12a). The percentage of positive milk samples is shown in Figure 4.

The control must be expanded to cover all the types of baby food which contain milk, the milk powders used in dairy and other food industries and dairy products such as local and imported cheeses, yogurts and whey cheeses.

Table 5: Incidence and levels of aflatoxin M1 in milk in Cyprus, 1995 - 2003

Sample	*No. of samples*	*No. of positives*[a]	*Range (avg of positives), μg/L*
Milk raw (cow, goat, sheep)[b]	198	10	0.003-0.04 (0.025)
Pasteurized milk, full[b] (cow)	48	10	0.002[c]-0.02 (0.011)
Pasteurized milk, light[b]	10	4	0.005-0.01 (0.008)
Pasteurized milk, skimmed[b]	14	6	0.01-0.04 (0.016)
Baby milk (imported)	15	-	-
Evaporated milk (imported)	4	-	-
Total	289	30	-

[a] Positive samples were those with AF M1≥0.002μg/L, by the IAC/HPLC method(2) and AOAC 2000.08 method (first action)

[b] Data were corrected for recovery

[c] Limit of detection

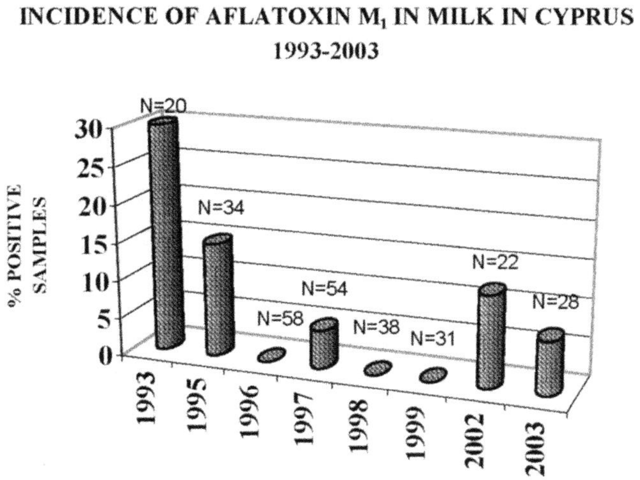

Figure 4: Incidence and control of aflatoxin M_1 in milk for the years 1993-2003

3.3 Ochratoxin A

All analyzed samples of locally produced and imported cereals (corn, wheat flour, barley etc) and wine (red and white) contained < 1μg/kg OTA, at least for the years 1997-2002 and were within the limit of 5 μg/kg OTA for cereals, the limit of 3 μg/kg OTA for cereal products, adopted by EU (11) and within the proposed E.U. limit (12b) of 2 μg/kg OTA for wines (Tables 6 - 7).

More specifically the average percentage of the positive samples (>0.2 μg/kg OTA) for the cereals and cereal products was 20% for the years 1997–2000 (Figure 5) and the levels of OTA ranged between 0.3 - 1.0 μg/kg OTA.

The levels of OTA in white and red wines (local and imported from Greece, France and Italy) were lower than that of cereals and they ranged between 0.01 to 0.8 μg/kg OTA (average 0.13 μg/kg) for white wines and between 0.03 to 1.0 μg/kg OTA (average 0.30 μg/kg) for red wines (Table 7 and Figure 6).

The levels of OTA in imported (green coffee) ranged between: 0.6 to 5.0 μg/kg and were within the proposed EU limit (12b). Generally, OTA levels in the examined foodstuffs are within the levels found in several E.U. countries (12c). Further monitoring and control is needed for these products as well as for dried vine fruit, grape juice, soluble coffee and beer.

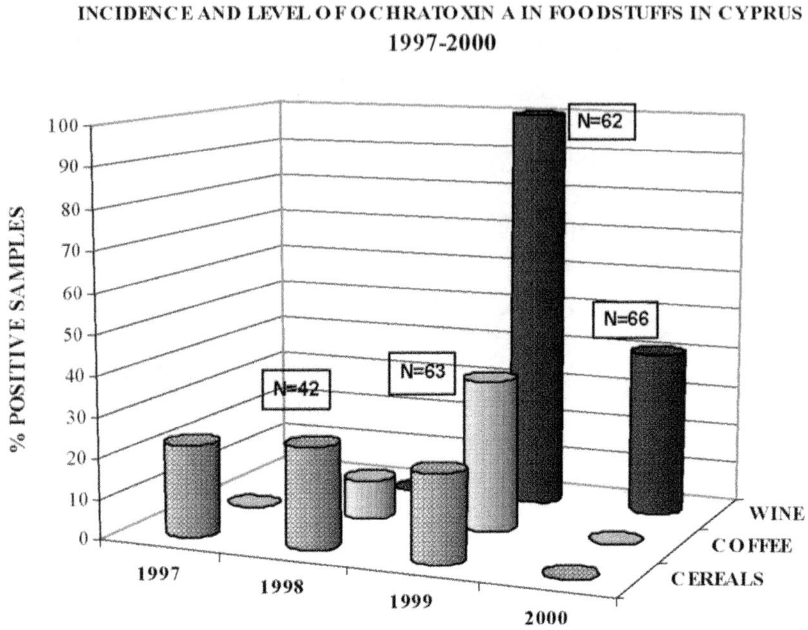

Figure 5: Occurrence of ochratoxin A in several foodstuffs, wine, coffee and cereals (Table 6)

3.4 Fumonisins (FB_1, FB_2) Zearalenone (ZON), Deoxynivalenol (DON).

During 1999 and 2002 some samples of cereals were analyzed for FB_1 and FB_2, ZON and DON and the results are presented in Table 8. All analyzed samples were within the relevant proposed EU limits (13). Positive samples were found (Table 8) and only one sample of imported maize was above proposed limits for FB_1 and FB_2 (ML: 100 μg/kg). Further investigation and monitoring is needed for cereals and cereal products for these mycotoxins, especially in the concept of the proposed new E.U. legislation (13).

Table 6: Incidence and levels of ochratoxin a in foodstuffs in Cyprus 1997 – 2000

SAMPLE	1997			1998			1999			2000		
	No. of samples	*No. of positives*[a]	*Range (avg of pos) μg/kg*	*No. of samples*	*No. of positives*[a]	*Range (avg of pos) μg/kg*	*No. of samples*	*No. of positives*[a]	*Range (avg of pos) μg/kg*	*No. of samples*	*No. of positives*[a]	*Range (avg of pos) μg/kg*
Wheat flour[a]	7	2	0.5-0.3 (0.4)	9	2	0.3-0.4 (0.3)	22	7	0.3-0.5 (0.4)	---	---	---
Corn[a]	7	0	---	9	3	0.3	5	0	---	---	---	---
Barley[a]	8	3	0.4-0.5 (0.4)	11	3	0.8-0.9 (0.8)	7	1	0.3	---	---	---
Biscuits/rusks[a]	9	2	0.8-1.0 (0.9)	3	0	---	3	0	---	---	---	---
Green coffee[a]	11	3	0.6-5.0 (2.8)	28	3	0.3	4	1	<0.2	---	---	---
Roasted coffee[a]	0	0	---	3	0	---	4	2	0.5-0.6	---	---	---
White wine[b]	---	---	---	---	---	---	10[c]	8	0.01-0.34 (0.12)	27[d]	10	0.02-0.8 (0.15)
Red wine[b]	---	---	---	---	---	---	7[c]	6	0.03-1.0 (0.2)	39[d]	17	0.01-0.2 (0.04)

a Positive samples were those with detectable OTA >0.2μg/kg by IAC/HPLC method(5)
b Positive samples were those with detectable OTA >0.01μg/kg by IAC/HPLC method(6)
c Local samples
d Local and Imported (France, Italy, Greece)

Table 7: Incidence and levels of ochratoxin A in wines in Cyprus, 1999 - 2000

		Import wines		*Local wines*	
Year	*Wines*	*Total no. of analyzed samples*	*Total no. of positive samples*	*Total no. of analyzed samples*	*Total no. of positive samples*
1999	Red	---	---	7	6 (85.7)
	White	---	---	10	8 (80)
	Total	---	---	17	14
2000	Red	11	3 (27.3)	28	14 (50)
	White	2	1 (50)	25	8 (32)
	Total	13	4 (30.7)	53	22 (41.5)

Positive samples were those with detectable OTA >0.01 µg/kg by IAC/HPLC method (6)

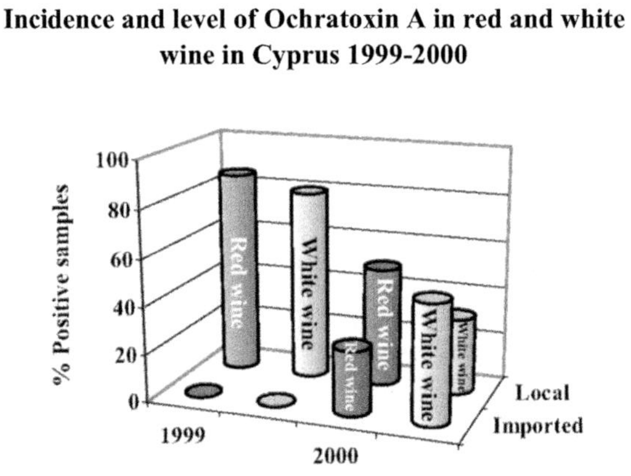

Figure 6: Occurrence of Ochratoxin A in local and imported red and white wines.

Table 8: Incidence and levels of deoxynivalenol, zearalenone & fumonisins (FB1 & FB2) deoxynivalenol – 2002 in Cyprus

Sample	*No. of samples*	*No. of positives*[a]	*Range (avg of positives), μg/Kg*
Barley	3	---	---
Wheat	3	---	---
Breakfast cereals	5	2	4.99 – 10.98 (7.98)
Total	11	2	

[a] : Positive samples were those with detectable DON > 3 μg/Kg analyzed by the IAC/HPLC method (8b)

Zearalenone – 2002 in Cyprus

Sample	*No. of samples*	*No. of positives*[b]	*Range (avg of positives), μg/Kg*
Barley	4	2	2.0 – 29 (11)
Wheat	4	3	2.0 – 3.1 (2.7)
Corn	10	10	10.8 – 89.8 (25.8)
Breakfast cereals	5	3	1.8 – 3.2 (2.66)
Total	23	18	

[b] : Positive samples were those with detectable ZON > 1 μg/Kg analyzed by the IAC/HPLC method (8a)

Fumonisins FB_1 & FB_2 - 1999 in Cyprus

Sample	*No. of samples*	*No. of positives*[c]	*Range (avg of positives), μg/Kg*
Corn	10	10	0.32 – 113 (85)
Total	10	10	

[c] : Positive samples were those with detectable FB_1 > 2 μg/Kg & FB_2 > 4 μg/Kg analysed by the IAC/HPLC method (7).

4. CONCLUSIONS

In general, the results for AFs and OTA occurrence are lower or comparable with those of other countries, e.g. E.U. countries (12c) United Kingdom (14), U.S.A. (15) , Greece (16) and Portugal (17), and indicate the effectiveness of the HACCP approach control applied, by preventing unfit products entering the Cyprus market and limiting the exposure of the population.

For all types of mycotoxins, the immunoaffinity column approach compared with conventional chromatographic clean-up methods, was significantly faster and more selective, sensitive enough and can be used for official and regulatory purposes (2,3,5-7,9).

5. AFFILIATIONS

E. Ioannou-Kakouri
Ministry of Health, State General Laboratory,
44 Kimonos Street, 1451 Nicosia, CYPRUS,
e-mail: elkakour@spedernet.com.cy, *http:/www.sgl.moh.gov.cy*

M. Aletrari
Ministry of Health, State General Laboratory,
44 Kimonos Street, 1451 Nicosia, CYPRUS,
e-mail: elkakour@spedernet.com.cy, *http:/www.sgl.moh.gov.cy*

E. Christou
Ministry of Health, State General Laboratory,
44 Kimonos Street, 1451 Nicosia, CYPRUS,
e-mail: elkakour@spedernet.com.cy, *http:/www.sgl.moh.gov.cy*

A. Ralli
Ministry of Health, State General Laboratory,
44 Kimonos Street, 1451 Nicosia, CYPRUS,
e-mail: elkakour@spedernet.com.cy, *http:/www.sgl.moh.gov.cy*

A. Koliou
Ministry of Health, State General Laboratory,
44 Kimonos Street, 1451 Nicosia, CYPRUS,
e-mail: elkakour@spedernet.com.cy, *http:/www.sgl.moh.gov.cy*

M. Christofidou
Ministry of Health, State General Laboratory,
44 Kimonos Street, 1451 Nicosia, CYPRUS,
e-mail: elkakour@spedernet.com.cy, *http:/www.sgl.moh.gov.cy*

6. REFERENCES

(a) Central Science Laboratory (2001), Zearalenone Collaborative trial No. 130.

(b) Central Science Laboratory (2002), Deoxinivalenone Collaborative trial No. 131.

Commission Directives: 98/53/EC of 16 July 1998 for Aflatoxins and 2002/26/EC of 13 March 2002 for Ochratoxin A.

Commission Regulations (EC) No. 466/2001 of 8 March 2001 for Aflatoxins and other contaminants.

Commission Reputation (EC) No. 472/2002 of 16 March 2002 for Ochratoxin A in foodstuffs.

(a) Draft Commission Reputation amending Reputation (EC) 466/2001 as regards aflatoxins and ochratoxin A in foods for infants and young children (SANCO/10983/2002 rev.6).

(b) Working Document of the European Commission for ochratoxin A SANCO/0063/2002 – rev. 2.

(c) Miraglia M. and Brera C. (as coordinators) (2002). "Assessment of dietary intake of Ochratoxin A by the population of EU Member States". Report on Task 3.2.7 for Scientific Cooperation. Directorate – General Health and Consumer Protection.

Ioannou-Kakouri E., Aletrari M., Christou E., Ralli A., Koliou A. & Akkelidou D. (1999) "Surveillance and Control of Aflatoxins B_1, B_2, G_1, G_2 and M_1 in Foodstuffs in the Republic of Cyprus: 1992-1996" J.AOAC International, vol. 82, p.883-892.

Ioannou-Kakouri E., Christodoulidou M., Christou E. & Constantinidou E., (1995) "Immunoaffinity column / HPLC Determination of Aflatoxin M_1 in milk", Food Agric. Immunogy, vol. 7, p. 131-137.

Ioannou-Kakouri E. Aletrari M., Ralli, A, Christou E., Koliou A., "Immunoaffinity Column / HPLC Determination of Ochratoxin A in Foodstuffs", poster presentation in the 5th International Conference Agri-food Antibodies 99, Norwich, UK., 1999.

Markari, P., & Melissari, E. (1997) "Occurrence of Aflatoxin M1 in Commercial Pasteurized Milk Determined with ELISA and HPLC" Food Additives and Contaminants 14, 451-456.

Martins, M. (2003), "Occurrence of Aflatoxins in Raw Material and in Feedstuffs in Portugal 2000-2002". Official Methods of Analysis, AOAC 999.31 (2000).

Sudamore K. and MacDonald S. (1998) "Automated Liquid chromatographic Determination of Ochratoxin A in Cereals and Animal Procucts Using Immunoaffinity Column Cleanup", Food Additives and Contaminants, vol. 15, p. 401-410.

Visconti A., Pascale M., and Centonze G., (2001) "Determination of Ochratoxin A in Wine and Beer by Immunoaffinity Column Cleanup and Liquid Chromatographic Analysis with Fluorometric Detection: Collaborative Study", Journal of the AOAC International, Vol. 84, No. 6, p.1818 – 1827.

Visconti A., Solfrizzo M., and De Girolamo A., (2000) "Determination of Fumonisins B1 and B2 in Corn and Corn Flakes by Liquid Chromatography with Immunoaffinity Column Cleanup: Collaborative Study", Journal of the AOAC International, Vol. 84, No. 6, p.1828 – 1837.

Wood, G.E. (1992) "Mycotoxins in Foods and Feeds in the United States", J. Anim. Sci. 70, 3941-3949.

Proceeding Book of Congress Food Protection, Monte da Caparica, Portugal, 12-13 Maio, 30.

Working Document of the European Commission for furasium toxins, SANCO/0062/2003 – rev. 2 (partim).

a) Ministry of Agriculture, Fisheries, and Food (1993) Mycotoxins: Third Report, Food Surveillance Paper No. 36, HMSO, London, UK.

b) Ministry of Agriculture, Fisheries and Food (1999) Survey No. 192 for Aflatoxins, Ochratoxin A, Fumonisins and Zearalenone in Raw Maize, MAFF, London, UK.

V. OSTRY, J. SKARKOVA, F. MALIR & S. SYCOROVA

ADVANCES ON THE OCCURRENCE OF TOXIGENIC FUNGI AND MYCOTOXINS IN THE CZECH REPUBLIC

1. INTRODUCTION

Mycotoxins are toxic metabolites produced by certain toxigenic microfungi inter alia in and on foodstuffs. Thus, the research and the control of the presence of toxigenic microfungi and mycotoxins in foodstuffs in the food chain – from the primary producer to the ultimate consumer – is an important part of the food safety strategy in the Czech Republic.

The food safety strategy is one of the priorities of the Czech government, in particular with regard to the accession of the Czech Republic to the European Union. In the context of the evaluation of the state of readiness for accession, the food safety system in the Czech Republic was reviewed by the European Commission Directorate-General for Health and Consumer Protection in May 2001. The entire organizational structure of the food safety system from farms to the consumer, including the division of the competencies and tasks of relevant state authorities was subject to review. The food safety system existing in the Czech Republic was concluded to be functional. However, it was recommended to accelerate the harmonization in some areas, to improve the coordination of some specific elements of the system and to introduce a clear division of responsibilities in the entire area of food safety, the competencies being shared by several ministries, mainly the Ministry of Health and the Ministry of Agriculture [1].

The occurrence of toxigenic microfungi and the production of mycotoxins in the Czech Republic have been studied for more than thirty years, since 1969. The state control of mycotoxins and toxigenic microfungi in the entire food chain – from the primary production to the points of sale – is assured by local institutes of Public Health Protection (Ministry of Health - MH), by local organs of the Veterinary Administration (Ministy of Agriculture - MAg), by the Czech Agriculture and Food Inspectorate (MAg) and to a certain degree also by the State Phytosanitary Administration (MAg) and by the Central Institute for Control and Testing in Agriculture (MAg) [1].

Currently, the content of mycotoxins in foodstuffs in the Czech Republic is regulated by the Ordinance of the Ministry of Health N°. 53/2002 Coll., which implements the Act N°. 110/1997 Coll. on Foodstuffs and Tobacco Products. The limits for aflatoxins, ochratoxin A and patulin are harmonized with the existing legislation of the European Union.

A. Logrieco and A. Visconti (eds.),
An Overview on Toxigenic Fungi and Mycotoxins in Europe, 67-81.

The research and monitoring of toxigenic fungi and mycotoxins in foodstuffs of food chains are carried out by the National Institute of Public Health in Prague (MH) and by the Institute of Public Health in Hradec Kralove (MH). The research and monitoring of mycotoxins in cereals from domestic crop are performed in the Research Institute of Crop Production in Prague-Ruzyne, furthermore in the Institute for Agriculture Research in Kromeriz, Ltd. and in the Research Institute for Fodder Crops, Ltd. in Troubsko.

The purpose of this article is to summarize results concerning toxigenic microfungi and mycotoxins in foodstuffs in the Czech Republic during the period of 2000-2002.

2. EXPERIMENTAL METHODS

2.1. Determination and Identification of Toxigenic Fungi in Foodstuffs

The study of the occurrence of toxigenic fungi (especially producers of aflatoxins and ochratoxin A) in foodstuffs was a part of the project „Monitoring of exposure to chemical substances from foodstuffs“ in the system of National Environment Health Monitoring in the Czech Republic in 1999. This study was called „MYCOMON”. Its objective was to obtain information on the degree of immediate contamination of foodstuffs with toxigenic micromycetes, which could serve as the basis for the evaluation of the dietary exposure and health risk assessment of afla- and ochra- toxigenic microfungi in the Czech foodstuffs, prediction on occurrence of „hidden mycotoxins“ in foodstuffs contaminated by toxigenic microfungi and protection of public health in this area. The study MYCOMON was conducted by the National Institute of Public Health in Prague [2-6].

2.1.1. Sampling of Foodstuffs in the Study MYCOMON

Twenty five different foodstuffs were purchased from retailers at 12 locations of the Czech Republic (Prague, Brno, Ostrava, Plzen, Hradec Kralove, Usti nad Labem, Ceske Budejovice, Zdar nad Sazavou, Znojmo, Jablonec nad Nisou, Benesov and Sumperk) in four terms (March, May, September, November), see Figure 1.

In total, the study involved three hundred randomly and not probabilistically collected samples of food. A survey of the whole sample collection is provided in Table 1.

2.1.2. Methods Applied in the Study MYCOMON

Aflatoxigenic microfungi *Aspergillus* Section *Flavi* (*Aspergillus flavus, Aspergillus nomius, Aspergillus parasiticus, Aspergillus tamarii*) and ochratoxigenic microfungi (*Aspergillus ochraceus, Aspergillus* Section *Nigri, Penicillium verrucosum, Penicillium nordicum*) were monitored in foodstuffs. The mycological examination norm CSN ISO 7954 (1994) for calculation of moulds in foodstuffs was applied. Classical mycological methods were used for identification [7-16]. AFPA medium (OXOID) and ADMB medium (HIMEDIA) were used for identification of *Aspergillus flavus/A. parasiticus.* DRYES agar (OXOID) and PRYES agar (HIMEDIA) were used for identification of *Penicillium verrucosum.* Chemotaxonomy, HPTLC determination of aflatoxins and cyclopiazonic acid in YES medium was carried out [17-20].

Figure 1. Sampling areas

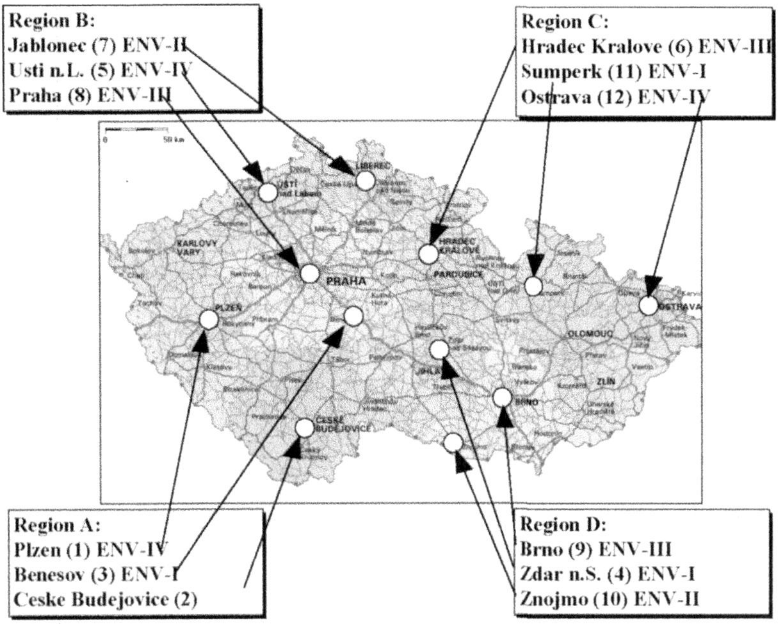

Aflatoxigenic microfungi were characterized by determining the total count (CFU/g) and the index of contamination (I_c), i.e. the ratio between the number of potentially toxigenic microfungi (CFU/g) and the total fungal count (CFU/g). It is originally an auxiliary parameter that was introduced for the study. The values of the index I_c were 0 – 1. The more the index approached number 1, the more serious the contamination of food with toxigenic microfungi was. If the index I_c = 1, the toxigenic microfungi in the foodstuffs occurred as „monoculture".

Subsequently, aflatoxigenic and ochratoxigenic ability of isolated strains was tested. Strains of fungi isolated from foodstuffs were cultivated on YES (Yeast Extract Sucrose) medium at 30 ^{0}C for 14 days. After the incubation the mycelium in

YES medium was homogenized with chloroform in order to extract mycotoxins. Purification of the extract was carried out by means of immuno-affinity chromatography. HPTLC determination of aflatoxins, cyclopiazonic acid and ochratoxin A in YES medium was used [21-27].

Validation of the methods was performed according to the principles of the ICH Guideline for planar chromatography and Food Mycology principles. The methods used were accredited to CSN EN ISO/IEC 17025. Laboratories successfully participated in FAPAS and FEPAS (Proficiency testing). The basic statistical processing of data was carried out according to the Report presented at the Workshop in the framework of GEMS/Food-Euro [28-30].

Table 1. Sample collection

Sampling period	*Foodstuffs*	*Number of samples*
1.	"Vysocina" salami, "Selsky" salami, semi-dry salami, "Polican" salami, raisins, pasta, rice	84
2.	Edam cheese	12
3.	sweet red pepper, caraway seeds, black pepper	36
4.	lentils, beans, peas, walnuts, peanuts, semi-granular flour, fine flour, granular flour, oat flakes, barley, fine semolina, instant semolina, fruit tea and black tea	168
Total		300

2.2. Determination of Mycotoxins in Foodstuffs

Determination of mycotoxins (aflatoxins, ochratoxin A, patulin and deoxynivalenol) in foodstuffs in 2000-2002 was performed by the Institute of Public Health in Hradec Kralove.

2.2.1. Sampling of Foodstuffs for Purposes of Mycotoxins Determination

Samples of different foodstuffs were purchased in the course of reference and expertise activity of the Institute of Public Health in Hradec Kralove.

2.2.2. Methods for Determination of Mycotoxins in Foodstuffs

Reversed-phase high-performance liquid chromatography (HPLC) with fluorescence detector and DAD detectors were employed for purposes of the detection and the quantification of mycotoxins. The cleaning-up of mycotoxins in the samples of foodstuffs was accomplished using standard, commercially available immunoaffinity columns [31,32]. All of these methods were validated and accredited to the Czech Accreditation Institute. The laboratory successfully participated in the FAPAS (Proficiency testing). The basic statistical processing of data was carried out according to the Report presented at the Workshop in the framework of GEMS/Food-Euro [28-30].

2.3. Determination of Deoxynivalenol in Cereals from the Domestic Crop

Investigation of deoxynivalenol was performed in the project of "Monitoring of deoxynivalenol occurrence in cereals" in the Czech Republic in the years 2000-2002. Determination of deoxynivalenol in cereals (wheat, barley, rye) from the domestic crop in 2000-2002 was performed in the Research Institute of Crop Production in Prague-Ruzyne.

2.3.1. Sampling of Cereals for Purposes of Mycotoxins Determination

Samples of cereals from the domestic crops were purchased in the context of the project „Monitoring of deoxynivalenol occurrence in cereals in the Czech Republic“ in collaboration with the Institute for Agriculture Research in Kromeriz.

2.3.2. Deoxynivalenol Determination in Cereals

The immunochemical method ELISA (Ridascreen® Deoxynivalenol, R-Biopharm GmbH, Darmstadt, Germany) was used for the determination of deoxynivalenol. Validation of the method was performed according to the protocol approved by the AOAC. The laboratory successfully participated in the FAPAS (Proficiency testing) [33,34]. The selected results were confirmed by GC/ECD methods. The basic statistical processing was carried out according to the Report presented at the Workshop in the framework of GEMS/Food-Euro [28-30].

3. RESULTS AND DISCUSSION

3.1. Determination and Identification of Toxigenic Fungi in Foodstuffs

Frequency data on the qualitative and quantitative occurrence of toxigenic microfungi in selected foodstuffs in the Czech Republic were obtained. The total count of microfungi (CFU/g) in selected foodstuffs was determined and the mycological profile was characterized. The results of total fungal count are shown in Table 2.

Qualitative mycological determination was specific and characteristic in some foodstuffs (e.g. black pepper, raisins, black tea and fruit tea).

Frequency of toxigenic strains *Aspergillus flavus* and *Aspergillus tamarii* is summarized in Table 3 and 4.

Total count of isolated *Aspergillus flavus* strains, index of contamination (I_c) and evaluation of the toxigenicity in the year 2002 are shown in Table 5.

Following the findings of aflatoxigenic Aspergillus flavus strains in foodstuffs samples (see Tables 5 and 6), analytical examination of selected samples was completed using the HPTLC method in order to determine the content of aflatoxins.

Aflatoxins were detected in black pepper and caraway seed samples in 1999 and in sweet red pepper in 2001 - 2002.

Frequency of occurrence of potential toxigenic strains *Aspergillus* Section *Nigri* is shown in Table 7.

Table 2. Total fungal count (CFU/g) in foodstuffs in 2002

Foodstuffs	*Number of samples n*	*Aritmetic mean (CFU/g)*	*Median (CFU/g)*	*Range min - max (CFU/g)*
Raisins	12	71353	160	<10 - 630000
Walnuts	12	4899	325	100 - 54000
Caraway seeds	12	2680	61	10 - 17000
Black pepper	12	2608	30	<10 - 20000
Fruit tea	12	793	235	20 - 5400
Black tea	12	658	380	86 - 3500
Rice	12	558	27	<10 - 4600
Fine flour	12	322	265	100 - 950
Semi-granular flour	12	191	64	15 - 1500
"Polican" salami	12	179	8	<10 - 2000
Semi-dry salami	12	127	5	<10 - 730
Sweet red pepper	12	91	30	<10 - 410
Fine semolina	12	71	62	10 - 150
Oat flakes	12	58	15	<10 - 270
Barley	12	44	5	<10 - 380
"Selsky" salami	12	44	15	<10 - 300
Granular flour	12	42	30	<10 - 130
"Vysocina" salami	12	41	8	<10 - 300
Edam cheese	12	34	5	<10 - 200
Instant semolina	12	18	8	<10 - 72
Pasta	12	11	5	<10 - 30
Peas	12	6	5	<10 - 10
Peanuts	12	5	5	<10
Lentils	12	5	5	<10
Beans	12	5	5	<10

The occurrence of ochratoxin A in raisins contaminated by *Aspergillus* Section *Nigri* is summarized in Table 8. Raisins containing high levels of ochratoxin A (OTA) were imported from Turkey (the country of origin).

Neither *Aspergillus parasiticus* nor *Aspergillus nomius* were isolated from the tested foodstuffs. *Penicillium verrucosum, Penicillium nordicum* and *Aspergillus ochraceus* were not isolated from the tested foodstuffs either. Potentially toxigenic strains of *Penicillium crustosum* were isolated from the tested sample of walnuts. Hard Edam cheese from retail was contaminated by *Penicillium roqueforti.* Dry salami from retail was contaminated by *Penicillium chrysogenum* and *Penicillium nalgiovense*.

Spores of aflatoxigenic microfungi have been stored in 30 % solution of glycerol in deep-freezer box at a temperature of – 71 °C for further research.

Data obtained in the course of the study MYCOMON represent the first step towards the conclusive evaluation of the dietary exposure and health risk of toxigenic microfungi in Czech foodstuffs. Trends in the occurrence of toxigenic microfungi in foodstuffs can be investigated and evaluated in the following years.

Table 3. Frequency of occurrence of toxigenic strains Aspergillus flavus *in foodstuffs in 1999-2002*

Year	*1999*	*2000*	*2001*	*2002*
Frequency (%) (strains/total samples)	28 (20/72)	17 (8/48)	18 (13/72)	17 (16/96)
Toxigenic strains (%)	75	63	77	56
Foodstuffs	black pepper black tea caraway seeds fine flour fruit tea oat flakes	black pepper black tea caraway seeds fine flour fruit tea	black pepper black tea caraway seeds fine flour fruit tea oat flakes	black pepper black tea caraway seeds fine flour fruit tea pasta raisins sweet red pepper

Table 4. Frequency of occurrence of toxigenic strains Aspergillus tamarii *in foodstuffs in 1999-2002*

Year	*1999*	*2000*	*2001*	*2002*
Frequency (%) (strains/total samples)	25 (3/12)	25 (6/24)	21 (5/24)	33 (4/24)
Toxigenic strains (%)	100	67	80	25
Foodstuffs	black pepper	black pepper black tea	black pepper black tea	black pepper black tea

Table 5. Total count of Aspergillus flavus *strains, index of contamination (I_c) and evaluation of the toxigenicity*

Number of isolates	*Identification of isolate*	*Isolated from foodstuffs*	*Aspergillus flavus total count (CFU/g)*	*Ic*	*Toxi-genicity*
1	*Aspergillus flavus* 19	raisin	100	0.001	-
2	*Aspergillus flavus* 62A	pasta	10	0.333	+
3	*Aspergillus flavus* 103	black pepper	10	0.250	+
4	*Aspergillus flavus* 105	caraway seeds	10	0.001	-
5	*Aspergillus flavus* 112A	black pepper	130	0.022	+
6	*Aspergillus flavus* 121A	black pepper	100	0.020	-
7	*Aspergillus flavus* 131	sweet red pepper	10	0.111	+
8	*Aspergillus flavus* 133A	black pepper	380	0.019	-
9	*Aspergillus flavus* 253	black tea	10	0.080	+
10	*Aspergillus flavus* 267A	black tea	10	0.020	-
11	*Aspergillus flavus* 281	black tea	20	0.010	+
12	*Aspergillus flavus* 308	fruit tea	10	0.050	+
13	*Aspergillus flavus* 316	fine flour	10	0.020	-
14	*Aspergillus flavus* 337	black tea	10	0.070	+
15	*Aspergillus flavus* 351	black tea	10	0.120	-
16	*Aspergillus flavus* 393	black tea	10	0.030	+

Table 6. Aflatoxins in indicated foodstuffs

Foodstuffs	*n*	*n+*	*Arithmetic mean (μg/kg)*	*Max. conc. (μg/kg)*
1999				
Black pepper	12	3	15.0	19.2
Caraway seed	12	1	-	8.1
2001				
Sweet red pepper	12	6	2.5	4.1
2002				
Sweet red pepper	12	3	1.0	1.4

Table 7. Frequency of occurrence of potential toxigenic strains Aspergillus *Section* Nigri *in food in 1999-2002*

Year	*1999*	*2000*	*2001*	*2002*
Frequency (%) *(strains/total samples)*	*67* *(40/60)*	*85* *(41/48)*	*32* *(27/84)*	*48* *(29/60)*
Foodstuffs	black pepper black tea fruit tea raisins sweet red pepper	black pepper black tea fruit tea raisins	black pepper black tea caraway seeds fine flour fruit tea raisins sweet red pepper	black tea fruit tea raisins peanuts fine flour

Table 8. Ochratoxin A in raisins contaminated by Aspergillus *Section* Nigri

Sampling areas	*1999*	*2000*	*2001*	*2002*
		OTA (μg/kg)		
01	< 1	9.3	< 1	8.0
02	< 1	3.0	4.7	< 1
03	< 1	< 1	5.1	6.5
04	1.9	< 1	< 1	< 1
05	< 1	< 1	< 1	< 1
06	< 1	< 1	1.6	< 1
07	< 1	2.4	3.1	< 1
08	< 1	< 1	6.1	< 1
09	12.0	< 1	< 1	< 1
10	< 1	< 1	< 1	< 1
11	2.0	3.2	63.6	17.5
12	52.5	< 1	5.2	< 1
Arithmetic mean [a]	6	1.8	7.7	3.0
Median [a]	0.5	0.5	2.4	0.5
90% percentile [a]	11	3.2	6	7.9

[a] The concentration of OTA <1 μg/kg considered as 1/2 limit of quantification (LoQ) = 0.5 μg/kg

3.2. Determination of Mycotoxins in Foodstuffs

The occurrence of mycotoxins in foodstuffs in the Czech Republic (2000–2002) is shown in Tables 9 - 13. The basic statistical processing (mean, median and 90% percentile) was carried out according to the approach of GEMS/Food-Euro [28]. Two types of results were calculated. The first result in brackets was with substitution of zero and the second result was computed with substitution of the limit of quantification (LoQ).

Table 9. Levels of aflatoxins B_1, B_2, G_1 and G_2 in foodstuffs in 2000-2002

Foodstuffs	*n*	*n+*	*Arithmetic mean (μg/kg)*	*Median (μg/kg)*	*Perc. 90% (μg/kg)*	*Min (μg/kg)*	*Max (μg/kg)*
Peanuts	432	19	(1.7) 5.5	(0) 4.0	(0) 4.0	7.0	161.0
Dried milk	120	2	(0.01) 0.4	(0) 0.4	(0) 0.4	0.6	0.9
Wheat Flour	70	2	(0.03) 0.4	(0) 0.4	(0) 0.4	0.9	1.4
Chocolate	22	2	(0.06) 0.4	(0) 0.4	(0) 0.4	0.6	0.8
Black tea	22	1	(0.02) 0.4	(0) 0.4	(0) 0.4	-	0.5
Spices	9	1	(0.08) 0.4	(0) 0.4	(0) 0.4	-	0.7
Cereal	75	0	-	-	-	-	< 0.4
Malt	47	0	-	-	-	-	< 0.4
Baby food	38	0	-	-	-	-	< 0.4
Green algae	7	0	-	-	-	-	< 0.4
Wafer	3	0	-	-	-	-	< 0.4

Table 10. Levels of aflatoxin M_1 in foodstuffs in 2000-2002

Foodstuffs	*n*	*n+*	*Arithmetic mean (μg/kg)*	*Median (μg/kg)*	*Perc. 90% (μg/kg)*	*Min (μg/kg)*	*Max (μg/kg)*
Milk	101	1	(0.005) 0.1	(0) 0.1	(0) 0.1	-	0.5
Milk porridge	35	0	-	-	-	-	< 0.1
Drink with milk	8	0	-	-	-	-	< 0.1

It is noteworthy that one sample of blueberries was extremely contaminated by patulin (3150 μg/kg).

Table 11. Levels of ochratoxin A in foodstuffs in 2000-2002

Foodstuffs	*n*	*n+*	*Arithmetic mean (μg/kg)*	*Median (μg/kg)*	*Perc. 90% (μg/kg)*	*Min (μg/kg)*	*Max (μg/kg)*
Cereals	64	7	(0.8) 1.1	(0) 0.35	(0.28) 0.35	0.4	37.0
Malt	35	2	(0.4) 0.7	(0) 0.35	(0) 0.35	0.7	13.4
Red wine	12	2	(0.03) 0.1	(0) 0.1	(0.16) 0.17	0.2	0.2
Baby Food	115	0	-	-	-	-	< 0.35
Milk	76	0	-	-	-	-	< 0.35
Cereal	56	0	-	-	-	-	< 0.35
Coffee	11	0	-	-	-	-	< 0.35
Herbal tea	6	0	-	-	-	-	< 0.35

Table 12. Levels of patulin in foodstuffs in 2000-2002

Foodstuffs	*n*	*n+*	*Arithmetic mean (μg/kg)*	*Median (μg/kg)*	*Perc. 90% (μg/kg)*	*Min (μg/kg)*	*Max (μg/kg)*
Baby fruit food	174	1	(0.08) 10	(0) 10	(0) 10	-	13
Fruit foodstuffs	76	6	(50.5) 59.7	(0) 10	(0) 10	15.0	3150
Fruit drink	23	4	(3.3) 11.6	(0) 10	(0) 14	14.0	24
Malt	14	4	(11.6) 18.8	(0) 10	(26.5) 26.5	22.0	90
Fruit tea	6	0	-	-	-	-	<10
Fruit porridge	21	0	-	-	-	-	<10

3.3. Determination of Deoxynivalenol in Cereals from the Domestic Crop

The occurrence of deoxynivalenol in cereals from the domestic crop in the Czech Republic (2000 – 2002) is shown in Tables 14 - 16.

Table 13. Levels of deoxynivalenol in foodstuffs in 2000 - 2002

Foodstuffs	*n*	*n+*	*Arithmetic mean (µg/kg)*	*Median (µg/kg)*	*Perc. 90% (µg/kg)*	*Min (µg/kg)*	*Max (µg/kg)*
Barley	79	21	(270.6) 417.5	(0) 200	(412) 412	210	6 010
Cereal	10	1	(20.0) 200	(0) 200	(20) 10	-	200
Malt	25	3	(42.4) 218.4	(0) 200	(126) 206	210	550
Wheat flour	6	0	-	-	-	-	< 200

Table 14. Levels of deoxynivalenol in cereals –domestic crop in 2000

Cereals Grain	*n*	*n+*	*Arithmetic mean (µg/kg)*	*Median (µg/kg)*	*Perc. 90% (µg/kg)*	*Min (µg/kg)*	*Max (µg/kg)*
Wheat	56	4	135	220	230	3	740
Barley	33	10	220	110	460	20	550
Rye	22	3	136	220	220	53	455

Table 15. Levels of deoxynivalenol in cereals –domestic crop in 2001

Cereals Grain	*n*	*n+*	*Arithmetic mean (µg/kg)*	*Median (µg/kg)*	*Perc. 90% (µg/kg)*	*Min (µg/kg)*	*Max (µg/kg)*
Wheat	55	9	177	220	320	10	2490
Barley	32	9	283	220	540	30	3770
Rye	15	1	124	220	220	20	330

Table 16. Levels of deoxynivalenol in cereals –domestic crop in 2002

Cereals Grain	*n*	*n+*	*Arithmetic mean (µg/kg)*	*Median (µg/kg)*	*Perc. 90% (µg/kg)*	*Min (µg/kg)*	*Max (µg/kg)*
Wheat	95	7	142	220	220	0	1450
Barley	30	5	237	110	391	0	1150
Rye	32	1	126	220	220	0	1100

4. AFFILIATIONS

Vladimir Ostry
National Institute of Public Health Prague, Centre for the Hygiene of Food Chains, NRC for Microfungi and Mycotoxins in Food Chains, Palackeho 3a, 612 42 Brno, The Czech Republic, e-mail: ostry@chpr.szu.cz ;

Jarmila Skarkova,
National Institute of Public Health Prague, Centre for the Hygiene of Food Chains, NRC for Microfungi and Mycotoxins in Food Chains, Palackeho 3a, 612 42 Brno, The Czech Republic, e-mail: ostry@chpr.szu.cz ;

Frantisek Malir
Institute of Public Health Hradec Kralove, CHL, Xenobiochemistry Dept., NRL for Biomarkers of Mycotoxins and Mycotoxins in Food, Nezvalova 958, 500 02 Hradec Kralove, The Czech Republic;

Jiri Ruprich
National Institute of Public Health Prague, Centre for the Hygiene of Food Chains, NRC for Microfungi and Mycotoxins in Food Chains, Palackeho 3a, 612 42 Brno, The Czech Republic, e-mail: ostry@chpr.szu.cz ;

Svetlana Sykorova
Research Institute of Crop Production, Drnovska 507, 161 06 Praha-Ruzyne, The Czech Republic.

5. REFERENCES

Food Safety Strategy in the Czech Republic. (2000). Ministry of Agriculture of the Czech Republic in cooperation with Ministry of Health of the Czech Republic.

Ostry, V., Ruprich, J., Prochazkova, I., Skarkova, J., & Kubatova, **A.** (2000). MYCOMON, project for the determination and the identification of toxigenic fungi in food in the Czech Republic. Bydgoszcz: In: 5th International Conference Mycotoxins and Dioxins and the Environment, Bydgoszcz University of Kazimierz Wielki, 191-196.

Ostry, V. (2000, August). Mycological analysis of foodstuffs. Kliment, V., (ed.), National Institute of Public Health, EnviTypo, Prague. In: System of monitoring the environmental impact on population health of the Czech Republic. Summary report-1999, ISBN 80-7071-144-2, p. 106.

Ostry, V., Ruprich, J., Skarkova, J., Prochazkova, I., & Kubatova, A. (2001). Occurrence of the Toxigenic Fungi (Producers of Aflatoxins and Ochratoxin A) in Foodstuffs in the Czech Republic 1999 - 2000. Mycotoxin Research, 17A, 2, 188 - 192.

Ostry, V., Ruprich, J., Skarkova, J., Prochazkova, I., & Kubatova, A. (2002). MYCOMON - monitoring project of toxigenic fungi in food in years 1999 - 2000. Mycotoxin Research, 18A, 2, 193 - 197.

Ostry, V., Ruprich, J., & Skarkova, J. (2002**).** Raisins, ochratoxin A and human health.. Mycotoxin Research, 18A, 2, 178 - 182.

Pitt, J. I., & Hocking, A. D. (1997). Fungi and Food Spoilage. London, UK: Chapman & Hall, Blackie Academic and Professional.

Samson, R. A., & Pitt, J. I. (2000). Integration of Modern Taxonomic Methods for *Penicillium* and *Aspergillus* Classification. Amsterodam, The Netherlands: Harwood Academic Publishers.

Samson, R. A., Hocking, A. D., Pitt, J. I., & King, A. D. (1992). Modern Methods in Food Mycology. Amsterdam, The Netherlands: Elsevier Science Publishers B. V.

Samson, R. A., & Reenen-Hoekstra, E. (1988). Introduction to Food-Borne Fungi. Baarn, The Netherlands: Centraalbureau Voors Schimmelcultures.

Gravesen, S., Frisvad, J. C., & Samson, R. A. (1994). Microfungi. Copenhagen, Denmark: High Tech PrePres A/S.

Kubatova, A., Vanova, M., Prasil, K., & Fassatiova, O. (1999). Microfungi contaminating foods with low water content. Novit. Bot. Univ. Carol., 13, 13-25.

Paterson, R. R. M., & Bridge, P. D. (1994). Biochemical Techniques for Filamentous Fungi. CAB International, UK: IMI Technical Handbooks No. 1.

Ostry, V., Ruprich, J., Skarkova, J., Prochazkova, I., & Kubatova, A. (2001). The System Approach to the Identification of Aflatoxigenic Fungi in Foodstuffs and Feedstuffs. Mycotoxin Research, 17A, 2, 178 - 182.

Skarkova, J., Ostry, V., Ruprich, J., & Kubatova, A. (2003). Chemotaxonomy of aflatoxigenic species of *Aspergillus* section *Flavi*. Czech Mycol., 55 (1-2), 145 - 146.

Skarkova, J., Ostry, V., & Ruprich, J. (2002). Application of Planar Chromatography to the Chemical Standardization of Aflatoxigenic Micromycetes. Nyiredy S (Ed.), Hévíz: In: Planar Chromatography 2002, International Society for Planar Chromatography, pp. 441 - 446.

Pitt, J. I., Hocking, A. D. & Glenn, D. R. (1983). An improved medium for the detection of *Aspergillus flavus* and *A. parasiticus*. J Appl Bacteriol., 54, 109-114.

Frisvad, J. C. (1983). A selective and indicative medium for groups of *Penicillium viridicatum* producing different mycotoxins in cereals. J Appl Bacteriol., 54, 409 - 416.

Frisvad, J. C. (1994). Evaluation of media based on yeast extract, sucrose or glycerol for enumeration of *Penicillium verrucosum* in flour. A collaborative study. Copenhagen: In: 3. International workshop on standardization of methods for the mycological examination of foods, 10-11.

Abarca, M. L., Bragulat, M. R., Castella, G., & Cabanes, F. J. (1994). Mycoflora and aflatoxin-producing strains in animal mixed feeds. J Food Protection, 57, (3), 256-258.

Paterson, R. R. M. (1985). Standardized one- and two-dimensional thin-layer chromatographic methods for the identification of secondary metabolites in *Penicillium* and other fungi. J Chromatogr, 368, 249-264.

Abarca, M. L., Bragulat, M. R., Castella, G., & Cabanes, F. J. (1994). Ochratoxin A production by strains of *Aspergillus niger* var. niger. Appl Environ Microbiol, 60, (7), 2650-2652.

Varga, J., Kevei, F., Rinyu, E., Teren, J., & Kozakiewicz, Z. (1996). Ochratoxin A production by *Aspergillus* species. Appl Environ Microbiol, 62, 4461- 4464.

Wicklow, D. T., Dowd, P. F., Alfatafta, A. A., & Gloer, J. B. (1996). Ochratoxin A: an antiisectan metabolite from sclerotia of *Aspergillus carbonarius* NRRL 369. Can J Microbiol, 42, 1100 - 1103.

Skarkova, J., Ostry, V., & Ruprich, J. (2001). Application of Planar Chromatography on Determination of Mycotoxins in the Czech Republic. CBS, 9, 2-3.

Skarkova, J., Ostry, V., & Ruprich, J. (2001). Application of Planar Chromatography on Determination of Mycotoxins in the Czech Republic. Nyiredy S, Kakuk A (Eds), Lillafüred: In: Planar Chromatography 2001 New Milestones in TLC, International Society for Planar Chromatography, pp. 343-349.

Ostry, V., Skarkova, J., & Ruprich, J. (2001). The Mycotoxin Research in Foodstuffs in the Czech Republic in the 90th Years. Mycotoxin Research, 17A, 2, 183 - 186.

GEMS/Food-EURO Second Workshop on Reliable Evaluation of Low-Level Contamination of Food. (1995, May 26 – 27). Kulmbach, Federal Republic of Germany: Report on a Workshop in the Frame of GEMS/Food-EURO, pp 8.

Vlachonikolis, I. G., & Marriott, F. H. C. (1995). Evaluation of censored contamination data. Journal of Food Additives and Contaminants, 12, 637-644.

Hecht, H., & Honikel, K. O. (1995). Assessment of data sets containing considerable values below the detection limits. Z Lebensm Unters Forsch, 201(6), 592-597.

Malir, F., & Ostry, V. (1999). Mycotoxins and their Health Hazard for Czech Population. Casopis Ceskych Lekarniku, 71, 8, 22-23 (in Czech).

Severa, J., Roubal, T., & Malir, F. (2003). The Levels of Mycotoxins in Biological Materials and Foodstuffs. Informacni zpravodaj Institutu ochrany obyvatelstva MV-GR HZS CR, 14 (1), 79 - 87 (in Czech).

Papouskova, L., & Sykorova, S. (2001, April 22 – 27). The survey of mycotoxin content determined by ELISA method in cereal samples obtained from different locations in Czech Republic in 2000. Impact and Exploitation "Univ. of Wales Swansea: In: Proceedings of International Symposium" Bioactive Fungal Metabolites, pp. 85.

Sykorova, S., Papouskova, L., Sip, V., Chrpova, J., Hysek, J., & Sychrova, E. (2002, August 11 – 17). The survey of spring barley and rye contamination by Fusarium mycotoxins (artificial and natural infection). Oslo: In: Book of Abstracts, The 7th International Mycological Congress, pp. 271.

T. YLI-MATTILA, S. PAAVANEN-HUHTALA, P. PARIKKA,
V. HIETANIEMI, M. JESTOI & A. RIZZO

TOXIGENIC FUNGI AND MYCOTOXINS IN FINNISH CEREALS

1. INTRODUCTION

In Finland barley, wheat, oats and rye are normally dried by heating already in the farm, which decreases the amount fungi in grains. The quality of dried barley is usually good and stable in storages, which means that *Aspergillus* and *Penicillium* cannot grow and form toxins. *Fusarium* fungi are monitored by food industry either visually or by *Fusarium*-specific media (e.g. CZID) before they decide to buy the barley for malting from the farmers. Random samples for mycotoxins in grains are also studied by the food industry companies, but they have not detected high mycotoxin concentrations in their raw material.

The variability of Finnish growing seasons was clearly observed during the years here presented data was collected. Heavy rainfalls during the cool growing season of 1998 caused abundant *Fusarium* mould infection in cereals in southern Finland, particularly in rye. Spring wheat, oats and barley were less damaged by the rain than winter rye and wheat. Due to the early rains in 1998 the *Fusarium* contamination level, especially *F. avenaceum*, was highest in rye (Eskola *et al.*, 2001), which resulted in high levels of beauvericin and enniatins in most of the rye samples studied (Logrieco *et al.*, 2002; in the original paper the crop was erroneously reported as wheat), but the levels of other mycotoxins were low (Eskola *et al.*, 2001). The growing season of the year 2000 was warmer, and abundant rainfall started only at the end of June, resulting in high *Fusarium* contamination levels in barley (Yli-Mattila *et al.*, 2002). The growing season 2001 was relatively warm and rainy, especially before the harvesting of late cereals in September, while the growing season 2002 was warm and very dry, except for a rainy period at the end of June and beginning of July.

The purpose of the present study was to collect data on the occurrence and abundance levels of different toxigenic fungi and their toxins in Finnish grains and imported food stuffs during the growing seasons 1998-2002 in Finland based on published (Eskola *et al.*, 2001, Yli-Mattila *et al.*, 2002, 2004, Jestoi *et al.*, 2004) and unpublished data (e.g. www.agronet.fi/cerveg).

A. Logrieco and A. Visconti (eds.),
An Overview on Toxigenic Fungi and Mycotoxins in Europe, 83-100.

2. MATERIALS AND METHODS

2.1. Grain samples and fungal strains

Grain samples were collected in Finland for toxin analyses during the years 1998-2002 (Eskola *et al.*, 2001; Yli-Mattila *et al.*, 2002, www.agronet.fi/cerveg; Tables 1,2); along with three Russian barley samples from the norhwestern Russia (Leningrad) region in 2000 and 2002. The contamination percentages of a part of the grain samples were studied in 1998 and 2000-2002 and the *Fusarium* isolates from the grain samples were identified by morphology (Eskola *et al.*, 2001). Fresh grain samples from the years 2001-2002 (Yli-Mattila *et al.*, 2004) were used for the analysis of *Fusarium* species, since drying at high temperatures may decrease the amount of living fungi. The morphological identification of ca. 1000 of the 3500 *Fusarium* isolates was confirmed by species-specific primers (Table 3, Yli-Mattila *et al.*, 2004). Several of the grain samples were collected from the same farms in 2000-2002, allowing comparison of the disease situation in the same locations.

2.2. Artificial inoculation

Artificial inoculation of barley and wheat in the field experiments was carried out during the growth season of 2002 in barley cv. Scarlett and wheat cv. Mahti in Marttila in southwestern Finland. Three strains per *Fusarium* species were used as a mixture. The cultures used for the inoculation were cultivated for three to four weeks on PDA. The cultures were washed with 200 ml of sterile water to obtain a concentration of 15000-50000 spores/ml, except for *F. arthrosporioides, F. tricinctum* and *F. graminearum*, which produced very few conidia (less than 2000 spores ml^{-1}). The strains from the year 2001 used for inoculation were 1, 2, 14 (*F. avenaceum*), 7, 11, 92 (*F. arthrosporioides*), 47, 67, 112 (*F. tricinctum*), 48, 65, 72 (*F. culmorum*), 5, 15, 16 (*F. poae*), 3, 4, 6 (*F. sporotrichioides*), 350, 364, 365 (*F. graminearum*). Wheat and barley were sown on 27 April and 2 May 2002, respectively. Plots of 2 m^2 were sprayed during flowering with 100 ml of the *Fusarium* spores per plot on 29 June 2002.

The heads from the plots were manually collected at the end of August and the grains were harvested from the dried heads. The final yields were 0.3-0.5 kg of grains/plot.

2.3. DNA extraction, molecular analyses and mycotoxin analyses

Mycelia were grown on potato dextrose agar (PDA) for 4-5 d at 25 °C under constant light before DNA extraction. DNA was extracted from the mycelia as described in Paavanen-Huhtala *et al.* (1999) and from naturally infected grain samples from the year 2001 (Konstantinova *et al.,* 2002). Another extraction method was modified from Taylor *et al.* (2001) for the grain samples from 2002 (Yli-Mattila *et al.*, 2004).

Table 1. Grain samples of the year 2001(to be continued)

No.	*Cereal/ Cv.*	*Origin*	*Fus%*	*JIA/ FA-ITS*	*Fp82/ Poae IGS*	*Fg-11*	*Fc-01*	*DON/NIV (μg/kg)*	*T-2/ HT-2 (μg/kg)*
1	w.wheat Ramiro	SW Finland	1	±/-	-/±	-	-	<10/nd	nd/nd
3	barley Rolfi	SW Finland	6.5	+/+	+/+	-	-	<10/nd	nd/nd
5	barley Inari	SE Finland	44	+/+	+/+	-	-	12/276	28/39
7	barley Inari	SE Finland	8	+/+	+/+	-	-	<10//nd	92/190
9	barley Scarlett	SW Finland	15.5	+/+	+/+	-	-	<10/85	35/83
11	barley Saana	Southern Finland	8.5	±/-	-/±	-	-	35/<30	nd/<20
13	s.wheat Reno	Southern Finland	16	+/+	+/+	-	-	119/<30	22/57
15	barley Scarlett	SW Finland	29.5	+/+	+/+	-	-	31/405	nd/<20
17	s.wheat Kruunu	SW Finland	6.5	+/-	±/+	-	-	<10/nd	nd/nd
19	barley Saana	SE Finland	7.5	+/-	-/+	-	-	14/<30	30/128
21	s.wheat Tjalve	SE Finland	18.5	+/+	-/+	±	+	702/nd	nd/38
23	barley Saana	SW Finland	4.5	-/-	+/+	-	±	81/107	nd/29
25	barley Scarlett	SW Finland	8.5	-/-	+/+	-	+	60/<30	37/86
27	barley Extract	SW Finland	23.5	+/-	-/+	-	-	120/94	nd/<20
29	barley Saana	Southern Finland	0.5	+/-	+/+	-	±	20/96	nd/nd
30	barley Scarlett	Southern Finland	5.5	-/+	+/+	-	+	48/<30	nd/46
31	w.wheat Ramiro	SW Finland	0	-/-	-/±	-	-	nd/<30	nd/<20
32	w.wheat Ramiro	SW Finland	0	-/-	-/±	-	-	<10/<30	nd/nd
33	barley Saana#	Western Finland	30.5	+/+	+/+	-	-	242/<30	nd/24
35/ 36	s.wheat Vinjett#	Western Finland	40.5	+/±	-/-	+	-	4301/nd	nd/21

37	barley Mette	SW Finland	16.5	-/-	-/±	-	±	12/nd	nd/68
441	s.wheat Tjalve	SW Finland	*	-/*	*/-	-	-	376/nd	nd/nd
881	oats Roope[#]	Western Finland	*	±/*	*/-	+	-	1510/nd	nd/nd
981	oats Leila[#]	Western Finland	*	±/*	*/-	+	-	7520/37	63/63
1052	s.wheat Mahti[#]	Western Finland	*	±/*	*/-	+	-	1110/nd	nd/nd
17b.	rye Amilo	SW Finland	2	-/*	*/-	-	-	<10/nd	nd/nd

F. avenaceum (JIA/FA-ITS), *F. poae* Fp82/PoaeIGS), *F. graminearum* (Fg11) and *F. culmorum* (Fc01)

SW = southwestern, SE = southeastern, + = positive reaction, - = negative reaction, ± = weak positive reaction, nd = not detected, * = not determined, # = harvested after a one-week-rain.

Table 2. Grain samples from 2002. Samples 18-31 are the same artificially inoculated samples as in Table 5(to be continued)

No	*Cereal/ Cv.& inocul.*	*Origin*	*Fus. %*	*JIA*	*Poae IGS*	*Fg11/ 175+ 430*	*Pfusf+ Fspor/ Pfusf+ Flanr*	*DON/ NIV/ 3Ac-DON (µg/kg)*	*T-2/ HT-2/ ZEN (µg/kg)*
1	barley Saana	SW Finland	15	±	±	-/+	-/-	941/ 1260/ 100	nd/nd/ 17
2	barley Saana	S Finland	3	-	-	-/±	-/-	<10/92 /nd	nd/nd/ nd
3	barley Inari	SE Finland	23	+	-	-/-	-/+	81/132 /41	71/27/ nd
4	barley Extract	SW Finland	16	+	-	-/±	-/+	37/ 1390/ 26	50/58/ nd
5	barley Extract	SW Finland	25	±	-	-/-	-/-	87/109 /34	30/nd/ nd
6	barley Cellar	SW Finland	15	+	-	-/+	-/+	594/ 492/59	nd/nd/ 22
7	barley Saana	W Finland	12	-	-	±/+	-/+	248/ 106/42	21/nd/ <6
8	barley Scarlett	SE Finland	15	(±)	-	-/-	-/+	<10/ <30/29	23/31/ nd
9	oats Belinda	SE Finland	5	-	±	-/-	-/+	<10/ 532/29	321/ 77/nd

10	s.wheat Mahti	SW Finland	17	-	±	-/+	-/-	1060/ 60/39	nd/nd/ nd
11	s.wheat Reno	S Finland	5	-	-	-/-	-/-	<10/nd /nd	24/nd/ nd
12	s.wheat Reno	S Finland	3	-	-	-/-	-/-	50/nd/ nd	27/nd/ nd
13	s.wheat Reno	S Finland	13	+	±	-/+	-/-	319/ <30/28	nd/nd/ nd
14	s.wheat Vinjett	SW Finland	5	-	-	-/±	-/-	350/79 /41	<20/ nd/nd
15	s.wheat Vinjett	W Finland	13	-	-	±/±	-/-	632/ <30/48	nd/nd/ nd
16	s.wheat Tjalve	SE Finland	6	(±)	-	-/+	-/-	<10/ <30/nd	<20/ nd/nd
18	barley Scarlett *F.tric.*	SW Finland	33	-	-	-/-	-/-	<10/ 2890/ nd	nd/nd/ nd
19	*F. arthr.*	SW Finland	52	±	+	±/±	-/-	94/ 4790/ 28	nd/nd/ nd
20	*F.aven.*	SW Finland	37	+	+	-/+	+/-	503/ 5510/ 72	nd/nd/ <6
21	*F. poae*	SW Finland	41	-	-	-/-	-/-	366/ 5460/ 47	nd/nd/ <6
22	*F. spor.*	SW Finland	31	+	±	-/+	-/-	513/ 6200/ 62	<20/nd /16
23	*F. culm.*	SW Finland	33	-	+	-/+	-/-	3060/ 1320/ 230	nd/nd/ 96
24	*F. gram.*	SW Finland	42	±	-	-/(±)	-/-	315/ 4970/ 28	nd/nd/ 8
25	s.wheat Mahti *F. tric.*	SW Finland	10	*	*	*/*	*/*	722/ 224/ 333	nd/nd/ nd
26	*F. arthr.*	SW Finland	18	-	-	±/±	*/-	649/ 114/32	nd/nd/ nd
27	*F. aven.*	SW Finland	25	-	-	-/-	-/-	592/ 553/37	21/nd/ nd
28	*F. poae*	SW Finland	19	-	-	-/(±)	-/-	196/ 333/nd	<20/nd /nd
29	*F. spor.*	SW Finland	25	-	-	+/-	+/-	33/98/ nd	466/62 /nd

30	*F. culm.*	SW Finland	54	-	±	-/+	-/-	22200/ 185/ 446	<20/nd /45
31	*F. gram.*	SW Finland	8	-	-	±/±	-/-	389/ 153/nd	nd/nd/ nd
32	wheat Lening-radka 97	Lenin-grad r.	5	+	-	*/-	-/*	<10/nd /nd	nd/nd/ nd
33	barley Suzdalets	Lenin-grad r.	7	±	-	-/-	-/±	<10/ 141/nd	nd/nd/ nd
34	barley Crinich-niy	Lenin-grad r.	14.5	(±)	-	-/-	-/-	<10/ 742/nd	nd/nd/ nd
35	wheat Inna	Lenin-grad r.	5	*	*	*/*	*/*	nd/nd/ nd	nd/nd/ nd
36	barley Crinich -niy	Lenin-grad r.	7	-	+	-/-	-/-	nd/nd/ nd	nd/nd/ nd
41	oats Borus	Lenin-grad r.	3	*	*	*/*	*/*	<10/ 241/nd	nd/nd/ nd
77	s.wheat Tjalve	SW Finland	*	±	+	-/+	*/-	1980/ nd/<1	nd/nd/ <1
94	oats Veli	SW Finland	*	-	+	+/+	*/+	1300/ 54/205	nd/80/ nd
102	oats Aarre	S Finland	*	-	+	-/-	*/+	188/nd /nd	159/26 8/nd
103	oats Aarre	S Finland	*	-	+	+/±	*/(±)	159/50 /nd	nd/nd/ nd
109	oats Aarre	C Finland	*	±	+	±/±	*/-	140/nd /5797	nd/nd/ nd
110	oats Roope	W Finland	*	-	+	±/+	*/+	1380/ nd/163	nd/nd/ nd
112	oats Veli	W Finland	*	-	+	+/±	*/+	3710/ 41/715	nd/112 /nd
115	oats Veli	W Finland	*	±	+	±/+	*/+	2990/ 891/340	nd/41/ nd

F. avenaceum (JIA), *F. poae* (PoaeIGS), *F. graminearum* (Fg11, 175/430), *F. sporotrichioides* (Pfusf/Fspor), *F. langsethiae* (Pfusf/Flanr) and *F. culmorum* (175/430)

SW = southwestern, S=southern, SE = southeastern, W=Western, C=Central, + = positive reaction, - = negative reaction, ± = weak reaction, (±) = very weak reaction, nd = not detected, * = not determined.

Table 3. Sequences of the Fusarium *species-specific primers.*

Name of the primer (primer size)	*Sequence 5′- 3′*	*Specificity of species-specific primers*
Fa5f Fa5r (1071 bp)[1]	GGGGTCTTGCCACTCAGCTTGT GGGGTCTTGCGGATCATGTGCT	Mainly in *F. arthrosporioides*
Fa8f Fa8r (604 bp)[1]	GTGACGTAGGGAAACTGCCTGG GTGACGTAGGACCAGAGATGTA	Mainly in *F. arthrosporioides* and *F. tricinctum*
Fa17f Fa17r (1550 bp)[1]	ACCGCTTGTACCGTACCGTCAAT GACCGCTTGTTGCATCTAGGTAG	Mainly in *F. arthrosporioides*
JIAf JIAr (300 bp)[2]	GCTAATTCTTAACTTACTAGGGGCC CTGTAATAGGTTATTTACATGGGCG	*F. avenaceum/ F. arthrosporioides*
FA-ITSf FA-ITSr (272 bp)[2]	CCAGAGGACCCAAACTCTAA ACCGCAGAAGCAGAGCCAAT	*F. avenaceum/ F. tricinctum/ F. arthrosporioides*
Fg11f Fg11r (400 bp)[3]	CTCCGGATATGTTGCGTCAA GGTAGGTATCCGACATGGCAA	*F. graminearum*
Fc01f Fc01r (300 bp)[4]	ATGGTGAACTCGTCGTGGC CCCTTCTTACGCCAATCTCG	*F. culmorum*
PoaeIGS CNL12 (306 bp)[5]	CAAGCTCTCCTCGGAGAGTCGAA CTGAACGCCTCTAAGTCAG	*F. poae*
Fp82f Fp82r (200 bp)[6]	CAAGCAAACAGGCTCTTCACC TGTTCCACCTCAGTGACAGGTT	*F. poae*
PulvIGS CNL12 (610 bp)[5]	GAACCGTCCGGCACCCATCC CTGAACGCCTCTAAGTCAG	*F. sporotrichioides/ F. langsethiae*
175f 430r (245 bp)[7]	TTTTAGTGGAACTTCTGAGTAT AGTGCAGCAGGACTGCAGC	*F.culmorum/F. graminearum*
Pfusf Fspor (300 bp)[8]	CCGCGCCCCGTAAAACG ACTGTGTTTGCACACAGATC	*F. sporotrichioides*
Pfusf Flanr (300 bp)[8]	CCGCGCCCCGTAAAACG CTGTCGGTAAGGACAGATC	*F. langsethiae* (+ a few *F. sporo-trichioides* strains)

[1]Paavanen-Huhtala (2000), [2]Turner *et al.* (1998), [3]Doohan *et al.* (1998), [4]Nicholson *et al.* (1998), [5]Konstantinova & Yli-Mattila (2004), [6]Parry & Nicholson (1996), [7]Mishra *et al.* (2003), [8]S. Klemsdal (2004, pers. comm.)

The species-and strain-specific primers (Table 3) were used as described in Paavanen-Huhtala (2000), Konstantinova *et al.* (2002) and Konstantinova & Yli-Mattila (2004) to identify *Fusarium* isolates from grain samples. Trichothecenes were extracted and determined from grain samples at EELA as described in Eskola *et al.* (2001), while zearalenone was extracted and determined as described in Eskola *et al.* (2002). Moniliformin (MON), enniatins (ENN) and beauvericin (BEA) were extractred and determined as described in Logrieco *et al.* (2002) and Jestoi *et al.* (2003, 2004). At MTT 25 g of grains were extracted with acetonitrile-water and filtered with suction. The filtrate was purified with the Romer Labs minicolumn.

The purified sample was evaporated to dryness. The trichothecenes were analysed by TMSI GC-MSD analysis (Hietaniemi & Kumpulainen, 1991).

3. RESULTS

3.1. Fusarium *species*

The most severe contamination in barley in Finland took place in 2000 (Eskola *et al.*, 2001; Yli-Mattila *et al.*, 2002, Table 4). In spring wheat the most severe contamination occurrec in 2001 (Table 4). In 1998 and 2000, *F. avenaceum* was clearly the dominant species in all cereals studied. In 2000 the contamination by *F. arthrosporioides* and *F. tricinctum* was greatest. The relative amount of *F. culmorum* started to increase in 2000 and the relative amounts of *F. sporotrichioides*, *F. poae* and *F. graminearum* increased in 2001. In 2002 *F. avenaceum* was still clearly the dominant species in barley, but in spring wheat *F. culmorum* was at the same level, although there was more variation between the samples (results not shown). *F. sporotrichioides* and *F. poae* were most abundant in 2001, while *F. graminearum* was most abundant in 2002. In northwestern Russia the contamination levels were similar to those in Finland, except for higher *F. poae* contamination levels in 2002 and the absence of *F. graminearum* isolates.

Fusarium culmorum contamination levels were highest in both wheat and barley plots inoculated with the spores of this species (Table 5). In wheat, inoculation with other *Fusarium* species reduced the *F. culmorum* contamination level compared to the control. Inoculation with *F. sporotrichioides* increased the *F. sporotrichioides* contamination level, especially in wheat, while *F. avenaceum*, *F. arthrosporioides*, *F. tricinctum* and *F. graminearum* inoculation on barley had no effect on the contamination level of these species as compared to the ones sprayed by other species. *F. graminearum* inoculation had only a slight positive effect on the *F. graminearum* contamination level in wheat. *F. avenaceum* inoculation increased the amount of this species on wheat, while *F. poae* inoculation slightly increased the amount of this species in both wheat and barley grains. The effect of *F. culmorum* inoculation could be observed already after three weeks on the heads of the wheat plot. Spraying with spores of any *Fusarium* species strongly increased *F. tricinctum* and less strongly *F. poae* contamination in barley, as compared to the unsprayed control.

3.2. Mycotoxins

In 2001 deoxynivalenol (DON) levels were highest in oats and spring wheat (late samples after a long rain, Table 2), while in 2002 high DON levels were also found in barley (Table 6). The highest T-2 and HT-2 levels were registered in one oats sample from 2002. The highest NIV levels were found in barley in south-western Finland and in north-western Russia in 2000-2002. *F. poae* was detected by species-specific primers in most barley samples with high NIV concentrations in 2001 and 2002, while *F. culmorum* and *F. graminearum* were connected to high DON concentrations. High enniatin levels were observed in several barley samples in 2001 and 2002 and in spring wheat in 2001, while high MON levels were found in spring wheat in 2001 and barley in 2002. In rye high BEA end enniatin levels were reported in 1998 (Logrieco *et al.*, 2002). After the year 1998 BEA levels were lower and the highest BEA concentration was 19 μg/kg in barley in 2002 (Jestoi *et al.*, 2004). In the rest of the grain samples (Tables 6-9) the *Fusarium* mycotoxin levels were low. NIV was not detected in rye during the years 1998-2002. The ochratoxin A concentrations in grains were also under the detection level (0.5 μg/kg) in 1998-2002, except in three samples (highest level 6 μg/kg in 2002).

According to the preliminary results mycotoxin levels were also low in barley and spring wheat in 2003, while in oats samples containing higher levels of DON (500-2000 μg/kg) and T-2 and HT-2 (500-1500 μg/kg) were found in Central Finland and Southwestern Finland, respectively.

In all inoculated barley plots the NIV levels were 2-5 times higher than in the neighbouring barley field (sample no. 1, Table 2), except for the barley plot inoculated with *F. culmorum*, which in turn had the highest DON and ZEN levels. The DON level of this plot was three-fold compared to the DON level in the neighbouring barley field, while in the other inoculated plots the DON levels were 40-90 % lower than in the neighbouring field. T-2 was only detected in the plot inoculated with *F. sporotrichioides*. The highest ZEN concentration occurred in the plot inoculated with *F. culmorum*.

In inoculated wheat plots DON levels were lower than in the non-inoculated sample from the neighbouring wheat field (sample no. 10, Table 2), except for the plot inoculated with *F. culmorum*. The DON level of this plot was 22-fold compared to the non-inoculated sample. The highest NIV levels (6-8-fold compared to the non-inoculated sample) were found (Table 2) in the plots with the highest *F. poae* contamination (Table 5), while the highest HT-2 and T-2 levels were found in the wheat plot inoculated with *F. sporotrichioides*. The ZEN concentration was highest in the wheat plot inoculated with *F. culmorum*, while no ZEN was detected in other inoculated wheat plots or in the neighbouring wheat field.

Table 4. The mean of the total Fusarium *contamination % and morphologically identified* Fusarium *species in 1998-2002. 200 seeds were studied per sample. In 1998 all samples were dried, in 2000-2002 the Finnish samples were not dried before isolation of* Fusarium *strains. In 2000 and 2002 Russian barley samples were also studied.*

	n	F%	range (%)	*F. avenaceum*	*F. arthro-sporioides*	*F. tricinctum*	*F. culmorum*	*F. sporo-trichioides*	*F. poae*	*F equiseti*	*F. graminearum*	*F. species*
1998												
rye	39	20.0	5-11	7.4	0.7	0.7	0.2	0.2	<0.1	-	<0.1	11.1
barley	15	11.0	4-16	3.1	0.5	0.4	0.3	0.3	<0.1	-	0	6.2
oats	5	9.0	7-14	2.4	0.6	0.4	0.2	0.2	0.2	-	0	5.6
wheat	4	12.0	7-20	4.7	0.3	0.7	0.3	0.3	0	-	0	6.5
2000												
barley	6	30.0	18-42	16.1	5.8	5.8	1.4	0.3	<0.1	0	0	2.7
barley Russia	3	7.7	2-15	5.2	1.2	<0.1	<0.1	0	0	<0.1	0	0.3
2001												
barley	10	18.0	5-44	3.5	3.2	2.1	0.8	2.6	1.8	0.2	0.2	5.0
s.wheat	4	20.4	7-41	6.6	3.6	0.1	0.8	1.1	0	0	0	4.9
w.wheat	1	1,0	1.0	0.5	0.5	0.5	0.5	0	0	0	0	0.5
2002												
barley	8	15.5	3-25	7.9	2.6	1.6	2.2	1.9	0.6	0.1	1.3	0
oats	1	5.0	5.0	5.0	0.5	0.5	0.5	0.5	0.5	0	0	0
wheat	7	8.9	3-17	2.9	0.5	0.1	3.0	0.1	0.1	0	1.1	0
barley Russia	3	10.8	8-15	1.5	0.3	0.9	0	2.4	5.1	0	0	0

Table 5. Mean of total Fusarium contamination % and morphologically identified Fusarium species in grains from inoculated plots of barley cultivar Scarlett and wheat cultivar Mahti in 2002. The mean of the controls for both cereals has been calculated from the six plots inoculated by one of the other six Fusarium species. In barley cultivar Scarlett there was no control not sprayed by any spores, but in barley cultivar Saana and wheat cultivar Mahti there was a control in the neighbouring field, which had not been treated at all by spraying. Three strains for each species were used for inoculation. The numbers of the samples are the same as in Table 2.

Table 5.

Cereal & inoculum	*No.*	*F%*	*F. avenaceum*	*F. arthro-sporioides*	*F. tricinctum*	*F. culmorum*	*F. sporo-trichioides*	*F. poae*	*F. graminearum*
Barley									
F. tricinctum	18	33	3	2	17	0.5	0.5	3	2
F. arthrosporioides	19	52	7	1	28	4	0	1	3
F. avenaceum	20	37	5	3	18	4	0	8	0
F. poae	21	41	7	2	20	5	0	5	0
F. sporotrichioides	22	31	7	0	9	11	4	3	0
F. culmorum	23	33	5	0	5	32	2	2	0.5
F. graminearum	24	42	2	2	23	5	2	1	0
Mean of controls			6.0	1.5	17.0	4.9	0.9	3.0	0.6
Control without spraying	1	15	7	0	2	7	0.5	0	2
Wheat									
F. tricinctum	25	20	2	0.5	2	3	0	0.5	0
F. arthrosporioides	26	18	5	3	3	8	1	0	0
F. avenaceum	27	25	9	3	7	2	0.5	2	3
F. poae	28	19	3	0	5	4	0	2	2
F. sporotrichioides	29	25	3	0	3	1	18	0	1
F. culmorum	30	54	1	0.5	2	47	0	0	0
F. graminearum	31	8	0.5	0.5	0.5	3	0.5	0	2
Mean of controls			2.4	0.7	3.4	3.5	0.3	0.4	1.0
Control without spraying	10	17	1	1	0	13	0.5	0	0.5

Table 6. Mean mycotoxin contents (µg/kg) in oats in Finland in 1998-2002. 2001b samples were harvested late after a rainy period of one week.

	DON		*NIV*		*T-2 + HT-2*		*ZEN*		*BEA/ENN*	
	All	*Pos.*	*All*	*Pos.*	*All*	*Pos.*	*All*	*Pos.*	*All*	*Pos.*
1998										
n=13	111	111	6	42	20	64	0.4	5.5	-	-
range	4-955		0-56		0-116		0-5.5			
1999										
n=10	629	697	0	0	0	0	-	-	-	-
range	0-2150									
2000										
n=25	544	890	38	119	120	336	-	-	-	-
range	0-5004		0-234		0-1369					
2001a										
n=30	237	273	21	70	21	104	-	-	-	-
	0-1560		0-131		0-184					
2001b										
n=7	1354	1680	11	37	116	162	-	-	-	-
range	0-7520		0-37		0-273					
2002										
n=31	609	609	73.5	134.0	71.7	130.7	0	0	18/<38 (n=1)	-
range	<10-3710		0-532		0-427					

Table 7. Mean mycotoxin contents (µg/kg) in feed barley (upper row) and malt barley in Finland in 1998-2002. If there is only one value, division into feed and malt has not been specified. In 2000 and 2002 Russian barley samples were also studied.

	DON		*NIV*		*T-2 + HT-2*		*ZEN*		*BEA+ ENN*	*MON*
	All	*Pos.*	*All*	*Pos.*	*All*	*Pos*	*All*	*Pos.*	*All*	*All*
1998										
n=22	12	18	3	20	5	22	0.2	4	-	-
range	0-43		0-20		0-41		0-4			
1999										
n=32	116	148	16	256	0	0	0	0	-	-
n=18	68	111	44	264	0	0	0	0		
range	0-278		0-351		0		0			
2000										
n=20	52	80	0	0	0	0	0	0	-	-
n=31	48	113	95	266	24	149	<1	6	<10(n=1)	
range	15-394		0-1496		0-388		0-6			

Russia										
n=3	100	100	1443	1443	0	0	-	-	-	-
range	15-169		209-3670							
2001										
n=59	71	85	40	80	21	90	8	31	3095 (n=14)	30 (n=14)
range	0-619		0-405		0-345		0-53		0-18437	0-42
2002										
n=53	355	373	10	50	12	34	9	60	2594 (n=8)	294 (n=8)
	158	193	214	247	23	62	<1	8		
range	0-941		0-1390		0-108		0-175		0-7359	0-747
Russia										
n=3	<10	<10	294	441	0	0	0	0	-	-
range	0-<10		0-742							

Table 8. Mean mycotoxin contents (µg/kg) in spring wheat (upper row) and winter wheat in Finland in 1998-2002.

	DON		*NIV*		*T-2 + HT-2*		*ZEN*		*BEA + ENN*	*MON*
	All	*Pos.*	*All*	*Pos.*	*All*	*Pos*	*All*	*Pos.*	*All*	*All*
1998										
n=31	36	43	2	17	0	0	0.3	5	-	-
range	0-190		0-17		0		0-5			
1999										
n=35	57	100	4	130	0	0	0	0	-	-
n=3	0	0	0	0	0	0	0	0	-	-
range	0-264		0-130		0		0			
2000										
n=25	224	250	0	0	0	0	<1	10	-	-
n=11	24	133	24	<30	0	0	0	0	<10 (n=1)	-
range	0-1026				0		0-12			
2001										
n=29	217	330	2	30	7	47	0	10	4122 (n=4)	280 (n=4)
n=13	2	10	2	30	2	20	0	0	88 (n=3)	<7 (n=3)
range	0-4301		0-30		0-79		0		0-24836	0-810
2002										
n=32	338	373	9	42	3	28	0	0	145 (n=7)	61 (n=7)
n=10	<25	40	8	38	0	0	0	0	-	-
range	0-1980		0-79		0-27		0			

Table 9. Mean mycotoxin contents (μg/kg) in rye in Finland in 1998-2002.

	DON		*T-2 + HT-2*		*ZEN*		*BEA*		*ENN*	
	All	*Pos.*	*All*	*Pos.*	*All*	*Pos.*	*All*	*Pos.*	*All*	*Pos.*
1998										
n=13	24	27	0	0	0	0	1895	1895	4900	7080
range	0-111		0		0		640-3500		0-13500	
1999										
n=10	0	0	0	0	0	0	-	-	-	-
range	0		0		0					
2000										
n=15	12	178	7	106	0	0	-	-	-	-
range	0-178		0-106		0					
2001										
n=11	7	25	0	0	0.3	3	0(n=1)	-	<62(n=1)	-
range	0-37		0		0-3		0			
2002										
n=10	4	43	8	42	0.7	7	-	-	-	-
range	0-43		0-56		0-7					

4. DISCUSSION

The growing seasons in Finland are highly variable, and great variation in the relative amounts of different *Fusarium* species in cereals appeared as well during the 1970s (Ylimäki *et al.*, 1979; Ylimäki, 1981) and 1980s (Hietaniemi & Kumpulainen, 1991). During the rainy and warm year of 1972 *Fusarium* levels were high (Ylimäki, 1981), as also during 1976 and 1977, especially in oats, in which they were almost at the same level as in 1972 (Ylimäki *et al.*, 1979). *F. avenaceum* has always been one of the most common *Fusarium* species, while the abundance of other *Fusarium* species has shown more variation. In 1998 *F. avenaceum* was the dominant *Fusarium* species, in 2000 *F. arthrosporioides* and *F. tricinctum* were also common and in 2001-2002 greater amounts of other *Fusarium* species were found. According to the inoculation experiments, the wheat cv. Mahti is most sensitive to *F. culmorum*, *F. sporotrichioides* and *F. avenaceum*, while the barley cv. Scarlett is most sensitive to *F. culmorum*. Spraying with the spores of any *Fusarium* species induced a natural infection of *F. tricinctum* in barley, while spraying with other *Fusarium* species inhibited the natural infection of *F. culmorum* in wheat. These findings confirm previous results of Ylimäki *et al.* (1979), according to which *F. avenaceum, F. culmorum* and *F. tricinctum* inocula in barley and *F. culmorum* inocula in oats clearly increased the contamination percentages of these species. The weak effect of *F. graminearum* inocula may be due to the low number of spores obtained.

Trichothecene concentrations were higher in Finland in 2000-2002 than during the rainy and cold growing season of 1998. Most of the highest toxin levels were found in western and central Finland. High levels of trichothecenes were also reported during the warm and rainy growing season of 1984 (Karppanen *et al.*, 1985) and the warm and drier growing season of 1988, as compared to the cold and rainy growing season of 1987 (Hietaniemi & Kumpulainen, 1991). In 1984 and 1988 *F. graminearum* was also more common in the grain samples than usually in the 1980s. In 2001 the highest DON levels in oats and spring wheat were registered in western Finland, where a heavy rain of one week delayed harvesting in September. In Norway high levels of DON have also been found in oats e.g. during 1988 and 1991-1994, when growing seasons were unusually rainy, especially during anthesis in July (Langseth *et al.*, 2001). At the end of the 1990s DON levels were lower in Norway, probably due to the dry July of 1996 and 1997, but the DON level also remained low during the cold and rainy growing season of 1998, probably because it favoured *F. avenaceum,* which does not produce DON (Langseth & Rundberget, 1999; Bottalico and Perrone, 2002).

In the oat and wheat samples of the years 2001-2002 the presence of *F. graminearum* correlated with high levels of deoxynivalenol, while Finnish *F. avenaceum* strains from rye samples of the year 1998 contained high levels of beauvericin and enniatins (Logrieco *et al.*, 2002). Enniatins are present in all grain samples in Finland (Jestoi *et al.* 2004), but the levels of beauvericin were low in 2001-2002 as compared to the study of Logrieco *et al.* (2002). A correlation has also been detected between the contamination levels by *F. avenaceum* and enniatins and moniliformin in Finnish grains in 2001-2002 (Jestoi *et al.*, 2003, 2004). Trichothecene concentrations were generally at the same level in the samples analysed at EELA and MTT. It seems that warm and rainy growing seasons, especially during anthesis, may result in higher levels of trichothecene-producing *F. graminearum* and *F. culmorum* strains in Finland, while during colder growing seasons *F. avenaceum* and its mycotoxins are more common in Finnish grains. Many more growing seasons and more samples per year from different regions of Finland are still needed before the toxin and *Fusarium* contamination levels can be considered representative of the whole country.

The nivalenol levels found in Finland and northwestern Russia may be due to the combination of exceptionally rainy and warm weather in certain parts of southern Finland and northwestern Russia. In Sweden, *F. poae* was the most common *Fusarium* species in NIV-contaminated grain samples (Pettersson *et al.* 1995). A slight connection was also observed between *F. poae* and high NIV levels in the present study. The capacity of each *F. culmorum*, *F. sporotrichioides* and *F. poae* isolate to produce nivalenol should be studied before it is possible to know by which species nivalenol was produced in the samples with the highest nivalenol contents. We are currently studying the toxin profiles of Finnish strains of *F. avenaceum, F. arthrosporioides, F. tricinctum, F. culmorum, F. sporotrichioides, F. poae* and *F. graminearum* grown on sterile rice at different temperatures.

Ochratoxin A was studied from almost all Finnish grain samples of the present study, but it was found only in three samples, which is probably due to the quick drying of grains after harvesting. Aflatoxins and fumosinins have not been detected in Finnish grains.

In 2001 243 (mainly different nuts, dried fruits and spices) and 244 (mainly grain products, dried fruits and coffee) imported food samples were studied for aflatoxins and ochratoxin A by the Finnish customs authorities, respectively (Tullilaboratorio, 2001). 17 % of the samples contained aflatoxins with highest levels in peanuts (0.38 mg/kg), hazelnuts (0.011 mg/kg), *Pistacia vera* (0.074 mg/kg), *Myristica fragrans* (0.029 mg/kg) and rice (0.037 mg/kg). 26 % of the samples contained ochratoxin A with highest levels 0.016 mg/kg in coffee and 0.017 mg/kg in biscuits. In addition, a few *Fusarium* toxin and patulin analyses were performed.

In 2002 251 (mainly different nuts) and 275 (mainly grain products, dried fruits and coffee) imported food samples were studied for aflatoxins and ochratoxin A by the Finnish customs authorities, respectively (Tullilaboratorio, 2002). 21 % of the samples contained aflatoxins with highest levels in *Pistacia vera* (0.169 mg/kg), hazelnuts (0.052 mg/kg), almonds (0.009 mg/kg) and curry (0.016 mg/kg). 26 % of the samples contained ochratoxin A, highest levels 0.011 mg/kg in grains and 0.025 mg/kg in coffee. 2 % in 2001 and 3.5 % in 2002 of the tested samples were rejected by the customs according to the regulations of EU. In addition, a few patulin analyses were performed.

5. CONCLUSION

In oats, samples containing high DON and ZEN concentrations were found in Finland, but the high ZEN concentrations were later identified as artefacts. The highest T-2 and HT-2 concentrations were found in oats in 2003 after an especially hot and dry summer. During the cold and rainy year 1998 trichothecene and ZEN concentration were low in all cereals studied, but BEA and ENN concentrations were high in rye. High ENN and MON concentrations were also registered in 2001 in barley and spring wheat after a long rainy period and in barley in 2002. In barley high concentrations of NIV were detected in a couple of samples in 2000, while in 2001 after a rainy period DON concentrations were highest in the oats of the western part of Central Finland. In winter wheat, rye and barley, which are harvested early, trichothecene and ZEN levels were generally low. In all domestic cereals ochratoxin A levels were low. At species level *F. avenaceum* was more dominant in 1998 and 2000 as compared to the year 2001, while *F. culmorum* was more common in 2000/2002 and *F. sporotrichioides, F. graminearum* and *F. poae* in 2001/2002. Quantitative PCR measurements of five *Fusarium* species from DNA extracted from grain samples of the years 2002-2003 are currently in progress.

6. ACKNOWLEDGEMENTS

This study was supported financially by the Academy of Finland (Grant nos 34375 and 52104 and travel grants to T. Yli-Mattila), National Technology Agency of Finland, COST 835 (travel grants to COST 835 meetings) and the Raisio Group and Turku University Foundations. The Russian grain samples and *Fusarium* isolates by Dr. T. Gagkaeva and the comments of Prof. M. Levitin and Dr. Gagkaeva on the manuscript are gratefully acknowledged. The English was checked by Ellen Valle of the Department of English, University of Turku.

7. AFFILIATIONS

Tapani Yli-Mattila & Sari Paavanen-Huhtala
Lab. of Plant Phys. and Mol. Biol., Dept of Biology, Univ. of Turku, FIN-20014 Turku, Finland; tymat@utu.fi

Päivi Parikka
MTT Agrifood Research Finland, Plant Production Research/Plant Protection, FIN-31600 Jokioinen, Finland
Veli Hietaniemi, MTT Agrifood Research Finland, MTT Chemistry Laboratory, FIN-31600 Jokioinen, Finland

Aldo Rizzo & Marika Jestoi
National Veterinary and Food Research Institute (EELA), Department of Chemistry, FIN-00581 Helsinki, Finland

8. REFERENCES

Bottalico, A. & Perrone, G. (2002) Toxigenic *Fusarium* species and mycotoxins associated with head blight in small-grain cereals in Europe. European Journal of Plant Pathology 108: 611-624.

Doohan, F.M., Parry, D.W., Jenkinson, P. & Nicholson, P. (1998) The use of species-specific PCR-based assays to analyse *Fusarium* ear blight of wheat. Plant Pathology 47: 197-205.

Eskola, M., Kokkonen, M. & Rizzo, A. (2002) Application of manual and automated systems for purification of ochratoxin A and zearalenone in cereals with immunoaffinity columns. Journal of Agricultural Food Chemistry 50: 41-47.

Eskola M, Parikka P and Rizzo A (2001) Trichothecenes, ochratoxin A and zearalenone contamination and *Fusarium* infection in Finnish cereal samples in 1998. Food Additives and Contaminants 18: 707-718.

Hietaniemi, V. & Kumpulainen, J. (1991) Contents of *Fusarium* toxins in Finnish and imported grains and feeds. Food Additives and Contaminants 8: 171-182.

Jestoi, M., Rokka, M., Rizzo, A., Peltonen, K., Parikka, P. & Yli-Mattila, T. (2003) Moniliformin in Finnish grains: analysis with LC-MS/MS. Aspects of Applied Biology 68: 211-216.

Jestoi, M., Rokka, M., Yli-Mattila, T., Parikka, P., Rizzo, A. & Peltonen, K. (2004) The presence and concentrations of the *Fusarium*-related mycotoxins, beauvericin, enniatins and moniliformin in Finnish grains. Food additives and Contaminants (manuscript submitted).

Karppanen, E., Rizzo, A., Berg, S., Lindfors, E. & Aho, R. (1985) *Fusarium* mycotoxins as a problem in Finnish feeds and cereals. Journal of of Agricultural Science in Finland 57: 195-206.

Konstantinova, P., Karadzhova, J., Yli-Mattila, T. & van den Bulk, R. (2002) Molecular detection of *Alternaria alternata* and *Fusarium* spp. In barley seeds and comparison with routine testing assays. Petria 12: 239-247.

Konstantinova, P. & Yli-Mattila, T. (2004) IGS-RFLP analysis and development of molecular markers for identification of *F. poae, F. pulverosum, F. sporotrichioides* and *F. kyushuense* International Journal of Food Microbiology (in press).

Langseth, W., Elen, O. & Rundberget, T. (2001) Occurrence of mycotoxins in Norwegian cereals 1998-1999. In: Logrieco, A (ed.) Cost Action 835 Occurrence of toxigenic fungi and mycotoxins in plants, food and feed in Europe. (pp. 105-110) European Communities, Belgium.

Langseth W and Rundberget T (1999) The occurrence of HT-2 toxin and other trichothecenes in Norwegian cereals. Mycopathologia 147: 157-165.

Logrieco, A., Rizzo, A., Ferracane, R. & Ritieni, A. (2002) Occurrence of beauvericin and enniatins in wheat affected by *Fusarium avenaceum* head blight. Applied and Environmental Microbiology 68: 82-85.

Mishra, P.K., Fox, R.T.V. & Culham, A. (2003) Development of a PCR-based assay for rapid and reliable identification of pathogenic Fusaria. FEMS Microbiology Letters 218: 329-332.

Nicholson, P., Simpson, D. R., Weston, G., Rezanoor, H. N., Lees, A. K., Parry, D. W. & Joyce, D. (1998). Detection and Quantification of *Fusarium culmorum* and *Fusarium graminearum* in cereals using PCR assays. Physiological and Molecular Plant Pathology 53: 17-37.

Paavanen-Huhtala, S. (2000) Molecular-based assays for the determination of diversity and identification of Fusarium and Gliocladium fungi. – Ph.D. Thesis, Annales Universitatis Turkuensis AII 139, University of Turku, Finland, pp. 198, Painosalama OY, Turku.

Parry, D. W. & Nicholson, P. (1996). Development of a PCR assay to detect *Fusarium poae* in wheat. Plant Pathology 45: 383-391.

Pettersson, H., Hedman, R., Engström, B., Elwinger, K. & Fossum, O. (1995). Nivalenol in Swedish cereals – occurrence, production and toxicity towards chickens. Food Additives and Contaminants 12: 373-376.

Tullilaboratorion vuosikertomus (2001) (report from Finnish customs, http://www.tulli.fi/tullihallinto/tullipiirit/tullilaboratorio).

Tullilaboratorio vuosikertomus (2002) (report from Finnish customs, http://www.tulli.fi/tullihallinto/tullipiirit/tullilaboratorio).

Turner, A.S., Lees, A.K., Rezanoor, H.N. & Nicholson, P. (1998). Refinement of PCR-detection of *Fusarium avenaceum* and evidence from DNA marker studies for phenetic relatedness to *Fusarium tricinctum.* Plant Pathology 47: 278-288.

Yli-Mattila, T., Paavanen-Huhtala, S., Parikka, P., Konstantinova, P., Gagkaeva, T. (2004) Molecular and morphological diversity of Fusarium fungi in Finland and northwestern Russia. European Journal of Plant Pathology (manuscript submitted).

Yli-Mattila, T., Paavanen-Huhtala, S., Parikka, P., Konstantinova, P., Gagkaeva, T., Eskola, M., Rizzo, A. (2002). Occurrence of Fusarium fungi and their toxins in Finnish cereals in 1998 and 2000. Journal of Applied Genetics 43A: 207-214.

Ylimäki, A. (1981) The mycoflora of cereal seeds and some feedstuffs. Annales Agriculturae Fenniae 20: 74-88.

Ylimäki, A, Koponen, H, Hintikka, E-L, Nummi, M, Niku-Paavola, M-L, Ilus, T, Enari, TM (1979). Mycoflora and occurrence of Fusarium toxins in Finnish grain. Technical Research Centre of Finland, Materials and Processing Technology publication 21, pp. 28, Valtion Painatuskeskus, Helsinki.

B. BARRIER-GUILLOT, M. DELAMBRE, A. MOREL,
C. MAUMENE, H. GOUET, F. GROSJEAN & M. LEUILLET

OCCURRENCE OF DEOXYNIVALENOL IN WHEAT (*TRITICUM AESTIVUM*) GROWN IN FRANCE AND IDENTIFICATION OF AGRONOMIC FACTORS INVOLVED IN CONTENT VARIATIONS

1. INTRODUCTION

The European Community has scheduled to reduce drastically sanitary risks for human and animals through foods and feeds. This decision is of importance since French agricultural production has the first place in Europe for many products. Consequently, it has been important for cereal producers to identify what contaminants could be present in grains, what are their contents, what are the factors that influence these contents and what are the possible ways to reduce the observed levels. This paper deals with the case of deoxynivalenol (DON), which is the most common mycotoxin found in cereals produced in France and Europe. DON belongs to the group of trichothecenes and is produced by some strains of *Fusarium graminearum* and *F. culmorum*. DON is considered to be a very stable compound which is not degraded, both during storage and further processing, including high temperatures during the processing-cooking food. Therefore, the best way to minimize the risk for DON to enter the food chain is to prevent its formation in the field.

2. MATERIAL AND METHODS

The study was carried out from 1999 to 2002. On a whole, this study involved 765 fields, derived either from different field trials carried out by ARVALIS-Institut du végétal, either from farm fields in a survey carried out by ARVALIS-Institut du végétal in collaboration with INVIVO cooperative union, and concerning wheat (*Triticum aestivum*). Among the 765 fields, 373 were used to assess between 1999 and 2002 the effect of different factors alone or in interaction, such as cultivation system, previous crops, varieties and fungicide protection. Samples involved originated from 16 different areas in France, but more than the half of them were from only 3 regions (Centre, Ile de France and Pays de Loire). Because several samples have been collected in each trial, 28 sites only were represented. The other 392 samples came from farm fields in the survey carried out with cooperatives in 2001 and 2002. This survey was performed in order to study the different agronomic factors related to DON contamination of wheat, their hierarchy, the relative weight

A. Logrieco and A. Visconti (eds.),
An Overview on Toxigenic Fungi and Mycotoxins in Europe, 101-108.

of each factor and the interactions between these factors. The survey was conducted in four main areas in France (Bretagne, Centre, Champagne-Ardennes and Lorraine). Farm fields involved were chosen in each area according to their agronomic characteristics: previous crop, cultivation system and wheat variety (susceptibility to *Fusarium* head blight). The aim of the experimental plan was to obtain, if possible, numbers of plots for each factor as similar as possible, and at least relatively balanced. The geographical origins on the 765 fields involved and their main agronomic characteristics are reported in table 1.

Table 1. Main characteristics of samples involved in the study (n=765)

French area	*(%)*	*Previous crop*	*(%)*	*Tillage system*	*(%)*	*Varietal susceptibility*	*(%)*
Centre	27.6	Maize	32.1	Ploughing	80	High	30
Bretagne	23.8	Rapeseed	23.0	No-ploughing	20	Medium	64
Champagne-Ardennes	9.5	Pea	10.6			Low	6
Ile de France	8.2	Wheat	7.5				
Lorraine	8.1	Sugar beet	6.9				
Normandie	6.2	Potato	5.5				
Pays de Loire	4.4	Sunflower	3.1				
Picardie	3.7	Ray grass	2.5				
Midi Pyrénées	3.0	Others	8.8				
Rhône-Alpes	2.2						
Others	3.3						

All agronomic parameters were recorded for each plot (experimental plots and farm fields). So, all agronomic practices were known in each field before cropping and sampling. Three different samples of a minimum of 1 kg were taken in the field at the emptying of the combine harvester. The 3 samples were then gathered, mixed to be homogeneous, divided and cleaned (laboratory cleaner Rohr, mini Petkus) to remove impurities (straw, glumes, small and broken grains, dusts, etc.) so that a minimum of 1 kg of cleaned and homogeneous grains remained. These obtained samples were entirely ground in a mill fitted with a 1 mm screen before analysis of trichothecenes (types A and B) by gas chromatography/mass spectrometry or DON by liquid chromatography (Qualtech laboratory, 7 rue du bois de la Champelle, BP 86, 54503 Vandoeuvre-lès-Nancy, France and INZO° laboratory, BP 19, Chierry, 02402 Château-Thierry, France).

Moreover, because of the major impact of climatic conditions, weather variables (daily rainfall, minimum, average and maximum daily temperatures...) around the flowering period were estimated for each field or experimental plot from nearby weather stations. The date of flowering was estimated for each field according to the date of sowing and to the variety (ARVALIS-Institut du végétal, unpublished model). Cultivar susceptibility was evaluated from initial score of *Fusarium* susceptibility assigned at the registration of each cultivar, and the results of DON accumulation observed in previous experiments in our Institute (unpublished data).

Indeed, each sample was characterized by agronomic and climatic variables with different modalities for each one and DON content. The median level was calculated for each modality of the different variables. Correlations between quantitative variables and DON content were calculated. The different factors involved in DON contamination were then evaluated by segmentation method, which consists of an explanation of a qualitative variable with a set of qualitative variables. DON content was converted into classes to establish a qualitative variable. Segmentation outlines a tree structure that allows delineating the hierarchy between the factors involved, and a ranking of the different modalities of each factor. Segmentation cuts the initial population at each level in two parts, in gathering modalities with the same DON content in a same branch of the tree structure.

3. RESULTS AND DISCUSSION

Trichothecene content was determined in wheat samples harvested between 1999 and 2001. The type A trichothecenes were not detected in any of all the samples, whereas among type B trichothecenes, nivalenol was exceptionally detected (less than 2% of positive results) and DON was detected more frequently. These results contrast with those recently reported by the European Commission with respectively 21%, 12% and 14% of positive samples for T-2 toxin, HT-2 toxin and nivalenol in wheat and wheat flour (SCOOP task, 2003). These results are slightly lower for the contribution of France in this SCOOP task, with respectively 10%, 11% and 11% of positive samples. Because of the lack or the very low level of occurrence of other trichothecenes than DON in our study on one hand, and the fact that the European Commission projects a regulation on *Fusarium* toxins for DON only in food and grains, we chose to analyse only DON content in 2002. The remainder of this paper will only deal with the case of deoxynivalenol (DON) between 1999 and 2002. The main results of DON contents are shown in table 2, in comparison with those of the SCOOP task ones for France contribution.

The median content *i.e.* the 50th percentile of the population is very low (60 µg/kg), close to the limit of quantification of DON. This result is in good agreement with the median contents reported in the SCOOP task for France with levels between the limit of detection and 190 µg/kg according to data sets for approximately the same years (2000 to 2002). However, it is important to notice a clear year-effect. Indeed 2001 was characterized by very low levels, with median values at the level of the limit of detection of methods in our study as well as in the SCOOP survey. On the opposite, 1999, 2000 and 2002 could be characterized as so-called "*Fusarium*-years" with higher median and mean concentrations of DON. However, if good conformity was observed between the two studies for median levels and year-effect, the results show higher maximums values for the present study (table 2).

Table 2. Main results concerning DON (μg/kg) in wheat in the present study compared with results of the SCOOP task for France.

	Present study	*French data** *(from SCOOP task, 2003)*
Number of data	765	844
Median value	60	lod-190
Mean value	275	177
% ≥ 1000 μg/kg	6.9	1.3
% ≥ 1500 μg/kg	3.7	-
% ≥ 2000 μg/kg	2.5	0.1
Median value in 1999	290	-
Median value in 2000	120	190
Median value in 2001	30 (=lod**)	lod
Median value in 2002	116	100

* Only data from chromatographic methods are selected
**lod : limit of detection

The DON results above 1000 μg/kg - a possible future threshold adopted by the European Commission for DON regulation (European Commission, 2003) -, 1500 or 2000 μg/kg reach 6.9, 3.7 and 2.5 % of the population involved whereas it concerns approximately 1% in the SCOOP task. This difference can be explained by at least two main reasons: it can be reminded here that part of fields in our study was selected according to the main agronomic factors well known to contribute to DON accumulation. Moreover, our data population was not representative of the real agronomic practices in the country (choice according to involved factors) and therefore overestimated the risk of DON contamination. The second reason emanates from the site of sampling: if most samples involved in general studies or surveys are taken in the silos or before the first processing stage after gathering of grains from several fields, our samples were directly taken from individual fields, so with a maximum of variability and subsequently more higher contents. Indeed, it is well established that crop residues play a role in the disease cycle, especially when decomposition is slow. Propagules from crop debris are the main source of initial inoculum (Xu, 2003), and populations of *Fusarium graminearum* on crop residues are affected by incorporation depth (Yi *et al.*, 2002). The nature of previous crop combined with non-ploughing system is of great influence, even with neighbored plots as shown in figure 1, with a trial carried out in 2000. These results confirm those of Dill-Macky and Jones (2000), Bauer, (2000) and Obst *et al.* (2000) who reported that absence of tillage after a previous maize crop was the highest risk factor leading to high levels of DON contamination.

However, *Fusarium* head blight epidemics are usually associated with abundance of inoculum. The crucial role of climatic conditions on the extent of *Fusarium* infection and subsequent mycotoxin formation was already reported, and is illustrated in Figure 1 with comparison of years. Regardless of the level of inoculum available, favourable weather conditions are critical for infection to occur in wheat heads (Hooker *et al*, 2002).

Figure 1: Effect of cultivation system and previous crop on DON results (µg/kg) in wheat grain at the Boigneville experimental station in 2000 and 2001

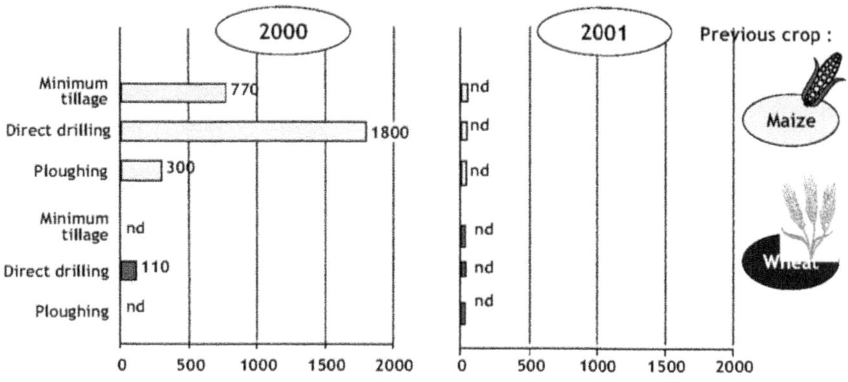

Table 3. Description of the different risk classes and the variables used in the model of prediction of risk of DON in wheat

	Weather at flowering	*Previous crop*	*Cultivation system*	*Variety susceptibility*	*Median level (µg/kg)*
Risk 1	No risk	No risk	All systems	All	30 (=lod)
Risk 2	No risk	High risk	All systems	All	30 (=lod)
Risk 3	Risk	No risk	All systems	Low, medium	60
Risk 4	Risk	No risk	All systems	High	120
Risk 5	Risk	High risk	Ploughing	All	315
Risk 6	Risk	High risk	No ploughing	All	1220

Figure 2. Allotment of DON results according to the 6 risks of the model

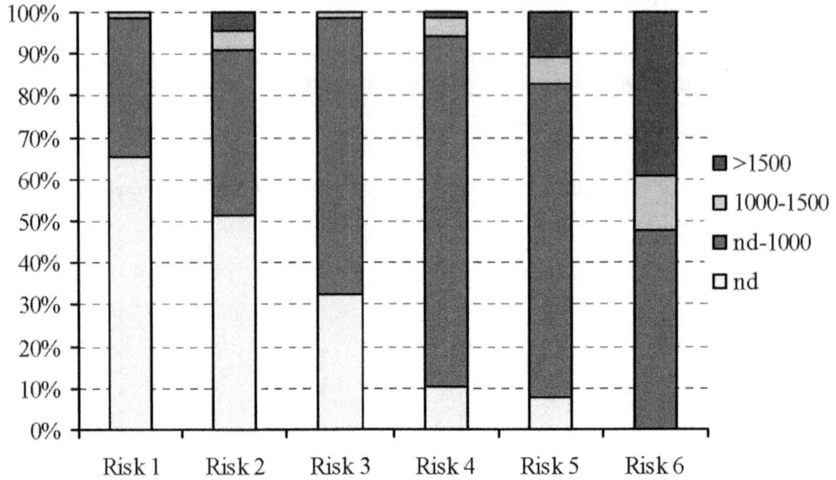

In particular, recent works on understanding *Fusarium* head blight development and mycotoxin production has been based on field observations (Obst *et al.*, 2000; Schaasma *et al.* 2001, Hooker *et al*, 2002) and these field-based studies confirmed that temperatures and moist conditions during heading and anthesis are the key factors (Xu, 2003).

In our study, we worked also during this period. After gathering all agronomic parameters and climatic variables, the description of the tree structure obtained by segmentation is shown in table 3, leading to 6 risk classes, and the allotment of DON results according to the 6 risks of the model are shown in figure 2.

Results of the segmentation lead to a tree structure with 6 different risks. The higher the risk number, the greater the DON content (median level). The results also confirm the essential impact of climate around flowering period since weather variables are the most important factor in the model. Without any climatic risks, the model leads to very low occurrence of DON, with 66% of negative results (under limit of detection) for risk 1, the probability being slightly higher for risk 2 with high risk previous crop, especially with some samples above 1000 µg/kg. The model cuts the modalities for previous crop in two populations. On the side of previous crop with high risk, maize was, as expected and also reported by several authors, effectively found in this category and was surprisingly joined with other crops as soyabean, sugar beet and potato. This result must be carefully taken into account because of the low number of data for these crops on one hand, and on another hand the fact that these crops were only present in our database some years, in particular years or sites with high DON contents. However, Obst *et al.* (2000) also reported slightly higher DON values in wheat grain after sugar beet and potato than after wheat or barley. It should also be noted that spring previous crops seem to induce higher DON contents in the following wheat than autumn ones.

In the case of climatic risk, the importance of available inoculum is essential. A high risk previous crop combined with a non-ploughing system leads to the highest DON contents (risk 6), with a median value of 1220 µg/kg for this population, *i.e.* more than half of this population (52%) is susceptible to surpass a possible threshold for future European regulation. It is once more demonstrated that the combination of different factors increased exponentially DON content, as already reported by Obst *et al.* (2000). In our study, the association of abundance of inoculum (high risk previous crop without ploughing) with unfavourable weather conditions around the time of flowering generates much higher levels than other situations. It represents approximately between 1 and 3% of surface areas of wheat grown in France. Ploughing under, thereby reducing the availability of crop residues at the soil surface, has greatly reduced DON contamination, with a median level of 335 µg/kg and 11% above 1000 µg/kg (risk 5). A last, with low risk previous crop, DON contamination depends of the sensitivity of cultivars against *Fusarium* (risks 3 and 4). Resistance of cultivars against *Fusarium* head blight and DON accumulation seems therefore to occur when the inoculum pressure is not too high.

4. CONCLUSION

The presence of DON in wheat grain cannot be completely avoided with preventive measures due to the critical influence of *weather* conditions at the time of flowering. Important weather variables and their timing for spore release and head infection need to be further examined for predicting DON contaminations in wheat grain. However, the hierarchy of the different agronomical and climatic factors has become better and better appreciated. If available models are not applicable to predict with a very good accuracy an absolute DON level in wheat grain field by field, they could be useful to help wheat storage industry to manage their crop sanitary qualities. Finally, in order to evaluate the DON occurrence in an area or a country with a good accuracy, the sampling plan has to take into account representatively agro-climatic factors and therefore needs to involve an important number of data.

5. REFERENCES

Bauer, G., 2000. Zur Analyse der Daten des Fuarium-Monitorings Bayern. In: Risiken durch den Ährenparasiten *Fusarium graminearum* – Ergebnisse eines LPB –Forschungsverbunds. Bodenkultur und Planzenbau, Schriftenreihe der Bayerische Landesanstalt für Bodenkultur und Pflanzenbau, 33-37.

Dill-Macky, R., Jones R.K., 2000. The effect of previous crop residues and tillage on *Fusarium* head blight of wheat. Plant disease, 84, 71-76.

European Commission, 2003. Consultation paper on *Fusarium*-toxins. European Commission, Health and Consumer Protection DG, October 2003, 1-6.

Hooker, D.C., Schaasma, A.W., & Tamburic-Ilincic, L., 2002. Using weather variables pre- and post-heading to predict deoxynivalenol content in winter wheat. Plant disease, june 2002, 611-619.

Obst, A., Lepschy, J., Beck, R., Bauer, G, & Bechtel, A., 2000. The risk of toxins by *Fusarium graminearum* in wheat – interactions between weather and agronomic factors. Mycotoxin research, 16A/1, 16-20.

Schaasma, A.W., Tamburic-Ilincic, L., Miller, J.D., & Hooker, D.C., 2001. Agronomic considerations for reducing deoxynivalenol in wheat grain. Canadian Journal of Plant Pathology, 23, 279-285.

SCOOP task, 2003. Collection of occurrence data of Fusarium toxins in food and assessment of dietary intake by the population of EU member states. Final Report SCOOP Task 3.2.10. Part A: Trichothecenes. European Commission, Health and Consumer Protection DG, October 2003, 1-237. http://europa.eu.int/comm/food/fs/scoop/task3210.pdf.

Xu, X., 2003. Effects of environmental conditions on the development of Fusarium ear blight. European Journal of Plant Pathology, 109, 683-689.

Yi, C., Kaul, H.-P., Klüber, E., & Aufhammer, W., 2002. Populations of Fusarium graminearum on crop residues as affected by incorporation depth, nitrogen and fungicide application. Journal of Plant Diseases and Protection, 109, 3, 252-263.

V. CURTUI, C. SEIDLER, E. SCHNEIDER & E. USLEBER

MYCOTOXINS IN FOOD IN GERMANY

1. INTRODUCTION

Mycotoxins are worldwide contaminants in food and feed, and they generate an increasing concern for human and animal health (Pittet, 1998). Cases of acute primary mycotoxicosis in humans have been reported, although with low frequency, while acute intoxication of animals is relatively common. However, chronic disease caused by low but constant dietary intake of mycotoxins seems to be the most important factor in aspects of public health. Long-term effects are well documented in animals, and epidemiological studies indicate that a correlation between chronic intake of some mycotoxins and the incidence of specific diseases may exist.

Therefore, many countries, including the European Union have adopted special regulations regarding the control of some mycotoxins in foods, and, undoubtedly, more regulations will follow in the years to come.

The control of mycotoxins is associated with economic costs, has implications on national and international trade, and therefore is the subject of not only scientific but also political discussions. An important factor influencing mycotoxin legislation is the knowledge of dietary exposure of the consumer. In this respect, there is the interest of each country to survey the occurrence of mycotoxins in foods, to assess the exposure of consumers, and ensure that food-producers apply all necessary measures to minimize the mycotoxin contamination in food. In Germany, research groups from federal or state laboratories, as well as from academic institutions, have spent considerable efforts during the last years to provide the necessary data basis.

The natural occurrence of about 20 mycotoxins has been investigated. Most attention has been given to aflatoxins, ochratoxin A (OA), deoxynivalenol (DON), zearalenone (ZEA), and fumonisins. This paper will therefore focus on recent studies regarding the occurrence of these mycotoxins, and on the exposure of the German consumer.

2. OCCURRENCE OF MYCOTOXINS AND CONSUMERS' EXPOSURE

2.1. Aflatoxins

Aflatoxins occur mostly in foods originating from tropical and subtropical countries. In Europe, they are considered as an imported problem, and therefore a strict control system is applied to the imported critical foods. Like in other EU countries, aflatoxins have been frequently detected in imported pistachio, figs, hazelnuts, and spices in the 1990s. Within a monitoring program developed between

A. Logrieco and A. Visconti (eds.),
An Overview on Toxigenic Fungi and Mycotoxins in Europe, 109-121.

1995 and 1999, about 50 to 70% of pistachio samples contained aflatoxins exceeding the maximum admissible level (4 μg/kg, sum of aflatoxins B_1, B_2, G_1, and G_2) (BGVV, 1999). Since 1999, the situation has gradually been improved, and less than 20% of samples exceeded the maximum admissible level (Landesanstalt für Umweltschutz Baden-Württemberg, 2003). In another study, only 6.3 % of food samples were reported with aflatoxin content exceeding the maximum admissible level (Bayerisches Landesamt für Gesundheit und Lebensmittelsicherheit, 2001). However, pre-processed material may also be of importance, as demonstrated by the high aflatoxin B_1 contamination of up to 415 μg/kg found in pistachio-paste for ice-cream production (Burgmaier-Thielert & Thielert, 2000).

The authors reported particular cases of uncontrolled imports from other EU countries. With low frequency, aflatoxins are still found in imported spices, like paprika, chilli, and nutmeg, but the incidence of samples with aflatoxin content exceeding the maximum admissible level seems to decrease with increasingly efficient control measures (Landesanstalt für Umweltschutz Baden-Württemberg, 2003).

Likewise, the control of imported feedstuffs for aflatoxin B_1 resulted in a marked decrease of aflatoxin M_1 (AFM_1) levels in milk. In the late 1980s, domestic milk contaminated with AFM_1 was relatively often identified; the levels occasionally exceeded the maximum admissible level (50 ng/kg; European Commission, 2001).

Due to stringent control of certain imported feedstuffs, the situation has been changed in the 1990s, when the incidence of AFM_1 in milk decreased yearly. Continuing monitoring studies of farm milk for AFM_1 performed in several states of Germany in recent years show that - in almost all samples - toxin levels are below 10 ng/L (Milchprüfring Baden-Württemberg, 2002).

2.2. Ochratoxin A

Ochratoxin A is a mycotoxin with nephrotoxic, teratogenic, and carcinogenic properties. It occurs mainly in foods and feeds of plant origin, but can also move up into the food chain. OA may accumulate in edible animal tissues, and therefore humans are exposed to OA not only by consuming contaminated vegetable foods but also by consuming food of animal origin.

Unlike aflatoxins, ochratoxins are considered a local problem in the European countries. In Germany, numerous surveys on the occurrence of ochratoxins in foodstuffs have been carried-out over the last 2 decades. The presence of OA was reported in different commodities, like cereals and cereal products (Wolff & Richter 1992; Majerus *et al.*, 1994; Rother & Taschan, 1998), coffee (Majerus *et al.*, 1993; Koch *et al.*, 1996), beer (Majerus *et al.*, 1993; Thellmann & Weber, 1997), animal edible tissues and meat products (Scheuer, 1989; Kipper *et al.*, 1991; Scheuer *et al.*, 1997; Gareis, 1996), and in human blood (Gareis *et al.*, 2000; Schwerdt *et al.*, 2001).

Although many data on the occurrence of OA have been available, the estimation of the exposure to this toxin from food surveys remained questionable, because the data were not representative for the whole country; most of the data originated from regional studies, and the analyses were performed by in-house laboratory methods.

In consequence, the German Federal Ministry of Health initiated a joint research project on the occurrence of OA and the consumers' exposure to this toxin.

The study was performed between 1995 and 1998, by seven institutes. Nearly 7,000 food samples and about 1,000 blood samples were examined for OA. Additionally, the nutritional habits of more than 2,500 persons were recorded and compiled along with the food contamination data.

These data enabled identification of the main food-groups with contribution to the OA intake, and calculation of the overall exposure of the consumer. The results of the project have been published in nine separate articles in *Archiv für Lebensmittelhygiene,* 51, 2000.

A short summary of the most relevant conclusions reported from this study is presented below.

(1) A total of 68.6% of cereals and cereal products were found positive for OA, but only in 1.4% of samples the contamination level exceeded 3 µg/kg (Wolff, 2000).

(2) About 50% of roasted and ground coffee samples, and up to 90% of instant coffee samples were found positive for OA. The mean contamination level was highest in instant coffee with caffeine (1.83 µg/kg). Among different sorts of teas, herbal teas for children contained the highest amount of OA. About 70-80 % of beer samples contained OA (maximum 0.29 µg/L) (Bresch *et al.*, 2000).

(3) Total incidence of OA in wine was 40%, but contamination was considerably higher in red wines and wines originating from southern Europe regions. Accordingly, in juices, the highest OA contamination was found in grape juice, particularly in red grape juices (Majerus *et al.*, 2000).

(4) Almost 100% of raisins and currants samples were found OA positive (maximum 21.4 µg/kg) (Engel, 2000).

(5) High incidence of OA was detected in pig kidneys and liver (44% and 17%, respectively), and in blood sausages and liver-type sausages (77% and 68%, respectively) (Gareis & Scheuer, 2000).

(6) Some between-region and within-region variations of OA level in human blood sera were observed, and were correlated with the different nutrition habits (Gareis *et al.*, 2000).

(7) Combining food data and consumption data, cereal products, like bread and rolls, were identified as the predominant sources of OA

intake (about 40% of the total OA intake). Other important sources are coffee and beer for adults, and juices and sweets for children, respectively (Cholmakov-Bodechtel *et al.*, 2000).

(8) The mean daily intake calculated from food data was approximately 0.5 ng/kg b.w.

(9) OA was detected in 98% of human blood serum samples with a median concentration of 0.23 ng/ml (Rosner *et al.*, 2000).

(10) The mean daily intake calculated from OA blood serum data was 0.46 ng/kg b.w.

(11) Daily intake calculations based on food data and blood serum data were highly consistent.

(12) The overall mean daily OA intake was reported as 0.5 ng/kg b.w. This is about tenfold lower than the Provisional Tolerable Daily Intake (PTDI) (5 ng/kg b.w.) estimated by the Scientific Committee on Food (European Commission, 1998).

As was concluded from this study by the study authors, exposure of the German consumer is not at a risk level. However, continued use of efficient preventive and control measures is needed to keep the OA contamination of foods as low as possible.

2.3. Deoxynivalenol

Deoxynivalenol is a type B trichothecene with high prevalence in cereals in central Europe. Cereal-based foods are of major importance in human nutrition, and thus high attention has been given in the last years to the monitoring of DON in these foods.

The incidence of DON in cereals and cereal products is influenced by seasonal weather conditions. In some years, with rain during bloom and harvest seasons, the contamination of cereals with DON, as well as with other *Fusarium*-toxins, can be very high. Such was the case in Germany in the year 1998, described as a "*Fusarium*-year", when concentrations up to 10 mg DON/kg were found in local wheat. Surveys on the occurrence of DON in German foodstuffs have been performed since the 1980s (Lepschy-v. Gleissenthal *et al.*, 1989), but the number of published studies increased considerably after 1998. DON was identified in cereals (wheat, rye, barley, maize, oats, and triticale) as well as in cereal products (flour, bread, noodles, semolina, malt, beer, and cereal-based baby foods) (Table 1).

Table 1. Occurrence of deoxynivalenol in German foods

Samples		*Collected*		*Positive samples (%)*	*Contamination level (µg/kg)*			*Reference*
Type	*n*	*Location*	*Year*		*Mean*	*Median*	*Maximum*	
Wheat (org.)	47	Rhineland and Eifel	1997	100	111§*	94§*	415§	Birzele *et al.*, 2000
Wheat (org.)	58		1998	100	280§*	229§*	1020§	
Wheat	98	Saxony	1998				2200	Horn *et al.*, 2000
Wheat (conv.)	150	Thuringia	1998	69	1540§*	320§**	11660§	Döll *et al.*, 2000
Rye (conv.)	50			54	760§*	270§**	4220§	
Wheat (org.)	46			34	490§*		3090§	
Rye (org.)	19			11	130§*		130§	
Wheat	115	Baden-Württemberg	1998, 2000	96	471*	196*	7730	Schollenberger *et al.*, 2002b
Wheat flour	60	Stuttgart	1999	98	292*	199*	1380	
Bread	141	Stuttgart	1998, 1999	90	134*	100*	690	
Noodles	67	Stuttgart	1998, 1999	91	322*	106*	4840	
Wheat, barley, triticale, oats	245	Germany	1999	45	317		10970	Ellner, 2000
Wheat, rye, barley, oats	60	Hesse	not specified		225	235	871	Taschan *et al.*, 2000
Flour	27				249	230	612	
Malt	32				119	96	451	
Beer	150	Germany and import			36	33	123	
Wheat flour type 405	42	Munich	1999	71	200*	200*	330	Usleber *et al.*, 2000
Wheat flour type 550	9			33	410*	400*	420	
Wheat flour type 1050	11			91	370*	340*	760	
Whole wheat flour	20			65	300*	270*	550	
Wheat grains	8			75	280*	270*	430	

continued ...

... continued (Table 1. Occurrence of deoxynivalenol in German foods)

Samples		*Collected*		*Positive samples (%)*	*Contamination level (µg/kg)*			*Reference*
Type	*n*	*Location*	*Year*		*Mean*	*Median*	*Maximum*	
Wheat bran	20	Munich	1999	85	830*	700*	2100	Usleber *et al.*, 2000
Rye flour and grits	7			29	120*	120*	120	
Wheat flour 405	28	Stuttgart	1999	100	239*	156*	965	Schollen-berger *et al.*, 2002a
Wheat flour 550	13			92	234*	157*	756	
Whole-grain wheat flour	19			100	404*	284*	1379	
Different foods	447	Baden-Württemberg	not speci-fied	33			500-900	Bassen & Thielert, 2000
Wheat, barley, triticale, oats	353	150 locations in Germany	2000	80	630**	110**		Ellner, 2001
Wheat, rye, barley and products thereof	48	Saxony-Anhalt	2001	33			400	Woese, 2001
Pasta	29	Berlin	2001	10% over 500 µg/kg			840	Bockhorn *et al.*, 2001
Rye flour	13	Hesse and Rhineland-Palatinate	2001	85	91*	86*	167	Majerus *et al.*, 2002
Wheat flour	81			73	161*	124*	497	
Wheat semolina	22			86	626*	587*	1260	
Flakes	23			30	194*	129*	598	
Bread	32			34	90*	87*	197	
Noodles	44			64	472*	297*	3020	
Cereal-based baby food	70			17	104*	104*	283	
Cereal-based baby food	110	Hesse	2001, 2002	79	79*	62*	517	Seidler *et al.*, 2002

§ = contamination expressed as µg/kg dry matter;
*= calculated based on positive samples; ** = calculated based on all samples; without superscript = not specified.

In particular, DON levels in baby foods (13% exceeding 100 µg/kg, maximum levels of 500 µg/kg) were alarming (Seidler *et al.*, 2002). It was clear that an unacceptable percentage of cereal foods contained DON at levels exceeding provisional maximum tolerable daily intake estimates. Meanwhile (February 2004), German food legislation is in the final stage of setting national maximum admissible levels for DON of 500 µg/kg in cereals and products thereof, 350 µg/kg in bread and rolls, and 100 µg/kg in cereals used in baby foods.

Regarding wheat and rye, principal cereals for bread-making in Germany, DON has been detected more frequently and in higher concentrations in wheat. Beer may also have a substantial contribution to the total intake of DON but no regulations are planned so far.

Another subject of interest was the distribution of DON in conventional food compared to organic food. It has to be mentioned that only few studies on this subject have been published so far, and in most of them, the number of organic samples was limited. However, a lower DON contamination in organic cereals and products thereof has been reported (Döll *et al.*, 2000; Usleber *et al.*, 2000; Schollenberger *et al.*, 2002a)

Although a substantial amount of data on the occurrence of DON exist in Germany, an estimate of the accurate value of consumers' exposure cannot be determined, partly because of inconsistencies in data presentation in published reports, partly because of discrepancies in sampling and analytical methodology.

Thus, a joint project on the occurrence of DON and ZEA and exposure of the German consumer has been initiated in 2001 by the German Ministry of Consumer Protection, Nutrition and Agriculture. Six institutes are involved in the project, which will provide representative analytical data of all critical foods. In parallel, the distribution of DON (and ZEA) during large-scale milling processes is investigated.

Another aspect is method evaluation and intercomparison. The report of the study is expected by the end of 2004.

2.4. Zearalenone

Zearalenone is an estrogenic mycotoxin. It is produced on cereals, by numerous species of *Fusarium*, wherefrom it is further transmitted to humans via cereal-based foods. The mode of action and adverse effects of low intake levels of ZEA are well known in animals, but toxicology of low ZEA intake in humans is only fragmentary documented. However, contribution of ZEA to the total load of exogenous estrogens should not be ignored. In the EU a provisional tolerable daily intake (PTDI) of 0.2 µg/kg b.w. has been recommended (European Commission, 2000). In Germany, tolerance levels for ZEA will become effective in 2004. The maximum admissible value for cereals, bread, and other cereal foods will be 50 µg/kg, whereas for cereals used for baby food the value will be 20 µg/kg.

Unlike DON, ZEA seems to have a lower incidence and it is usually detected only in trace amounts in cereals and cereal products, with the exception of maize (Usleber *et al.*, 1998). However, concentrations of >50 µg/kg in cereal samples have been registered in wheat, rye, maize, and barley (Table 2). Low incidence and levels of ZEA has been reported for baby food (Seidler *et al.*, 2002).

The exposure of the German consumer to ZEA will also be determined within the *Fusarium* toxin project described for DON.

Table 2. Occurrence of zearalenone in German foods

Samples		*Collected*		*Positive samples (%)*	*Contamination level (µg/kg)*			*Reference*
Type	*n*	*Location*	*Year*		*Mean*	*Median*	*Maximum*	
Wheat	20	Munich	1996, 1997	30	9*	10*	12	Usleber *et al.*, 1998
Barley	10			30	5.7*	5.5*	6.4	
Rye	10			40	7.6*	6.5*	12	
Maize	10			80	55*	48*	92	
Wheat (conv.)	135	Thuringia	1998	7	74§*	-	250§	Döll *et al.*, 2000
Wheat (org.)	46			4	47§*	-	55§	
Wheat grains	115	Baden-Württemberg	1998, 2000	71	15*	8*	158	Schollenberger *et al.*, 2002b
Wheat flour	60	Stuttgart	1999	38	5*	3*	24	
Wheat flour 405	28	Stuttgart	1999	11	1*	1*	2	Schollenberger *et al.*, 2002a
Wheat flour 550	13			31	4*	4*	8	
Whole-grain wheat flour	19			79	6*	4*	24	
Wheat, barley, triticale, oats	245	Germany	1999	5	2		67	Ellner, 2000
Different foods	359	Baden-Württemberg		10			59 (in maize)	Bassen & Thielert, 2000
Wheat, barley, triticale, oats	353	150 locations in Germany	2000	80% below 50 µg/kg				Ellner, 2001
Wheat grains	79	Schleswig-Holstein	2001	42		19	379	Reutter, 2002
Cereal-based baby-food	105	Hesse	2001, 2002	4	9*	7*	15	Seidler *et al.*, 2002

§ = contamination level expressed as µg/kg dry matter;
*= calculated based on positive samples; mean values without superscript = not specified.

2.5. Fumonisins

Fumonisins are a group of potential carcinogenic mycotoxins produced by certain *Fusarium* species. Usually fumonisin B_1 is the predominant compound found in naturally contaminated samples, followed by fumonisin B_2 and B_3. Fumonsins have been found to be worldwide contaminants in maize. In Germany, fumonisins have been identified in maize, maize meal, maize semolina, popcorn, cornflakes, snacks (Thielert & Reusch, 1998; Bresch *et al.*, 1998; Usleber & Märtlbauer, 1998) and rice (Usleber *et al.*, 1998). The average level of contamination was not very high, but in some cases amounts of milligrams/kg have been found (Table 3). Thielert and Reusch (1998) reported that 17.6% of maize and maize-products samples contained fumonisins at levels >500 µg/kg. Usleber and Märtlbauer (1998) reported the highest contamination in imported maize products, but fumonisins were also detected in maize originating from Germany (Thielert & Reusch, 1998).

The existing data on the occurrence of fumonisins in German food demonstrate that the highest contribution to fumonisins dietary intake in humans is represented by maize and maize products, other cereals were free of fumonisins (Meister & Symmank, 1996). Germany will regulate fumonisins (FB_1 + FB_2) in foods at 500 µg/kg in maize for direct consumption, and 100 µg/kg in corn flakes and infant foods.

A large study on the occurrence of fumonisins in foods and exposure of the German consumer was recently completed. Results will be published in the near future.

Table 3. Occurrence of fumonisins in German food.

Samples		*Collected*		*Positive samples (%)*	*Contamination level (µg/kg)*			*Reference*
Type	*n*	*Location*	*Year*		*Mean*	*Median*	*Maximum*	
Maize	10	Munich	1996, 1997	80	530*	52*	2600	Usleber *et al.*, 1998
Maize grit and maize meal	79	Germany	1993 - 1995				16000	Usleber & Märtlbauer, 1998
Popcorn	40						1600	
Cornflakes	108						130	
Maize-containing baby food	100			0				
Maize and maize-products	227	Germany	1995-1997	30			>1000	Thielert & Reusch, 1998

*= calculated based on positive samples

3. FURTHER RESEARCH

An enormous amount of data on the occurrence of some major mycotoxins in foods has been obtained in recent years. Exposure of the consumer in Germany to ochratoxin A has been well characterized, similar data for deoxynivalenol, zearalenone and fumonisins will soon be available. The validity of such data is based on the following aspects:

(1) Representative sample collection for all sorts of food which potentially may contribute to the mycotoxin intake, considering regional and year-to-year variability.

(2) Use of validated analytical methods (intra- and inter-laboratory) to ensure the reliability of the results.

(3) An as accurate as possible estimate of food consumption, based on food consumption surveys.

(4) Consideration both of average and high consumption (and high risk) subgroups within the population, in particular baby and infant foods.

Still urgently needed is a better data basis for chronic effects of these toxins. The amount and quality of toxicological information about most mycotoxins is still far from being adequate. In the long term, the costs of such studies would be negligible compared with those arising from efforts to fulfil regulatory requirements.

Along with that, further surveys of mycotoxins which may have significant negative effects on health but for which the data basis is poor (e.g. nivalenol, T-2 toxin) should be conducted.

4. CONCLUSIONS

Mycotoxins have been identified in different foods in Germany, mostly in cereals and cereal-based foods. Control of cereals immediately after harvest is necessary to decide the future use for food industry. In some years, only selected cereals may be suitable for certain products. Since only a relatively small percentage of the total cereal production in Germany is used directly for food processing, such a selection should be feasible.

Although the mean exposure of the German consumers is not considered to be at level of risk, particular attention should be given to some consumer groups, like babies and children.

Although the limited toxicological data basis has to be acknowledged, the proposed maximum tolerated limits for fumonisins, deoxynivalenol, and zearalenone seem to be scientifically justified and realistic in aspects of public health, and will have only a moderate impact on cereal production in Germany.

5. AFFILIATIONS

Valeriu Curtui,
Institute of Veterinary Food Science, Justus Liebig University, Ludwigstrasse 21, 35390 Giessen, Germany. Milchwissenschaften@vetmed.uni-giessen.de

Caroline Seidler,
Institute of Veterinary Food Science, Justus Liebig University, Ludwigstrasse 21, 35390 Giessen, Germany. Milchwissenschaften@vetmed.uni-giessen.de

Elisabeth Schneider
Institute of Veterinary Food Science, Justus Liebig University, Ludwigstrasse 21, 35390 Giessen, Germany. Milchwissenschaften@vetmed.uni-giessen.de

Ewald Usleber
Institute of Veterinary Food Science, Justus Liebig University, Ludwigstrasse 21, 35390 Giessen, Germany. Milchwissenschaften@vetmed.uni-giessen.de

6. REFERENCES

Bassen, B. & Thielert, G. (2000). Fusarientoxine (DON und ZEA) in Lebensmitteln. Mycotoxin Research, 16 A, 75-78.

Bayerisches Landesamt für Gesundheit und Lebensmittelsicherheit, LGL (2001). Lebensmittelüberwachung 2001. Retrieved January 23, 2004, from www.vis-ernaehrung.bayern.de/de/left/ueberwachung/ statistiken/kurz_lebueb_2001.pdf

BGVV (Bundesinstitut für gesundheitlichen Verbraucherschutz und Veterinärmedizin, now BfR). (1999). Lebensmittel-Monitoring 1999. Gemeinsamer Bericht des Bundes und der Länder. Berlin.

Birzele, B., Prange, A. & Krämer, J. (2000). Deoxynivalenol and ochratoxin A in German wheat and changes of level in relation to storage parameters. Food Additives and Contaminants, 17, 1027-1035.

Bockhorn, I., Bockhorn, A. & Pohler, S. (2001). Deoxynivalenol (DON) in raw and cooked pasta. Mycotoxin Research, 17A, 67-70.

Bresch, H., Majerus, P. & Urbanek, M. (1998). Fumonisinkonzentrationen in diversen Maisprodukten im Raum Karlsruhe und Rheinland-Pfalz in den Jahren 1993-1996. Deutsche Lebensmittel Rundschau, 94, 80-83.

Bresch, H., Urbanek, M. & Hell, K. (2000). Ochratoxin A in coffee, tea and beer. Archiv für Lebensmittelhygiene, 51, 89-94.

Burgmaier-Thielert, E. & Thielert, G. (2000). Aflatoxine in Verarbeitungserzeugnissen von Pistazien. Mycotoxin Research, 16 A, 142-145.

Cholmakow-Bodechtel, C., Wolff, J., Gareis, M., Bresch, H., Engel, G., Majerus, P., Rosner, H. & Schneider, R. (2000). Ochratoxin A: Representative food consumption survey and epidemiological analysis. Archiv für Lebensmittelhygiene, 51, 111-115.

Döll, S., Valenta, H., Kirchheim, U., Dänicke, S. & Flachowsky, G. (2000). *Fusarium* mycotoxins in conventionally and organically grown grain from Thuringia/Germany. Mycotoxin Research, 16 A, 38-41.

Ellner, F. M. (2000). Occurrence of *Fusarium* toxins in the 1999's harvest. Mycotoxin Research, 16 A, 21-25.

Ellner, F. M. (2001). *Fusarium* toxins in cereals – results from eight German Federal States in 2000. Mycotoxin Research, 17A, 41-44.

Engel, G. (2000). Ochratoxin A in sweets, oil seeds and dairy products. Archiv für Lebensmittelhygiene, 51, 98-101.

European Commission. (1998). Opinion of the Scientific Committee on Food on Ochratoxin A (expressed on 17 September 1998). Retrieved January 14, 2004, from www.mycotoxin.de/Dokumente/ochr_eu.pdf

European Commission. (2000). Opinion of the Scientific Committee on Food on *Fusarium* toxins. Part 2: Zearalenone (ZEA). SCF/CS/CNTM/MYC/22 Rev 3 Final.

European Commission. (2001). Verordnung (EG) Nr. 466/2001 der Kommission vom 8. März 2001 zur Festsetzung der Höchstgehalte für bestimmte Kontaminanten in Lebensmitteln. Amtsblatt der Europäischen Gemeinschaft, L 77/1.

Gareis, M. & Scheuer, R. (2000). Ochratoxin A in meat and meat products. Archiv für Lebensmittelhygiene, 51, 102-104.

Gareis, M. (1996). Fate of ochratoxin A on processing of meat products. Food Additives and Contaminants, 13, Suppl., 35-37.

Gareis, M., Rosner, H. & Ehrhardt, S. (2000). Blood serum levels of ochratoxin A and nutrition habits. Archiv für Lebensmittelhygiene, 51, 108-110.

Horn, F., Hanschmann, G., Gebhart, G., John, A. & Mietke, H. (2000). Untersuchung von Weizen sächsischer Herkunft auf Fusarien und die Toxine Deoxynivalenol (DON) und Zearalenon (ZEA). Mycotoxin Research, 16 A, 42-45.

Kipper, L., Schweigmann, A. & Vater, N. (1991). Untersuchungen zum Carry-over von Ochratoxin A in der Nahrungskette Futtermittel – Schlachtschwein – Fleischerzeugnis. Tierärztliche Umschau, 46, 538-542.

Koch, M., Steinmeyer, S., Tiebach, R. & Weyerstahl, P. (1996). Bestimmung von Ochratoxin A in Röstkaffee. Deutsche Lebensmittel-Rundschau, 92, 48-51.

Landesanstalt für Umweltschutz Baden-Württemberg. (2003). Lebensmittel und Verbraucherschutz. Retrieved January 23, 2004, from www.lfu.baden-wuerttemberg.de/lfu/abt2/umweltdaten2003_pdf/pdf/LFU_189_206.pdf

Lepschy-v. Gleissenthal, J., Dietrich, R., Märtlbauer, E., Schuster, M., Süß, A. & Terplan, G. (1989). A survey on the occurrence of *Fusarium* mycotoxins in Bavarian cereals from the 1987 harvest. Zeitschrift für Lebensmittel-Untersuchung und Forschung, 188, 521-526.

Majerus, P., Bresch, H. & Otteneder, H. (2000). Ochratoxin A in wines, fruit juices and seasonings. Archiv für Lebensmittelhygiene, 51, 95-97.

Majerus, P., Curtui, V., Otteneder, H. & Usleber, E. (2002). Deoxynivalenol in Lebensmitteln: Ergebnisse einer Pilotstudie. Mycotoxin Research, 18A, 32-34.

Majerus, P., Cutka, I., Dreyer, A., El-Desouki, S., Eyrich, W., Reusch H., Schurer, B. & Waiblinger, H. U. (1993). Zur Belastungssituation von Ochratoxin A in Lebensmitteln pflanzlichen Ursprungs. Deutsche Lebensmittel-Rundschau, 89, 112-114.

Majerus, P., Weber, R. & Wolff, J. (1994). Nachweis und Bestimmung von Ochratoxin A in Getreide und Getreideprodukten. Bundesgesundheitsblatt, 37, 454-458.

Meister, U. & Symmank, H. (1996). Fumonisine in Getreide. In M. Gareis & R. Scheuer (Eds.), Proceedings of 18. Mykotoxin-Workshop (pp. 26-30). Kulmbach.

Milchprüfring Baden-Württemberg e. V. (2002). Geschäftsbericht. Stuttgart.

Pittet, A. (1998). Natural occurrence of mycotoxins in foods and feeds – and update review. Revue de Médicine Vétérinaire, 149, 469-478.

Reutter, M. (2002). Mykotoxingehalte in Getreide und Futtermitteln. Mycotoxin Research, 18 A, 35-38.

Rosner, H., Rohrmann, B., Peiker, G. (2000). Ochratoxin A in human serum. Archiv für Lebensmittelhygiene, 51, 104-107.

Rother, B. & Taschan, H. (1998). Ochratoxin A in Lebensmitteln auf Getreidebasis. In J. Wolff & T. Betsche (Ed.), Proceedings of 20. Mykotoxin-Workshop (pp. 25-28). Detmold.

Scheuer, R. (1989). Untersuchungen zum Vorkommen von Ochratoxin A. Fleischwirtschaft, 69, 1400-1404.

Scheuer, R., Dietrich, R., Märtlbauer, E. & Gareis, M. (1997). Nachweis von Ochratoxin A in Fleisch und Fleischerzeugnissen. In E. Märtlbauer & E. Usleber (Eds.) Proceedings of 19. Mykotoxin-Workshop, (pp. 142-146). München.

Schollenberger, M., Jara, H. T., Suchy, S., Drochner, W. & Müller, H. M. (2002a). *Fusarium* toxins in wheat flour collected in an area in southwest Germany. International Journal of Food Microbiology, 72, 85-89.

Schollenberger, M., Müller, H. M. & Drochner, W. (2002b). *Fusarium* toxins in different food samples. Mycotoxin Research, 18A, 39-42.

Schwerdt, G., Schuster, C., Silbernagl, S., Gareis, M. and Gekle, M. (2001). Preliminary results on ochratoxin A concentrations in blood of patients with various kidney diseases in Germany. Mycotoxin Research, 17 A, 146-149.

Seidler, C., Curtui, V., Schneider, E., Dietrich, R., Märtlbauer, E. & Usleber, E. (2002). Nachweis von Fusarientoxinen in Säuglings- und Kleinkindernahrungsmitteln. In 43. Arbeitstagung des „Arbeitsgebietes Lebensmittelhygiene“ (pp. 411-416). Giessen, DVG Service GmbH.

Taschan, H., Puchtinger, G., Waller, U. & Puchtinger, T. (2000). Deoxynivalenol in grain products. Mycotoxin Research, 16 A, 26-29.

Thellmann, A. & Weber, W. (1997). Bestimmung von Ochratoxin A in Getreide, Malz und Bier nach Anreicherung und Separation an Immunaffinitätssäulen und anschließender Hochleistungsflüssigkeitschromatographie mit Fluoreszenzdetektion. Deutsche Lebensmittel-Rundschau, 93, 1-3.

Thielert, G. & Reusch, C. (1998). Untersuchung von Lebensmitteln auf Fumonisine. Deutsche Lebensmittel-Rundschau, 94, 153-159.

Usleber, E. & Märtlbauer, E. (1998). Occurence of fumonisins in foods in Germany. In M. Miraglia, H. P. van Egmond, C. Brera & J. Gilbert (Eds.), Mycotoxins and phycotoxins – Developments in Chemistry, Toxicology and Food safety (pp. 81-86). Alaken, Colorado, USA.

Usleber, E., Lepschy, J. & Märtlbauer, E. (2000). Deoxynivalenol in Mehlproben des Jahres 1999 aus dem Einzelhandel. Mycotoxin Research, 16 A, 30-33.

Usleber, E., Schneider, E. and Märtlbauer, E. (1998). Untersuchungen zum Vorkommen von Deoxynivalenol, Zearalenon und Fumonisinen in Speisegetreide. In J. Wolff & T. Betsche (Eds.), Proceedings of 20. Mykotoxin-Workshop (pp. 131-135). Detmold.

Woese, K. (2001). Monitoring of deoxynivalenol and zearalenone in cereals and cereal products from the state of Saxony-Anhalt, Germany. Mycotoxin Research, 17A, 45-48.

Wolff, J. & Richter, W. I. S. (1992). Ochratoxin A in Getreide. Getreide, Mehl und Brot, 46, 355-360.

Wolff, J. (2000). Ochratoxin A in cereals and cereal products. Archiv für Lebensmittelhygiene, 51, 85-88.

J. VARGA, B. TÓTH, Á. MESTERHÁZY, J. TÉREN & B. FAZEKAS

MYCOTOXIGENIC FUNGI AND MYCOTOXINS IN FOODS AND FEEDS IN HUNGARY

1. INTRODUCTION

Mycotoxins are a diverse group of fungal secondary metabolites, which are harmful to animals and humans. Several fungi are able to produce mycotoxins. Most of the mycotoxin-producing species are filamentous ascomycetes, basidiomycetes or deuteromycetes. *Aspergillus*, *Fusarium* and *Penicillium* are usually considered as the most important mycotoxin-producing genera. Post-harvest spoilage caused by these and other fungi is the most important source of mycotoxins in foods and feeds. However, many fungi produce toxins while interacting with living plants as pathogens or as endophytes (pre-harvest spoilage).

The occurrence of mycotoxigenic fungi and their mycotoxins have been examined since the late sixties in Hungary (Hornok & Tóth, 2001). The reports on occurrence of mycotoxin producing fungi and mycotoxins published between 1990 and 1999 in Hungary have recently been dealt with (Hornok & Tóth, 2001). In this review, we will focus on data available for the period 1998-2003. We will also review data published in this period not covered in the previous report.

2. FOOD CONTROL AND MYCOTOXIN ANALYSIS IN HUNGARY

Three food control authorities exist in Hungary: Veterinary and Food Control Stations (under the supervision of Ministry of Agriculture and Rural Development) are responsible for products of plant origin, the National Institute of Food Investigation is responsible for products of animal origin, while the National Public Health and Medical Officer Service (under the supervision of Ministry of Health, Social and Family Affairs) is responsible for prevention and investigation of food-borne diseases. The Inspectorate for Consumer Protection (under the supervision of Ministry of Economic Affairs) is another independent authority dealing with food control.

In the period 1994-2003, altogether 819 (1994), 1015 (1995), 1635 (1996), 1259 (1997), 2061 (1998), 1715 (1999), 1185 (2000), 1015 (2001), and 2211 (2002) plant-derived samples were analyzed at the Veterinary and Food Control Stations for the presence of various mycotoxins. The investigated mycotoxins include aflatoxins (ATs) B_1, B_2, G_1, G_2, M_1 and M_2, ochratoxin A (OTA), zearalenone (ZON), deoxynivalenol (DON), T-2 toxin and patulin. For screening, ELISA is used in most cases after sample-pretreatment and clean-up. Immunochemical kits suitable for the detection of aflatoxin B_1, ochratoxin A, zearalenone, T-2 toxin and fumonisin B_1 have recently been developed in the Agricultural Biotechnology Center in Gödöllő (Barna-Vetró *et al.*, 1994, 1996, 1997, 2000a). In addition, immune affinity column

A. Logrieco and A. Visconti (eds.),
An Overview on Toxigenic Fungi and Mycotoxins in Europe, 123-139.

chromatography and high performance liquid chromatography (HPLC) with UV or fluorescent detection are also used in some cases. For confirmation, GC/MS, HPLC/MS are applied mostly. The results are summarized in statistical reports every year and sent to the Ministry of Agriculture and Rural Development, and to international organizations.

3. MYCOTOXIN REGULATIONS

For foods and food ingredients, Codex Alimentarius Hungaricus 2000, Ministry of Health, Social and Family Affairs Regulations 40/2000 (XII.20.) and 9/2003 (III.13.) and Ministry of Agriculture and Rural Development Regulation 10/2000 (I.23.) are in agreement with EU Commission Directive 98/53/EC. Maximum tolerable levels of mycotoxins in foodstuff in Hungary are detailed in Table 1. While most countries regulate mycotoxin levels in foods by law or codex standards, such regulations are usually nonexistent for feeds or feed ingredients (with the exceptions of aflatoxins and ochratoxin A in some countries). In Hungary, the advisory limits listed in Table 2 have been recommended in Codex Pabularis Hungaricus (Barna-Vetró *et al.,* 2000b). Ministry of Agriculture and Rural Development Regulation 44/2003. (IV.26.) regulates aflatoxin levels in Hungarian feeds and feed ingredients. However, *Fusarium* toxin levels in feeds are not regulated by law. In a recent statement, the Veterinary Committee of Hungarian Academy of Sciences has emphasized the need for changing the recent limits for feeds, and suggested to distinguish between mycotoxin levels leading to depressive and toxic effects in animals (Mátrai *et al.,* 2003; Table 3). These suggested limits are to be included in the III. Hungarian Feed Codex. These limits are subject to change as analytical techniques become more sensitive, and as data accumulate on the effects of these toxins on animals.

4. OCCURRENCE OF MYCOTOXIN-PRODUCING FUNGI IN HUNGARY

Fusarium species are responsible for *Fusarium* head blight (scab, FHB), one of the most important fungal diseases of cereals. The average *Fusarium* contamination of Hungarian wheat seeds is shown in Fig. 1 (Aponyi *et al.,* 1998). As expected, *Fusarium* contamination of wheat was high in so-called "epidemic" years when weather conditions were favorable for *Fusarium* growth. In these years (1970, 1975, 1979-1981, 1985, 1996-1999) *Fusarium* head blight caused serious yield losses in Hungary. Such data are not available for 2000-2003 (in these years the relatively dry summer resulted in low *Fusarium* contamination levels; Mesterházy Á., personal communication).

Table 1. Maximum tolerable mycotoxin levels in foodstuffs in Hungary [Ministry of Health, Social and Family Affairs Regulation 9/2003 (III.13.)].

Commodity	*Mycotoxin*	*Level* $\mu g\ kg^{-1}$
Groundnut, hazelnut, walnut, almond, chestnut, dried fruits for direct human consumption or as ingredients	Aflatoxin B_1	2
	$\Sigma (B_1 + B_2 + G_1 + G_2)$	4
Groundnuts if they are selected or cleaned by other physical treatment before human consumption	Aflatoxin B_1	8
	$\Sigma (B_1 + B_2 + G_1 + G_2)$	15
Walnut, hazelnut, almond, chestnut, dried fruits if they are selected or cleaned by other physical treatment before human consumption	Aflatoxin B_1	5
	$\Sigma (B_1 + B_2 + G_1 + G_2)$	10
Cereals (including buckwheat) and milled products for human consumption or as ingredients	Aflatoxin B_1	2
	$\Sigma (B_1 + B_2 + G_1 + G_2)$	4
Confectionery products	Aflatoxin B_1	1
	$\Sigma (B_1 + B_2 + G_1 + G_2)$	1
Spices (red pepper, chilli, white pepper, black pepper, ginger, turmeric, nutmeg	Aflatoxin B_1	5
	$\Sigma (B_1 + B_2 + G_1 + G_2)$	10
Milk and milk products	Aflatoxin M_1	0.05
Raw (unroasted) coffee beans	Ochratoxin A	15
Roasted coffee, other food of plant origin	Ochratoxin A	10
Cereals (including rice and buckwheat)	Ochratoxin A	5
Cereal products	Ochratoxin A	3
Raisin	Ochratoxin A	10
Cereal bran for human consumption	Deoxynivalenol	1200
Flour, milled products, cereals for muesli	Deoxynivalenol	1000
	Zearalenone	100
	T-2 toxin	300
Fruit and vegetable products	Patulin	50

Table 2. Acceptable mycotoxin limits in feeds in Hungary (Barna-Vetró et al., *2000b).*

Mycotoxins	*Animal feed*	*Regulation $\mu g\ kg^{-1}$*
Zearalenone	Raw materials	10000
	Grown-up ruminants	2000
	Breeding (cattle, swine, turkey)	80
	Laying hens, broilers, swine	500
T-2/HT-2 toxins	Grown-up ruminants	1000
	Laying hens, broilers, swine	300
DON	Grown-up ruminants	2000
	Swine, goose, duck, turkey	400
	Laying hens, broilers, swine	2000
Ochratoxin A	Grown-up ruminants	100
	Laying hens, broilers, swine	10
	Other feeds	25
	Raw materials	50
Aflatoxin B_1	Breeding (cattle, swine, turkey)	5
	Laying hens, broilers, swine	20
	Other feeds	10

The most frequently observed *Fusarium* species on cereals are listed in Table 4. The main causative agents of FHB are members of section *Discolor* of the *Fusarium* genus. Aggressiveness tests proved that isolates of *Fusarium graminearum*, *F. culmorum* and *F. avenaceum* are the most aggressive to wheat among Fusaria in Hungary (Mesterházy, 1984). The genetic variability and toxin producing abilities of Hungarian *F. culmorum* and *F. graminearum* isolates have recently been examined (Tóth *et al.*, 2003a, 2003b). According to these studies, most Hungarian isolates of both species produce deoxynivalenol and therefore belong to chemotype I. These studies have also clarified that the Hungarian *F. culmorum* population reproduces clonally, while the *F. graminearum* population is a recombining one indicating that outcrossing is frequent among different isolates. Cereals can also be infected with Penicillia and other fungi which may produce mycotoxins during storage (Veres, 1999; Table 4).

Hungarian onions have been found to be infected by *Aspergillus niger* belonging to mitochondrial DNA groups 1b and 1c (Varga *et al.*, 1993, 1994). Some of these isolates have also been found to produce ochratoxin A. Further studies are needed to clarify whether *A. niger* contaminates onions during storage or in the field as recently it has been proposed that *A. niger* has an endophytic life style in onions (Tuffley & Lorbeer, 2002).

The mycobiota of green coffee beans of different origin have been examined recently (Téren *et al.*, 1997; Varga, J. & Téren, J., unpublished data). Altogether, 24 green coffee bean samples were analysed. Among the *Aspergillus* strains examined,

Table 3. Recommended depressive and toxic mycotoxin levels in feedstuffs (Mátrai et al., 2003).

Toxin/feedstuff	*Depressive level* $\mu g\ kg^{-1}$	*Toxic level* $\mu g\ kg^{-1}$
ZEARALENONE		
Cow	150	300
Heifer	-	-
Beef cattle	-	-
Calf (preruminant)	250	-
Swine (breeding)	150	250
Piglet (breeding)	50	*
Porker	200	400
Broiler, hen	500	-
Turkey-hen, goose, laying duck	200	-
Other feedstuff	500	1000
T-2 TOXIN		
Horse	-	-
Cattle	1000	2000
Swine	250	600
Broiler	300	600
Laying poultry	250	800
Other feedstuff	1000	2000
DEOXYNIVALENOL		
Cattle	5000	-
Calf	200	-
Swine	400	1000
Chicken	2000	-
Other poultry	500	-
ALL TRICHOTHECENES		
Cattle	2000	4000
Swine	500	1200
Broiler-chicken	600	1200
Laying poultry	300	1600
Other feedstuffs	2000	4000
FUMONISIN B_1		
Horse	5000	-
Cattle	50000	-
Swine	5000	10000
Poultry	30000	-
Other feedstuffs	30000	-
OCHRATOXIN A		
Swine, poultry	200	-
Other feedstuffs	200	-
AFLATOXIN B_1	50	

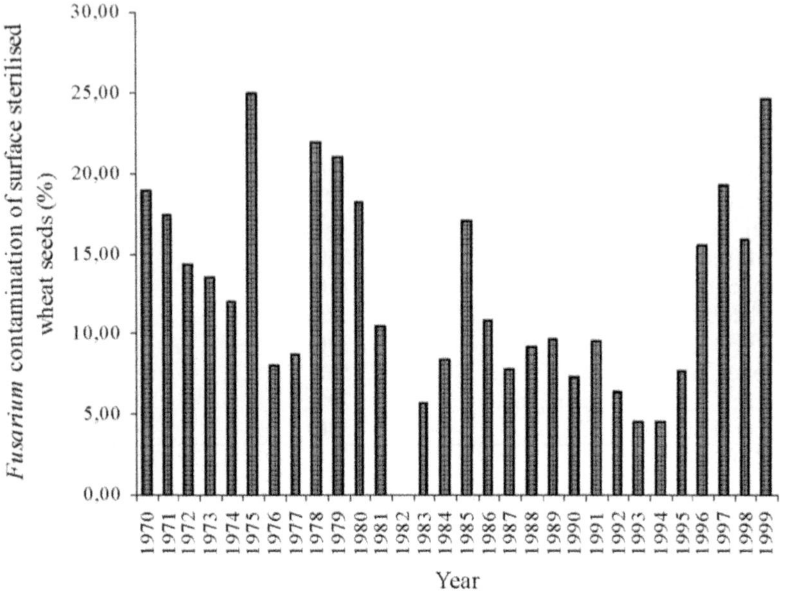

Figure 1. Fusarium *contamination of Hungarian wheat samples in 1971-1999 (modified after Aponyi* et al.*, 1998; data were not available for the year1982).*

some *A. ochraceus, A. niger* and *A. carbonarius* strains were found to produce OTA. This observation indicates that *A. ochraceus* is not the only source of OTA contamination in green coffee beans as observed previously by other authors (Téren *et al.*, 1996; Bucheli & Taniwaki, 2002). Other *Aspergillus* species found in green coffee beans include *A. flavus*, *A. fumigatus*, *A. tamarii* and *A. parasiticus*, none of which produced ochratoxins. Similarly, none of the isolates representing other fungal genera (*Cladosporium, Alternaria, Trichothecium, Penicillium*) were found to produce OTA in YES media. African robusta type coffee beans were in general more heavily contaminated than American coffee beans in accordance with previous findings (Bucheli & Taniwaki, 2002). Other OTA contaminated plant products including spices (chilli powder, pepper, red pepper, sage, dried parsley) and nuts have also been examined; black *Aspergilli* and *A. ochraceus* have been found to be responsible for OTA contamination in these products (data not shown). In another survey, the mycobiota of Hungarian red peppers was examined (Halász *et al.*, 2002). Among the 191 identified fungi, 38.2%, 34.0% and 14.7% were found to belong to the genera *Penicillium, Fusarium* and *Alternaria*, respectively. Some *Aspergillus, Cladosporium, Botrytis, Scopulariopsis, Phomopsis* and *Trichoderma* isolates have also been identified. In the same survey, *Aspergillus* and *Scopulariopsis* species could be isolated from ginger samples (Halász *et al.*, 2002).

Table 4. Occurrence and mycotoxin producing abilities of Fusarium *species in cereals in Hungary.*

Fusarium sp.	*Occurrence on cereals*[1]	*Aggressiveness on heads*	*Mycotoxins produced*[3]
F. graminearum	+++++	High	DON, AcDON, NIV, ZON, DeDON
F. culmorum	++	High	DON, AcDON, NIV, ZON
F. avenaceum	+++	Medium	MON
F. crookwellense	++	High	NIV,FUS, ZON
F. poae	+++	Medium	NIV, FUS
F. sporotrichioides	++	Medium	T-2, HT-2, NEO
F. tricinctum	+	Low	MON
F. chlamydosporum	+	Low[2]	MON
F. semitectum	+	Saprophyte	BEA
F. equiseti	++	Saprophyte	FUCH, ZON
F. acuminatum	+	Low	T-2, MON
F. verticillioides	++++[2]	Medium[2]	FB_1, FB_2, FB_3, FU-C
F. subglutinans	+++++[2]	Medium[2]	MON, BEA, FUP
F. proliferatum	++[2]	Low[2]	FB_1, FB_2, MON, BEA, FUS
F. anthophilum	+[2]	Saprophyte	MON
F. solani	++	Saprophyte	MON
F. oxysporum	++	Saprophyte	MON
Microdochium nivale (F. nivale)	+++	Low	-

1: occurrence: +++++ very frequent, ++++ frequent, +++ medium occurrence, ++ rare, + very rare; 2: on maize; 3: DON: deoxynivalenol, AcDON: acetyl deoxynivalenol, NIV: nivalenol, ZON: zearalenone, DeDON: dideoxynivalenol, MON: moniliformin, FUS: fusarenon, NEO: neosolaniol, BEA: beauvericin, FUCH: fusarochromanon, FB: fumonisin B, FU-C: fusarin C, FUP: fusaproliferin, T-2: T-2 toxin, HT-2: HT-2 toxin.

5. OCCURRENCE OF MYCOTOXINS IN FOODS IN HUNGARY

5.1. Aflatoxins

Among the economically most important mycotoxins, aflatoxins have only very rarely been identified in Hungarian foods and feeds due to climatic factors. In the 1980's, about 250 milk and milk products were analysed, none of which contained aflatoxin M_1 in detectable quantities (Varga *et al.,* 2000). However, aflatoxins can be present in imported products such as peanuts, pistachio, red and black pepper, chocolate, sunflower seeds or rice (Varga *et al.,* 2000; Table 5). Aflatoxin levels in some of these products were above 100 μg kg^{-1}, and reached 3 mg kg^{-1} in a peanut batch in 1995 (Varga *et al.,* 2000). Regarding aflatoxin contamination of peanut samples, about 30% of the samples were contaminated with 1-2 μg kg^{-1} aflatoxins, and about 2% of the examined imported samples contained aflatoxins above the maximum allowable levels between 1995 and 1999 (Varga *et al.,* 2000). Aflatoxin content of imported foods raised slightly in 2000 and 2001, while dropped significantly in 2002, partly due to the fact that different statistical analyses were introduced in that year (Table 5).

Table 5. Aflatoxin (AT) and patulin in food products in Hungary

Year	*Rice AT μg kg⁻¹*	*Muesli AT μg kg⁻¹*	*Peanuts AT μg kg⁻¹*	*Red pepper AT μg kg⁻¹*	*Apple juice concentrate Patulin μg kg⁻¹*
1994	6.1	6.0	n.d.	n.d.	28.5
1995	n.d.	n.d.	n.d.	5.6	22.6
1996	4.6	6.6	5.6	3	70.3
1997	2.6	4.1	3.4	1.5	33.7
1998	2.5	1.5	2.4	3.7	19.5
1999	3.6	3.8	3.3	4.7	20.5
2000	6.1	8.1	5.4	7.6	122.8
2001	5.2	5.9	6.9	5.1	14.0
2002	0.5	0.6	0.3	0.3	0.0

n.d., no data available

5.2. Patulin

At the beginning of the 1980's, 30-100 μg kg-1 patulin were detected in about half of the Hungarian apple juice concentrates. In the nineties, about 0-25% of the concentrates were found to contain patulin above the allowable limit (Varga *et al.,* 2000; Table 5). These findings are in agreement with those found in other European countries (data not shown). The average patulin content of apple juice concentrates was above 100 μg kg-1 in 2000, when one of the examined samples contained 440 μg kg-1 patulin, 9 times higher than the maximum allowable limit.

5.3. Ochratoxin

Ochratoxin content of agricultural products are regularly examined in Hungary at Veterinary and Food Control Stations. Average OTA content of the examined products were usually below the maximum tolerable levels (Figure 2). Ochratoxin A was detected at 1-4 μg kg^{-1} levels in about half of the examined coffee, cocoa, muesli, lentil, rice, red pepper, cereal, biscuit and red wine samples in the period 1995-1999 (Varga *et al.,* 2000). Ochratoxin concentrations were higher than the maximum allowable level only in coffee samples. In 1995, several coffee batches contained 50-60 μg kg^{-1} OTA. Since than, the situation has improved considerably partly due to extensive sampling and analysis of imported coffee loads. Consequently, the rate of coffee loads containing more than 10 μg kg^{-1} OTA decreased in the late nineties (Varga *et al.,* 2000, Figure 2). Most of the coffee loads containing more OTA than the maximum allowable limit came from Africa (Burundi, Ivory Coast, Uganda). Today only 0-2 coffee loads are disapproved due to OTA levels exceeding the official limit every year (Figure 2).

Ochratoxin A content of 36 baking wheat, 16 wheat flour and 6 maize coarse meal samples derived from cereal processing plants in counties Hajdú-Bihar and Békés were analysed at the Veterinary Institute of Debrecen at the end of storage period in 2001 (May-June; Table 6). The samples originated from the harvest of 2000. Ochratoxin A was determined by HPLC analysis coupled with fluorescent detection following immunaffinitiy cleanup of the samples (Fazekas *et al.*, 2002). The results are presented in Table 6. The low OTA content of cereal samples prepared for human consumption is presumably due to favourable weather conditions during harvest and optimal storage conditions. Ochratoxin content of coffee samples examined in the same survey was also lower than the maximum tolerable limit (Fazekas *et al.*, 2002; Table 6).

The ochratoxin content of Hungarian wines and beers has been examined recently in our laboratory using an immunochemical technique (Varga, J., Téren, J., & Kiss, R., unpublished observations). Altogether 50 Hungarian wine and 25 Hungarian beer samples were analyzed. The presence of OTA was confirmed by HPLC in positive samples. OTA was detected in 97.7 % of wines, and ranged from 0 to 0.533 μg l^{-1}. The mean OTA concentration in wines was 0.104 μg l^{-1}. Only one of the Hungarian wines examined contains more than 0.5 μg l^{-1} OTA, the previously suggested EU limit for wine. Our data indicate that red wines are more frequently contaminated, and have higher mean OTA contamination (0.116 μg l^{-1}) than white wines (0.0915 μg l^{-1}), in accordance with previous observations. These values are close to those observed in German white (0.07 μg l^{-1}) and red wines (0.19 μg l^{-1}; Majerus & Otteneder, 1996), but much lower than those found in Italian wines (data not shown). Evidence for a North-South gradient in wine OTA concentrations is not evident from our data.

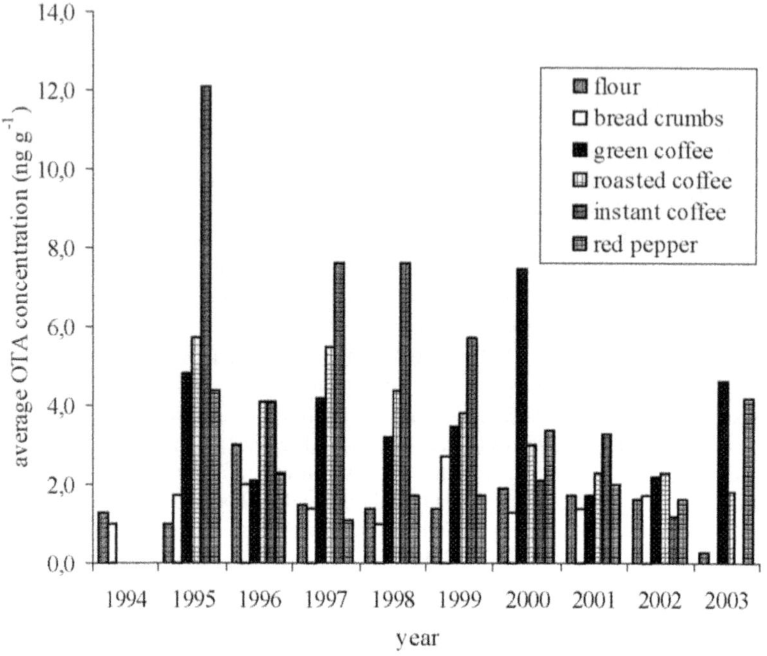

Figure 2. Ochratoxin contamination of foods in Hungary in 1994-2003.

Table 6. Ochratoxin content of cereal products and coffee samples in 2001 in Hungary (Fazekas et al.*, 2002).*

Product	*Positive samples/total samples (n)*	*Positive samples (%)*	*OTA concentration $\mu g\ kg^{-1}$*	*Average $\mu g\ kg^{-1}$*
Wheat	3/36	8.3	0.12 – 0.50	0.29
Wheat flour	2/16	12.5	0.11 – 0.15	0.13
Maize coarse meal	1/6	16.7	0.40	0.40
Roasted coffee*	22/38	58.0	0.17 – 0.91	0.50
Coffee blend**	11/12	92.0	0.34 – 1.30	0.72
All coffee samples	33/50	66.0	0.17 – 1.30	0.57

* products containing 100 % real coffee

** products containing real coffee and roasted barley

For beers, all samples were found to be contaminated with small amounts of OTA with a mean concentration of 0.135 µg l^{-1} (range: 0.030- 0.250 µg l^{-1}).

None of the beers contained OTA above 0.5 µg l^{-1}, the anticipated European Community maximum allowable limit in beer (data not shown).

We could not detect correlation between the type or origin of beer and OTA contamination.

5.4. Fusarium *toxins*

Hungarian cereals and cereal products have been extensively studied for the presence of trichothecenes, zearalenone and fumonisins in the period 1994-2002. In general, zearalenone and trichothecene contamination levels were higher in "epidemic" years, when weather conditions were favorable for *Fusarium* species to infect plant tissues.

5.4.1. Zearalenone

Regional laboratories of the Ministry of Agriculture and Rural Development examined about 2300 cereal samples for zearalenone contamination in 1994-2002. Zearalenone levels exceeding the maximum allowable limits were observed only in about 0.5% of the samples originated from the 1998 harvest. The average values were around 20 µg kg^{-1}. Zearalenone concentrations were even lower in wheat, flour and wheat meal samples in the period 2000-2002 (Table 7).

Table 7. Mycotoxins in Hungarian cereal products in 1994-2002.

	Flour			*Wheat*			*Wheat Meal*		
Year	*DON* µg kg^{-1}	*ZON* µg kg^{-1}	*T-2* µg kg^{-1}	*DON* µg kg^{-1}	*ZON* µg kg^{-1}	*T-2* µg kg^{-1}	*DON* µg kg^{-1}	*ZON* µg kg^{-1}	*T-2* µg kg^{-1}
1994	80.6	n.d.	65.5	n.d.	n.d.	n.d.	230.0	14.5	35.4
1995	132.7	16.8	42.2	n.d.	n.d.	n.d.	75.9	9.7	36.8
1996	255.6	17.0	40.0	n.d.	n.d.	n.d.	174.9	11.3	45.6
1997	498.0	19.0	20.5	903.0	18.0	36.0	311.0	18.7	36.5
1998	407.0	17.2	32.2	852.0	14.0	16.9	316.0	21.8	28.3
1999	425.4	9.9	12.8	490.9	8.3	12.5	332.2	10.9	12.8
2000	239.1	9.0	19.1	186.4	5.0	10.4	164.3	14.7	21.1
2001	230.7	6.8	29.8	216.7	7.1	13.5	233.0	6.4	18.2
2002	122.5	4.7	27.2	93.7	3.5	11.0	115.2	1.4	6.8

5.4.2. Trichothecenes

Deoxynivalenol content of cereals were examined in more than 1300 cereal and flour samples in 1995-1999 at Veterinary and Food Control Stations. Altogether 79 samples (6%) were contaminated with DON above the maximum allowable level (Table 7). In the period 2000-2002, more than 700 samples were analysed. Contamination levels were much lower than in the late nineties due to dry weather conditions (Table 7).

The occurrence of T-2 toxin was examined in 1995-1999 in about 1500 food samples. None of the samples contained T-2 toxin above the maximum allowable limit (average: 13-84 $\mu g\ kg^{-1}$). In the period 2000-2002, 850 samples were analysed, and similar contamination levels were detected (Table 7).

5.4.3. Fumonisins

The fumonisin content of maize flour and maize coarse meal samples have been examined in 1997 (Fazekas, 2001). Fumonisins could be detected in 67% of the samples, although the contamination levels were very low (16-58 $\mu g\ kg^{-1}$).

6. MYCOTOXINS IN FEEDS IN HUNGARY

6.1. Fusarium *toxins*

In Hungarian feeding cereals, zearalenone was detected in 18-66% of the samples in 1992-1998 at Veterinary and Food Control Stations, and the concentrations were 100-300 $\mu g\ kg^{-1}$ (Varga *et al.,* 2000). In feeding cereals, DON was detected in 62-78% of the samples in 1991-1998, with average concentrations between 188-313 $\mu g\ kg^{-1}$ (Varga *et al.,* 2000).

The DON content of 99 feeding wheat samples harvested in three counties in 1998 was determined by HPLC at the Veterinary Institute of Debrecen (Table 8). Eighty-eight percent of the feeding wheat samples were contaminated with 0.94 mg kg^{-1} DON on average. Contamination levels were higher in an eastern county (Békés) than in the north-eastern part of Hungary. The relatively high contamination levels were due to the weather conditions (rainy summer; Fazekas *et al.,* 2000).

These data also indicate that cereals grown in different regions could be contaminated with drastically different levels of mycotoxins due to differences in the susceptibilities of the cultivars used, to different agricultural practices and to different weather conditions.

Table 8. Deoxynivalenol content of feeding wheat samples in three regions of Hungary in 1998 (Fazekas et al., *2000).*

	Positive (%)	*Minimum* $\mu g\ kg^{-1}$	*Maximum* $\mu g\ kg^{-1}$	*Average* $\mu g\ kg^{-1}$
Szabolcs-Szatmár-Bereg county (n=29)	82	50	1300	600
Hajdú-Bihar county (n=45)	84	50	2800	700
Békés county (n=26)	100	270	4300	2000

In another study, the presence of *Fusarium* toxins in 2657 feed samples has also been examined in 1995-2001 at the Veterinary Institute of Debrecen (Fazekas, B., Tar, A.K., & Fazekas, G., unpublished data; Table 9). About 65% of the samples were feed mixtures prepared for swine and poultry, while 35% were feeding cereals (maize, wheat, barley, rye, oat, etc.). The relatively high occurrence of positive samples and toxin levels could partly be due to the fact that most of these samples were examined because they were suspected to cause yield loss or veterinary problems in animals.

Table 9. Fusarium *toxins in feed mixtures and feeding cereals in Hungary in 1995-2001.*

	Zearalenone		T–2 toxin		DON	
Year	Positive (%)	Min – Max Average $\mu g\ kg^{-1}$	Positive (%)	Min – Max Average $\mu g\ kg^{-1}$	Positive (%)	Min – Max Average $\mu g\ kg^{-1}$
1995	30.1	8 – 140 22.3	8.3	50 – 480 75.7	45.1	50 – 650 200.0
1996	64.5	6 – 525 33.6	5.3	50 – 500 104.0	83.4	50 – 3100 441.0
1997	41.5	5 – 140 30.3	4.1	100 – 140 113.0	72.3	50 – 2500 327.0
1998	21.1	10 – 1180 63.5	3.5	45 – 150 65.0	71.0	50 – 6200 547.8
1999	37.0	10 – 200 25.7	4.3	48 – 310 98.0	64.6	50 – 3500 371.0
2000	9.8	10 – 120 40.0	3.4	50 – 100 62.0	14.3	50 – 870 283.0
2001	46.8	10 – 400 45.4	5.3	60 – 310 143.0	48.2	70 – 4100 438.9

Fusarium toxin levels of feeding cereal samples have also been examined independently in another institution in 1991-1998 (Rafai *et al.*, 2000). Altogether 760 maize, 367 wheat, 119 soybean, 222 barley, 85 bran, 32 triticale, 60 oat, 14 rye and 22 sunflower samples were analysed for the presence of T-2 toxin, zearalenone, deoxynivalenol, nivalenol, diacetoxyscirpenol, HT-2 toxin, fusarenone X and ochratoxin A. Based on the contamination data, the samples were assorted into three categories: A, either free of any mycotoxins, or contaminated at levels which are considered harmless (can be used safely in feeds); B, containing one or more mycotoxins in considerable concentrations (may be used for compounding with care, e.g. should be diluted with non-contaminated cereals); C, highly contaminated with one or more mycotoxins (cannot be used for feeding). The proportion of group C samples was 3.0, 2.2, 2.3, 1.7, 7.1, 6.7 and 6.3% for maize, wheat, barley, soybean, bran, oat and triticale samples, respectively (Rafai *et al.*, 2000).

Fumonisin content of Hungarian feeding maize samples have been examined in 1993-1999 (Fazekas, 2001). The fumonisin contamination levels were relatively low in average maize samples, while heavy contamination was detected in mouldy samples (Table 10).

In a recent study, Fazekas *et al.* (1998) clarified that a disease of pig described for the first time in 1941 in Hungary is caused by fumonisins.

Table 10. Fusarium *toxins in mouldy and average maize samples in 1993-1999 in Hungary (Fakezas, 2001).*

Year	*Mouldy maize samples*				*Average maize samples*			
	FUMO[1] $\mu g\ kg^{-1}$	*DON* $\mu g\ kg^{-1}$	*T-2* $\mu g\ kg^{-1}$	*ZON* $\mu g\ kg^{-1}$	*FUMO* $\mu g\ kg^{-1}$	*DON* $\mu g\ kg^{-1}$	*T-2* $\mu g\ kg^{-1}$	*ZON* $\mu g\ kg^{-1}$
1993	2600	4330	200	1260	100	220	100	20
1994	6640	90	165	30	1520	50	50	8
1995	8650	2400	430	360	1600	370	180	100
1996	5520	3200	380	230	1380	420	160	150
1997	5940	3400	480	560	1170	470	220	190
1998	3960	n.d.	n.d.	n.d.	n.d.	n.d.	n.d.	n.d.
1999	n.d.	n.d.	n.d.	n.d.	1190	480	210	n.d.

[1] fumonisins; n.d., not determined

6.2. Ochratoxins

The OTA content of 30 feeding wheat, 32 feeding maize and 20 feeding barley samples were also analysed, the data are presented in Table 11. The feeding cereal samples were heavily contaminated with OTA possibly due to bad storage conditions.

Table 11. Ochratoxin contamination of feeding cereals in 2001 in Hungary.

Product	*Positive samples/total samples (n)*	*Positive samples (%)*	*OTA concentration* $\mu g\ kg^{-1}$	*Average* $\mu g\ kg^{-1}$
Feeding wheat	8/30	26.7	0.3 – 62.8	12.2
Feeding maize	5/32	15.6	1.9 – 8.3	4.9
Feeding barley	7/20	35.0	0.14 – 212.0	72.0

The ochratoxin content of several feeding cereals have been examined in 1991-1998 (Rafai *et al.,* 2000). The average OTA contamination of maize, rye and soybean samples were above 200 μg kg^{-1} (320, 250 and 350 μg kg^{-1}, respectively), while those of wheat, barley, oat, triticale and sunflower samples were lower. However, the rate of incidence was below 10% with the exception of sunflower samples, where 4 of the examined 22 samples were contaminated. Ochratoxins were not detected in bran samples (Rafai *et al.,* 2000). Some maize samples contained more than 1.0 mg kg^{-1} OTA.

7. CONCLUSIONS

In this review, we attempted to summarise our knowledge on the occurrence of mycotoxigenic fungi and mycotoxins in Hungarian foods and feeds in the period 1995-2002. Current Hungarian mycotoxin regulations for foods and food products are in agreement with EU directives. The situation is less satisfying for feeds and feed ingredients. Mycotoxin regulations for feeds and feed ingredients are subject to change in the near future in agreement with the recommendations set of the Veterinary Committee of the Hungarian Academy of Sciences (Mátrai *et al.,* 2003).

Plant-derived agricultural products are regularly screened for mycotoxins in Veterinary and Food Control Stations. *Fusarium* toxins contaminate agricultural products at high levels in "epidemic" years when weather conditions are favorable for *Fusarium* infections of cereals. Ochratoxin concentrations above the tolerable limit have mostly been detected in imported products such as peanuts and coffee. Ochratoxin levels close to the tolerable limit have been observed in Hungarian red peppers. Aflatoxins are usually detected in considerable quantities only in imported agricultural products in Hungary, while patulin concentrations were usually below the allowable limit in Hungarian apple juice concentrates. In the future, continuous sampling and analysis of foods and feeds are required to ensure consumer safety in Hungary.

8. ACKNOWLEDGEMENTS

Preparation of this review was supported by OTKA grants (D38486, T037217). J. Varga was in receipt of a Széchenyi fellowship grant. We thank Professor Pál Rafai (Department of Animal Hygiene, Szent István University, Faculty of

Veterinary Sciences, Budapest, Hungary) and Dr. Anna Makáry (Animal Health and Food Control Station, Kecskemét, Hungary) for contributing data on contamination levels of Hungarian foods and feeds.

9. AFFILIATIONS

János Varga - Department of Microbiology, Faculty of Sciences, University of Szeged, P.O. Box 533, H-6701 Szeged (Középfasor 52, H-6726 Szeged), Hungary. Tel.: 36-62-544-515, Fax: 36-62-544-823, e-mail: jvarga@bio.u-szeged.hu

Beáta Tóth, Ákos Mesterházy - Cereal Research non-Profit Company, P.O. Box 391, H-6701 Szeged, Hungary

József Téren - Animal Health and Food Control Station, P.O. Box 446, H-6701 Szeged, Hungary

Béla Fazekas - Veterinary Institute of Debrecen, H-4031 Debrecen, Bornemissza u. 3-5, Hungary

10. REFERENCES

Aponyi, I., Nagy, G., Princzinger, G., & Kajati, I. (1998). *Fusarium* infection of wheat seeds in Hungary between 1970 and 1997. Cereal Research Communications, 26, 253–258.

Barna-Vetró, I., Balázs, E., & Solti, L. (2000b). Immunodiagnostics for food and feed safety. American Laboratory News, 2000 August, 64-65.

Barna-Vetró, I., Gyöngyösi, A., & Solti, L. (1994). Monoclonal antibody-based enzyme-linked immunosorbent assay of *Fusarium* T-2 and zearalenone toxins in cereals. Applied and Environmental Microbiology, 60, 729-731.

Barna-Vetró, I., Solti, L., Szabó, E., & Gyöngyösi-Horváth, A. (1997). Sensitive enzyme-linked immunosorbent assay for screening aflatoxin B_1 in food- and feedstuffs. Cereal Research Communications, 25, 291-292.

Barna-Vetró, I., Solti, L., Téren, J., Gyöngyösi, A., Szabó, E., & Wolfling, A. (1996). Sensitive ELISA test for determination of ochratoxin A. Journal of Agricultural and Food Chemistry, 44, 4071-4074.

Barna-Vetró, I., Szabó, E., Fazekas, B., & Solti, L. (2000a). Development of a sensitive ELISA for the determination of fumonisin B_1 in cereals. Journal of Agricultural and Food Chemistry, 48, 2821-2825.

Bucheli, P. & Taniwaki, M.H. (2002). Research on the origin, and on the impact of post-harvest handling and manufacturing on the presence of ochratoxin A in coffee. Food Additives and Contaminants, 19, 655-665.

Fazekas, B. (2001). Occurrence and economic effect of mycotoxins in North-Eastern Hungary. In F. Kovács (Ed.), Moulds and mycotoxins in the food chain (pp. 69-83). Budapest: Section of Agricultural Sciences, Hungarian Academy of Sciences [in Hungarian].

Fazekas, B., Bajmoczy, E., Glavits, R., Fenyvesi, A., & Tanyi, J. (1998). Fumonisin B1 contamination of maize and experimental acute fumonisin toxicosis in pigs. Journal of Veterinary Medicine B, 45, 171-181.

Fazekas, B., Hajdu, E.T., Tar, A.K., & Tanyi, J. (2000). Natural deoxynivalenol (DON) contamination of wheat samples grown in 1998 as determined by high-performance liquid chromatography. Acta Veterinaria Hungarica, 48, 151-160.

Fazekas, B., Tar, A.K., & Zomborszky-Kovács, M. (2002) Ochratoxin A contamination of cereal grains and coffee in Hungary in the year 2001. Acta Veterinaria Hungarica, 50, 177-188.

Halász, A., Baráth, Á., & Németh, E. (2002). Selection of lactic acid bacteria for the inhibition of aflatoxin producing mold. In L. Sobotova, J. Stroka, & E. Anklam (Eds.) Minutes and results of the workshop on mycotoxin analysis: The need for scientific and technical support in implementation of European legislation (pp. 95-99). Ispra: European Commission Joint Research Centre (EUR 20410 EN).

Hornok, L., & Tóth, A. (2001). Reports on the natural occurrence of toxigenic fungi and mycotoxins in Hungary between 1990-1999. In A. Logrieco (Ed.), Occurrence of toxigenic fungi and mycotoxins in plants, food and feed in Europe (pp. 63-68). Luxembourg: European Commission Directorate-General for Research (EUR 19695).

Majerus, P., & Otteneder, H. (1996). Detection and occurrence of ochratoxin A in wine and grape juice. Deutsche Lebensmittel Rundschau 92, 388-390.

Mátrai, T., Rafai, P., & Varga, J. (2003). Statement of the Veterinary Committee of Hungarian Academy of Sciences on mycotoxin limits in mixed feedstuffs. Magyar Allatorvosok Lapja, 125, 506-508 [in Hungarian].

Mesterházy, Á. (1984). *Fusarium* species of wheat in South Hungary. Cereal Research Communications, 12, 167-170.

Rafai, P., Bata, A., Jakab, L., & Ványi, A. (2000). Evaluation of mycotoxin-contaminated cereals for their use in animal feeds in Hungary. Food Additives and Contaminants, 17, 799-808.

Téren, J., Palágyi, A., & Varga, J. (1997). Isolation of ochratoxin producing *Aspergilli* from green coffee beans of different origin. Cereal Research Communications, 25, 303-304.

Téren, J., Varga, J., Hamari, Z., Rinyu, E., & Kevei, F. (1996). Immunochemical detection of ochratoxin A in black *Aspergillus* strains. Mycopathologia, 134, 171-176.

Tóth, B., Horváth, Z., Téren, J., Mesterházy, A., & Varga, J. (2003a). Examination of molecular variability of Hungarian *Fusarium graminearum* isolates. 14th International Congress of the Hungarian Society of Microbiology, Balatonfüred, Hungary, 2003

Tóth, B., Mesterházy, A., Nicholson, P., Téren, J., & Varga, J. (2003b). Mycotoxin production and molecular variability of European and American *Fusarium culmorum* isolates. European Journal of Plant Pathology (in press).

Tuffley, J.W., & Lorbeer, J.W. (2002). Endophytic activity of *Aspergillus niger* in onion. American Phytopathological Society Annual Meeting, 2002, Milwaukee, Wisconsin, USA, Abstract No. 115.

Varga, I., Matyasokszky, K., & Sohár J. (2000). Mycotoxin contamination of food items based on investigations in Hungary. Egészségtudomány, 44, 224-241 [in Hungarian].

Varga, J., Kevei, F., Debets, F., Kozakiewicz, Z., & Croft, J.H. (1994). Mitochondrial DNA restriction fragment length polymorphisms in field isolates of the *Aspergillus niger* aggregate. Canadian Journal of Microbiology, 40, 612-621.

Varga, J., Kevei, F., Fekete, C., Coenen, A., Kozakiewicz, Z., & Croft, J.H. (1993). Restriction fragment length polymorphisms in the mitochondrial DNAs of the *Aspergillus niger* aggregate. Mycological Research, 97, 1207-1212.

Veres, E. (1999). Mycotoxin content and *Fusarium* contamination of autumn wheat. Tessedik Napok '99 (Tiszántúli Mezőgazdasági Tudományos Napok '99, Debrecen, 1999. október 28-29) [in Hungarian].

A. MORETTI, A. LOGRIECO, A. VISCONTI & A. BOTTALICO

AN OVERVIEW OF MYCOTOXINS AND TOXIGENIC FUNGI IN ITALY

1. INTRODUCTION

The continental and Mediterranean climates in Italy have a great influence on variability of the growing season and, in particular, on the kind of toxigenic fungal contamination on various crops. The main concerns for wheat head blight and maize ear rot (both diseases caused by *Fusarium* species) are recorded in Northern and Central Italy, while the grape black rot (caused by *Aspergillus* species) is more common in Southern Italy.

An examination of the Italian literature clearly indicates that most of the surveys carried out on the occurrence of toxigenic fungi and mycotoxins in Italy are cognitive studies, performed for preliminary information about the occurrence of specific mycotoxins in specific products.

This overview is mainly focused on toxigenic fungi and mycotoxins more frequently occurring in agricultural and food products in Italy. The major mycotoxins, such as aflatoxins (AFBs and AFGs), ochratoxin A (OTA), patulin (PAT) and *Fusarium* toxins [deoxynivalenol (DON), fumonisins (FBs), zearalenone (ZEA)] have been considered.

Part of the data has been obtained from investigations on infected plants and pre-harvest products, to stress the mycotoxicological importance of the attacks of toxigenic fungi in the field. Data on the occurrence of toxigenic fungi and mycotoxins are referred to the last decade literature and reported separately.

2. OCCURRENCE OF TOXIGENIC FUNGI AND THEIR TOXIGENICITY

Investigations carried out in Italy on the occurrence of toxigenic fungi and their capability to produce mycotoxins are mainly related to the representatives of the genus *Fusarium*, which are widespread in Italy. In particular, several surveys have been performed on the occurrence of the toxigenic *Fusarium* species on cereals, namely wheat and maize. However, recently, the increasing importance of OTA contamination of wines elicited in Italy a proliferation of investigations on the OTA producing fungi on grape. Main data regarding this section are summarized in Table 1.

A. Logrieco and A. Visconti (eds.),
An Overview on Toxigenic Fungi and Mycotoxins in Europe, 141-160.

2.1. Occurrence of toxigenic Fusarium *species in maize*

Bottalico *et al.* (1995) reported *F. moniliforme* (syn. *F. verticillioides*) as the main species isolated from selected samples of preharvest maize ear rot, from different localities in Sardinia in 1992. All samples were affected almost exclusively by *F. moniliforme*, which was isolated from 100% of kernels from infected ears. In two samples, in addition to *F. moniliforme*, also *F. proliferatum* was present but in a reduced percentage of kernels (up to 34%).

Logrieco *et al.* (1995) examined the relative incidence of *Fusarium* species in forty-two samples of pre-harvest maize ear rot collected in 1992-1993 from different maize fields throughout Italy. *F. proliferatum* (up to 78% of samples in the Northern area), together with *F. moniliforme* (54%), were the predominant species in kernels from infected ears. Less frequently isolated were *F. equiseti* (8%) and *F. graminearum* (2%). This was the first evidence that toxigenic *F. proliferatum* strains cause ear rot in Italy. These results strongly suggested a more significant role of *F. proliferatum* in maize ear rot and in associated fusariotoxicoses in Italy.

Ritieni *et al.* (1997), investigating the occurrence of mycotoxins on preharvest maize samples collected from 22 fields in Italy during 1994 in the Southern and Northern areas, showed that *F. moniliforme* (up to 81% of samples) and *F. proliferatum* (up to 73% of samples) were the two main species isolated from the kernels since they occurred in all the maize fields analyzed, while few more species, *F. equiseti* and *F. compactum*, were isolated with a very low frequency.

More recently, Avantaggiato *et al.* (2002), analyzing 25 maize hybrids samples from Central Italyfor *Sesamia nonagrioides* attacks and fumonisin accumulation in maize kernels, confirmed *F. verticillioides* and *F. proliferatum*, isolated from all maize samples, as the main *Fusarium* species occurring on maize.

2.2. Occurrence of toxigenic Fusarium *species in wheat*

Data on the incidence of *Fusarium* head blight (FHB) on wheat in Italy in the year 2000 (Pasquini *et al.*, 2001), showed an incidence of the disease in the northern regions (Lombardia, and Emilia-Romagna), which was lower than that in central regions (Umbria, Toscana, Lazio and Abruzzo). The disease was absent in the southern regions (Molise, Puglia and Sicily).

The higher FHB incidence in central than in northern regions can be explained by the more pronounced influence of maritime warmer weather in Central Italy. In addition, the data obtained in the year 2000 showed a slightly higher FHB incidence on soft wheat than on hard wheat. But other data indicate that the hard cultivars were more susceptible than the soft ones (Balmas *et al.* 1999). The species predominantly found associated with FHB of both soft and hard wheat in Italy were *F. graminearum*, *F. culmorum* and *F. avenaceum.*

A prevalence of *F. graminearum* on *F. culmorum* was observed when weather conditions around the time of flowering were wetter and warmer than normal, while drier weather appears more conducive to *F. culmorum* and less to *F. graminearum.*

In particular, *F. graminearum* (32%), *F. avenaceum* (31%), *F. culmorum* (25%) and, to a lesser extent, *F. cerealis*, *F. poae* and *F. tricinctum* were mostly isolated from FHB of soft wheat (Balmas and Corazza, 1994; Casulli *et al.*, 1995). Among the other less frequently isolated species there are several indications according to regions and to crop year. In 1994 and 1995, Lops *et al.* (1998) showed that durum and soft wheat were both highly contaminated by *F. avenaceum* (94% and 65%), *F. graminearum* (78% and 48%) and *F. culmorum* (58% and 27%) in Northern Italy, while in Southern Italy there was no contamination by *Fusarium* species.

During three years of observations (1994-1996) of FHB in hard wheat all over Italy Pancaldi and Torricelli (1999) found *F. culmorum* and *F. graminearum*, associated with *F. avenaceum*, predominant in all three years, and also *F. verticillioides*, *F. tricinctum* and *F. cerealis* in 1994. In 1996, *F. poae* was also isolated with high frequency. Balmas *et al.* (1999) confirmed the predominance of *F. graminearum* and *F. culmorum* from blighted heads of durum and bread wheat and stressed the prevalence of the former in Northern Italy and of the latter in Central and Southern Italy. Moreover, among the other occurring species involved in FHB of hard wheat were reported *F. avenaceum*, *F. acuminatum*, *F. cerealis*, *F. chlamydosporum*, *F. semitectum*, *F. compactum*, *F. equiseti*, *F. poae*, *F. solani*, *F. oxysporum*, *F. proliferatum*, *F. subglutinans* and *F. sporotrichioides* (Balmas *et al.*, 1999; Pasquini *et al.*, 2001).

2.3. Occurrence of toxigenic Fusarium *species in other crops*

In a survey over 32 localities spread all over Italy, Moretti *et al.* (1995) investigated several asparagus fields in 1993 spring, collecting samples of asparagus plants showing symptoms of *Fusarium* crown and root rot. The data showed that *F. proliferatum* (71% of samples) and secondarily *F. oxysporum* (28% of samples) were the two most prevalent fungi isolated from the asparagus plants. Other species occurring at a lower concentration were *F. acuminatum*, *F. scirpi* and *F. equiseti*.

The data obtained confirmed previous worldwide reports showing *F. proliferatum* as the main species causing *Fusarium* crown and root rot.

2.4. Occurrence of ochratoxin A producing fungi on grapes

Battilani *et al.* (2003) performed a wide investigation on grape for two years in 1999 and 2000, showing that *Aspergillus carbonarius* and *A. niger* aggregate are the predominant fungal species colonizing the berries at early veraison and ripening, and, at a lesser extent, *A. japonicus*. Moreover, they isolated also fungal strains of *Penicillium* spp. although at very low incidence. From these results authors conclude that the OTA occurrence on wines is mostly related to the *Aspergillus* section *Nigri* species.

3. OCCURRENCE OF MYCOTOXINS IN PREHARVEST INFECTED PLANTS AND PRODUCTS

Since the first finding of ZEA in maize in Italy in 1974 (Bottalico *et al.*, 1977), there have been several studies on the occurrence of mycotoxins in infected plants and in preharvest products in Italy. The main aim of these observations was to stress the mycotoxicological importance of the colonization of toxigenic fungi on agricultural commodities, especially during the ripening stage and the harvest. However, most investigations regarded infection of cereals (maize and wheat) by toxins produced by *Fusarium* species that are common fungal contaminants of these crops (Table 2, for summary of this section data).

Table 1. Occurrence of toxigenic fungi in preharvest plants, in Italy

Region	*Plant*	*Main fungal species*	*Year*
North	Asparagus	*F. proliferatum* (51%) *F. oxysporum* (64%)	1993
Centre	Asparagus	*F. proliferatum* (82%) *F. oxysporum* (19%)	1993
South	Asparagus	*F. proliferatum* (88%) *F. oxysporum* (14%)	1993
Sardinia	Maize ear	*F. verticillioides* (81%) *F. proliferatum* (34%)	1992/1993
North	Maize ear	*F. verticillioides* (56%) *F. proliferatum* (78%)	1992/1993
Centre	Maize ear	*F. verticillioides* (76%) *F. proliferatum* (35%)	1992/1993
South	Maize ear	*F. verticillioides* (81%) *F. proliferatum* (47%)	1992/1993
North	Maize ear	*F. verticillioides* (81%) *F. proliferatum* (47%)	1994
North	Wheat soft/durum	*F. graminearum* +++ *F. culmorum* ++	various
North	Maize ear	*F. verticillioides* (100%) *F. proliferatum* (100%)	2000
Centre	Wheat soft/durum	*F. graminearum* ++ *F. culmorum* +++	various
South	Wheat soft/durum	*F. graminearum* + *F. culmorum* +++	various
South	Grape berries	*Aspergillus carbonarius* *A. niger* aggregate *A. japonicus*	1999/2000

3.1. Occurrence of mycotoxins in preharvest Fusarium *infected plants and products*

Bottalico *et al.* (1995) recorded the occurrence of FB_1 in preharvest *F. moniliforme* maize ear rot in Sardinia. Selected samples of preharvest maize ear rot, from different localities in Sardinia, were examined for causal *Fusarium* species and associated mycotoxins. All samples were found to be contaminated by FB_1 (up to 250 mg/kg).

Logrieco *et al.* (1995) reported the natural occurrence of FB_1 (up to 250 mg/kg), in preharvest selected maize ears mostly colonized by *F. proliferatum*. These results strongly suggested a more significant role of *F. proliferatum* in maize ear rot and in associated fusariotoxicoses.

Ritieni *et al.* (1997) analyzed the natural occurrence of FB_1 in twenty-two samples of preharvest maize ear *Fusarium* rot, collected in 1994 in Italy. The most frequently recovered *Fusarium* species were *F. moniliforme* and *F. proliferatum*. Chemical analysis identified FB_1 (up to 300 mg/kg) in eighteen samples mostly infected by *F. proliferatum*.

Recently, maize hybrids were investigated for the occurrence of FB_1 and FB_2 (Avantaggiato *et al.* 2002). All the samples resulted contaminated by both toxins with a range for the total toxin content from 0.01 to 20 mg/kg.

Investigations on toxin contamination in wheat have been mostly focused on DON occurrence, showing that this toxin is a common contaminant in wheat cultivated in the Northern Italy. To this respect, Lops *et al.* (1998) showed that 100% of both durum and soft wheat samples (36 and 13 samples, respectively) were contaminated by DON (range 0.095 – 3.1 and 0.06 – 0.92 mg/kg, respectively). Such widespread contamination was confirmed by Pascale *et al.* (2000) in 26 samples of durum wheat and 77 samples of soft wheat collected in 1998 in Northern Italy, showing a contamination of 24 out 26 samples of durum wheat (ranging from 0.06 to 1 mg/kg) and 23 out of 77 samples of soft wheat (ranging from 0.05 – 0.33 mg/kg).

Recently, Pascale *et al.* (2002) carried out a two-year investigation on soft and durum wheat samples collected in several regions throughout the country, showing a contamination of DON up to 6 mg/kg for durum wheat and 3 mg/kg for soft wheat. The investigation demonsstrated that durum wheat was more contaminated than soft wheat and that the samples collected in North were more contaminated than those collected in Central and Southern Italy.

Logrieco *et al.* (1998) reported the natural occurrence of fumonisins in asparagus plants infected by *F. proliferatum*. Toxigenic strains of *F. proliferatum* are the causal agent of a destructive disease of asparagus plants called *Fusarium* crown and root rot. Asparagus plants heavily infected by *F. proliferatum* led to detection of FB_1 at levels of 7.4 and 0.83 mg/kg, and FB_2 at levels of 0.46 and 0.06 mg/kg in crown and stem samples, respectively.

4. OCCURRENCE OF MYCOTOXINS IN FOODS AND FEEDS OF PLANT ORIGIN

Several surveys were performed in Italy on the occurrence of mycotoxins in foods and feeds. Extensive investigations regarded essentially *Fusarium* toxins, i.e., ZEA, DON, and more recently FBs; but also AFs and OTA were surveyed in different Italian foods and feeds.

Some representative surveys are here briefly recalled, but a more complete account of the data obtained from surveys carried out in Italy on the contamination of foods and feeds of plant origin by AFBs, AFGs, OTA, PAT, ZEA, DON, and FBs, is illustrated in Tables 3-8.

Table 2. Mycotoxin occurrence in preharvest Fusarium *infected plants and products, in Italy*

Region	*Product*	*Year*	*No. positive/ total samples*	*Mycotoxin content (mg/kg)*
Fumonisins B_1				
South	Asparagus	1994	2/2	0.5 - 7.4
Sardinia	Maize ear	1992-93	6/6	125 - 250
North	Maize ear	1992-93	2/2	150 - 210
South	Maize ear	1994	8/9	Traces - 175
North	Maize ear	1994	12/13	Traces - 300
Deoxynivalenol				
North	Soft wheat	1994-1995	13/13	0.06 – 0.9
North	Durum wheat	1994-1995	36/36	0.095 – 3.1
North	Soft wheat	1998	23/77	0.05 - 0.3
North	Durum wheat	1998	24/26	0.06 - 1
North	Soft wheat	1999-2000	52/72	0.05-3.0
Centre	Soft wheat	1999-2000	18/34	0.06-0.2
South	Soft wheat	1999-2000	8/32	0.08-0.1
North	Durum wheat	1999-2000	60/63	0.09-6.5
Centre	Durum wheat	1999-2000	36/40	0.05-0.4
South	Durum wheat	1999-2000	9/34	0.08-0.1

4.1. Occurrence of aflatoxins in foods and feeds

Pietri *et al.* (in press) collected 503 maize samples during five years (1995-1999) from storage bins and feed mills in Northern Italy. They investigated the occurrence of AFB_1 and other toxins, finding that the incidence and the mean content of AFB_1 were generally low. However, two samples resulted highly contaminated (109 and 158 ng/g), and five of them exceeded 20 ng/g.

Finoli & Ferrari (1994) screened for the presence of AFB_1, AFB_2, AFG_1, and AFG_2, forty samples of spices, 20 samples of herbs, and 12 samples of food products prepared with Cayenne pepper. Analyses carried out using high-pressure liquid chromatography showed 75% of the spices containing detectable levels of aflatoxins with the highest contamination (234 ng/g of AFB_1) in samples of Cayenne pepper powder. Moreover, 7 out of 20 herb samples gave results over the detection limit of the analytical method (0.2 ng/g), as were all the 12 samples of Cayenne pepper food products where AFB_1 reached 6.4 ng/g in a sample of pasta.

Minervini *et al.* (1998a) analysed AFB_1 in 142 samples of farm feeds for cow, and in 71 feeds for buffalos. Aflatoxin B_1 was found in 94% of compound feed samples for both animal species at levels lower than 5 ppb; the remaining samples were aflatoxin free or contaminated at lower levels.

Miraglia *et al.* (1998a) reported a summary of the results obtained from the various approaches for aflatoxin contamination in several foods in Italy, over the last decade, mostly derived from the detailed reports of the European SCOOP programmes. Samples analyzed included cereals and derived products (whole cereals, whole wheat, wheat flour, bran, breakfast cereals); peanut and peanut products; pistachio nuts; other nuts (brazil nuts); dried mixed fruits; dried figs; spices (red chilli pepper); confectionery products; beverages, beer, olive oil, milk and milk products. They evaluated the incidence and level of aflatoxin contamination of Italian food, similar to that of other European countries.

Barbagallo & Russo (1999) examined for aflatoxin contamination by HPLC thirthy-eight pistachios samples from Sicily (11 samples), California (7), Greece (7), Iran (7), and Turkey (6). Aflatoxins were found in 27% of the samples from Sicily (up to 45.07 ng/g, of AFB_2 and AFG_2), in 29% of the samples from California (up to 87.9 ng/g, of AFB_2 and AFG_1), in 43% of the samples from Greece (up to 87,5 ng/g, of AFB_1, AFB_2, AFG_1 and AFG_2), in 57% of the samples from Iran (up to 183,6 ng/g, of AFB_1, AFB_2, AFG_1 and AFG_2), and in 50% of the samples from Turchey (up to 102.4 ng/g, of AFB_1, AFB_2, AFG_1 and AFG_2).

Finally, it is important to report the exceptionally high level of AFB_1 contamination of maize-based feeds in Northern Italy during the last season (2003), due to a widespread level of *A. flavus* contamination of maize in the field. This contamination has been due mainly to very high temperatures during the whole growing season of maize and the extraordinary drought from early May to early September, with consequent high levels of water stress for maize cultivation. Data kindly provided by Pietri (Piacenza University, personal communication), showed that among the maize-based feeds samples investigated, a high percentage of samples (around 13%) contained AFB_1 levels higher than 50 ng/g, which is the EU maximum permitted level for such products.

4.2. Occurrence of ochratoxin A in foods and feeds

The surveys on the occurrence of OTA in food and feed in Italy mostly regarded cereals, green coffee and wine. Wine is receiving increased attention because it appears, after cereals, to be a major source of daily intake of this toxin for both developed and developing countries.

Brera *et al.* (1998) screened twenty-one samples of the major coffee brands distributed at national level and found 16 of them contaminated by OTA. In particular, the maximum level (13 ng/g) was found in soluble coffee, and a marked presence of OTA remained in roasted samples.

Solfrizzo *et al.* (1998) employing a more rapid and reliable analytical method, detected OTA in cereal samples collected in Italy, i.e, in 42 rye samples (up to 105.9 ng/g); and in 4 out 13 trout feed (0.4 to 1.9 ng/g).

Table 3. Occurrence of aflatoxins in foods and feeds, in Italy

Region	*Product*	*Year*	*No. positive/ total samples*	*Aflatoxin B_1 content (ng/g)*
		Foods		
North/Import	Spices	--	30/40	0.5 - 231.6
North/Import	Herbs	--	7/20	0.2 - 1.5 (a)
North/Retail	Foods (b)	--	12/12	0.5 - 6.4 (a)
North/Center	Beverages	1992	0/91	<0.1
North/Center	Breakf.cereals	1995-96	15/50	0.1- 12.1 (a)
North/Center	Bran	1995-96	10/30	0.1 - 24.6 (a)
North/Center	Cereals	1995-96	10/35	0.1 - 9.3 (a)
North/Center	Chilli pasta	1995-96	32/40	0.1 - 150.2 (a)
North/Center	Chilli pepper	1995-96	39/43	0.1 - 504.0 (a)
North/Center	Confectionery	1990-92	3/18	0.1 - 0.8
North/Center	Figs (dried)	1990-94	4/119	0.10 – 120.0
North/Center	Figs (dried)	1995-96	3/15	0.1 - 121.1 (a)
North/Center	Fruit mix-dried	1989-95	75/584	0.1 - 52.0
North/Center	Fruit mix-dried	1995-96	10/33	0.1 - 102.0 (a)
North/Center	Marzipan	1995-96	7/25	0.1 - 7.2 (a)
North/Center	Nuts	1989-95	0/15	<0.1
North/Center	Olive oil extra	1995-96	1/15	0.1 - 2.3 (a)
North/Center	Olive oil	1995-96	3/20	0.1 - 11.2 (a)
North/Center	Peanuts	1989-95	39/457	0.1 - 789.0
North/Center	Peanuts	1995-96	17/30	0.1 - 1870.3 (a)
North/Center	Pistachio nuts	1991-94	2/11	0.1 - 2.4
North/Center	Pistachio nuts	1995-96	15/46	0.1 - 360.4 (a)
North/Center	Spices	1993-95	16/35	0.1 - 204.0
North/Center	Wheat flour	1995-96	9/40	0.1 - 18,5 (a)
North/Center	Wheat	1995-96	12/38	0.1 - 23.0 (a)
Sicily	Pistachio nuts	--	3/11	<0.1-45.3 (a)
Imported	Pistachio nuts	--	12/27	<0.1-183.6 (a)
		Feeds		
South	Cow feed	1997-98	89/95	0.10 - 4.8
South	Buffalo feed	1996/97	30/32	0.20 - 4.0
North	Maize feed	1995-99	503/503	0.05 - 158

(a)-Total aflatoxins. (b)-Different kind of pasta and sausage prepared with cayenne pepper.

Visconti *et al.* (1999) analyzed for OTA contamination 56 samples of red (38 samples), rosè (8), white (9) and dessert (1) Italian wines, applying a new method, based on a commercial immune-affinity column for clean-up and HPLC with fluorescence detection for quantification. Ochratoxin A was found in 37/38 red wine (traces to 7.63 ng/ml), in 7/8 rosè wine (traces to 1.15 ng/ml), in 4/9 white wines (traces to 0.97 ng/ml), and in the sample of dessert wine (0.29 ng/ml).

Visconti *et al.* (2000) reported the contamination of imported and domestic beer in 30 out of 61 samples examined by using immunoaffinity clean-up and liquid chromatography. The survey showed no substantial difference between strong and pale beers.

Pietri *et al.* (2001) investigated the incidence of OTA in 96 and 15 red and white dessert Italian wines, respectively, from 19 Italian regions. They showed that OTA occurred in both kinds of wines with a range from <0.01 to 3.86 ng/ml. 82 out of 96 samples of red wines were contaminated with a mean of 0.419 ng/ml, while 9 out of 15 samples of dessert wines resulted contaminated by OTA with a mean of 0.74 ng/ml. However, the survey showed that the geographic region of origin had a strong influence on OTA contamination both for red and dessert wines. Indeed, wines produced in Southern Italy were markedly more contaminated.

Miraglia *et al.* (1998a) also reported a summary of the results obtained from various approaches for OTA contamination in several foods in Italy, referred to the 1988-1998 decade, mostly derived from detailed reports of the European SCOOP programmes. Samples analyzed included cereals and derived products (durum and soft wheat, rice; maize, wheat flour, bran, extruded foods; breakfast cereals, maize flakes); pulses; green and roasted coffee; barley drinks; beer, wine. They concluded that the incidence and levels of OTA contamination of Italian foods is considerably higher than that of other European countries.

Screening the presence of OTA in green coffee beans of different origin and type, Romani *et al.* (2000) showed that of the 162 samples analyzed originated from various countries (Africa, America and Asia), 106 were positive for OTA, with a concentration up to 48 ng/g.

Recently, Tafuri *et al.* (in press), reported OTA contamination of nine out of eighteen pure cocoa powder samples marketed in Italy (from 0.22 ng/g to 0.77 ng/g). The high incidence (50%) of positive samples is an alert signal for consumers, since cocoa is widely used in the production of child snack foods and may contribute to OTA daily intake.

4.3. Occurrence of patulin in foods and feeds

Patulin is a mycotoxin produced by certain species of *Penicillium* and *Aspergillus*, often detectable in mouldy fruits and their derivatives. A limit of 50 ng/ml has been fixed for fruit and their derivatives. Beretta *et al.* (2000) analyzed 26 samples of rotten apples, 23 baby-food products, 22 samples of apple juices, and 11 commercial juices with pulp.

While none of the baby-food samples was over the safe limit (the highest concentration was 6.39 ng/ml), nine samples of commercial juices with pulp were positive with a dramatic variability (range from 0.68 to 1150 ng/ml).

Table 4. Occurrence of ochratoxin A in foods and feeds, in Italy

Region	*Product*	*Year*	*No. positive/ total samples*	*Ochratoxin A content (ng/g)*
		Foods		
Import/Retail	Green coffee	--	9/40	0.5 - 23.0
Imported	Green coffee	--	17/29	0.2 - 15.0
Imported	Green Coffee	1999	106/162	0.1 - 48.0
n.d	Coffee (a)	--	16/21	up to 13.0
Import/Retail	Cocoa powder	2003	9/18	0.2 - 0.8
Import/Retail	Beer	--	30/61	0.01 – 0.1
North/Center	Barley drinks	1996	3/8	0.1 - 7.0
North/Center	Beer	1996	9/55	0.1 - 2.6
North/Center	Bran	1990-93	5/35	0.1 - 11.0
North/Center	Bran	1996	8/20	0.1 - 12.0
North/Center	Breakf.cereals	1996	7/25	0.1 - 6.3
North/Center	Cereals	1996	5/12	0.1 - 10.2
North/Center	Coffee green	1989	17/29	0.1 - 15.0
North/Center	Coffee roasted	1996	14/22	0.1 - 7.0
North/Center	Food extruded	1990-93	8/15	0.1 - 15.0
North/Center	Maize	1982-84	52/202	0.1 - 2.0
North/Center	Maize	1996	3/10	0.1 - 6.0
North/Center	Maize flakes	1996	15/23	0.1 - 11.2
North/Center	Pulses	1996	4/20	0.1 - 1.2
North/Center	Rice	1990-93	8/15	0.1 - 1.0
North/Center	Wheat durum	1990-93	9/10	0.1 - 5.6
North/Center	Wheat soft	1990-93	10/10	0.1 - 2.6
North/Center	Wheat flour	1996	13/58	0.1 - 6.1
North/Center	Wine	1996	4/41	0.1 - 0.6
n.d	Rye kernels	1997-98	13/42	1.3 - 105.9
South	Red wine (b)	1998	26/27	0.01 - 7.6
South	Red wine (c)	1998	11/11	0.5 - 4.7
Various areas	Red wine	1995-97	82/96	0.01 – 3.2
South	Rosè wine (b)	1998	5/6	0.01 - 1.2
South	Rosè wine (c)	1998	2/2	0.4 - 0.6
South	White wine (b)	1998	2/7	0.01 - 0.06
South	White wine (c)	1998	2/2	0.1 - 1
South	Marsala wine	1998	1/1	0.3
Various areas	W. dessert wine	1995-97	9/15	0.01-3.9
		Feeds		
North	Trout feed	1997-98	4/13	0.7 - 1.9

(a)-roasted (14 samples) and soluble (7 samples) coffee brands (9,6 mg/kg). (b)-samples from local retail (bottled wine). (c)-sample from home-made wine.

On the other hand, the apple juices samples were all positive, but with a highest concentration of 28.24 ng/ml, below the safe limit. Finally, 21 out of 26 rotten apples resulted contaminated.

The analyses were performed on the rotten area, unaffected area with peel and affected area without peel.

The range of the concentration of PAT for each of the three areas were: from 2 to 113343 ng/g, 0.73 to 1166 ng/g, and 0.44 to 93 ng/g, respectively.

Recently, Ritieni (2003) analyzed, for PAT occurrence, forty samples of four apple products marketed in Italy and purchased from supermarkets, herbalists and retail shops. The eleven samples infected had a PAT concentration ranging between 1.4 and 74.2 ng/ml. In particular, while all vinegar samples were negative for PAT, two samples of apple-based baby food out of ten analyzed were positive (from 13.1 to 17.7 ng/ml) and both labelled as "organic food"; three samples industrial puree to prepare commercial juice were positive with the "organic" one showing the highest concentration (74.2 ng/ml) of PAT; finally six samples of apple juice were contaminated from 1.4 to 56.4 ng/ml. Significant differences between organic and conventional products were not found. However, particular attention should be given in this report on the data regarding baby food contamination since the daily intake of apple juice or similar products together with the low baby weight could be a dangerous combination.

4.4. Occurrence of zearalenone in foods and feeds

Visconti and Pascale (1998), using a rapid method to quantify ZEN in maize, by application of immunoaffinity clean-up and HPLC with fluorescence detection for quantification of the toxin, examined 15 maize samples of 1997 crop, collected from North Italy. The analyses revealed the presence of ZEN in 13 out of 15 samples (from 4 to 140 ng/g).

A five-year survey of Pietri *et al.* (in press) from 1995 to 1999 investigated also the incidence of ZEA in 503 maize samples collected from storage bins and feed mills. The mean of ZEA occurrence varied from 13 ng/g in 1998 to 453 ng/g in 1996, when the percentage of samples contaminated by more than 1000 ng/g of ZEA was 14.4%.

4.5. Occurrence of deoxynivalenol in foods and feeds

Visconti *et al.* (1997) applying the immuno-affinity columns for the clean-up and pre-concentration of DON, combined with quantitative toxin determination by reversed phase HPLC/diode array UV detector (DAD), analysed 70 samples of durum (58 samples) and soft (12) wheat collected in 1995 in different parts of Italy.

The results showed a widespread contamination of DON in samples originating from Emilia Romagna, with levels of contamination higher in durum wheat (up to 3085 ng/g) than in soft wheat (up to 920 ng/g). No DON was found in samples of durum wheat from Southern Italy.

Table 5. Occurrence of patulin in foods and feeds, in Italy

Region	*Product*	*Year*	*No. positive/ total samples*	*Patulin content (ng/g)*
Retail	Apple baby-foods	1999	21/23	0.03-6.4
Retail	Apple puree	1999	9/11	0.7-1150
Retail	Apple-juice	1999	22/22	0.1-28.2
Retail	Apple baby-foods	2002	2/10	13.1-17.7
Retail	Apple-juice	2002	6/21	1.4-56.4
Retail	Apple puree	2002	3/6	15.9-74.2
Retail	Vinegar	2002	0/3	-

Cirillo *et al.* (2003) performed a survey to obtain the frequency and the level of contamination by DON in Italian marketed foods. They analyzed 202 samples including raw materials and processed cereal foods (bread, pasta, breakfast cereals, biscuits, baby and infant foods), showing that 84% of samples were contaminated with the toxin at levels ranging from 7 to 930 ng/g. The highest level of DON was detected in raw cereals and wholemeal flours, and a widespread DON contamination was found in baby and infant foods at levels ranging from 7 to 166 ng/g.

Pietri *et al.* (in press) investigated also the DON occurrence in 503 maize samples collected from storage bins and feed mills from 1995 to 1999 and registered a level of DON contamination ranging from 194 ng/g (1995) to 2716 ng/g (1996) as mean. Moreover, of particular interest were the data of 1996, showing DON levels higher than 1000 ng/g in 76% of total samples.

Table 6 – Occurrence of zearalenone in food and feed, in Italy

Region	*Product*	*Year*	*No. positive/ total samples*	*Zearalenone content (mg/kg)*
North	Maize Food	1997	13/15	0.004 - 0.1
North	Maize Feed	1995-97	232/503	0.05 - 2.5

4.6. Occurrence of fumonisins in foods and feeds

Doko and Visconti (1994) investigated the occurrence of FB_1 and FB_2 in maize and maize-based human foodstuffs in Italy. Samples of maize kernels from regional distributors and maize genotypes from breeding stations, as well as samples of commercially available maize-based food products were included. All maize samples were positive for FB_1 (up to 5310 ng/g) and for FB_2 (up to 1480). In maize-based foods, the highest FB contamination levels were recorded in puffed (extruded) maize containing up to 6100 ng/g of FB_1 and up to 520 ng/g of FB_2. Levels ranging from 420 to 3760 ng/g FB_1 and from 80 to 910 ng/g FB_2 were detected in maize grits, and in maize flour or polenta. All sweet maize samples examined were positive for FB_1, at levels varying from 60 to 790 ng/g, but negative for FB_2. Lower levels of fumonisins were found in other commodities, such as popcorn (up to 60 ng/g FB_1 and 20 ng/g FB_2), tortilla chips (up to 60 ng/g FB_1 and 10 ng/g FB_2), and maize flakes (10 ng/g FB_1).

Table 7. Occurrence of deoxynivalenol in foods and feeds, in Italy

Region	*Product*	*Year*	*No. positive/ total samples*	*Deoxynivalenol content (mg/kg)*
Center/North	Durum wheat	1995	38/38	0.095 – 3.1
Center/North	Soft wheat	1995	10/12	0.060 – 0.9
South	Durum wheat	1995	0/20	--
Center/South	*Triticum spelta*	1995	7/100	up to 0.06
North	Maize-based Feed	1995-99	468/503	0.02 - 9.4
Retail	Cereals, Flours	2002	111/111	0.07 - 0.97
Retail	Bread	2002	20/24	0.007 - 0.3
Retail	Durum wheat pasta	2002	8/17	0.009 - 0.1
Retail	Breakfast cereals	2002	9/14	0.012 - 0.5
Retail	Biscuits	2002	20/24	0.016 - 0.2
Retail	Baby/infant foods	2002	7/12	0.007 - 0.2

Pascale *et al.* (1995) analyzed 20 samples of maize meal (polenta) for the occurrence of FBs. All samples were contaminated by FB_1 (150 to 3760 ng/g) and by lower concentrations of FB_2 (60 to 910 ng/g). The contamination levels were not reduced by cooking, confirming previous data on FB stability at high temperature.

Cirillo *et al.* (2003) performed a survey to obtain the frequency and the level of contamination by FBs in Italian marketed foods. They analyzed 202 samples including raw materials and processed cereal foods (bread, pasta, breakfast cereals, biscuits, baby and infant foods), showing that 26% of samples were contaminated by the FB_1 with a range from 10 to 2870 ng/g, and FB_2 with a range from 10 to 790 ng/g. The highest level of FB_1 was detected in raw cereals and wholemeal flours, while the highest level of FB_2 was detected in durum wheat pasta.

Finally, Pietri *et al.* (in press) investigated also the FB_1 occurrence in 503 maize samples collected from storage bins and feed mills from 1995 to 1999. The FB_1 incidence recorded was very high in all the years and ranged from 1324 ng/g (1996) to 5173 ng/g (1999). Among the 503 maize samples, 69.6% resulted contaminated over 1000 ng/g and 16.9% over 5000 ng/g.

Table 8. Occurrence of fumonisins in foods and feeds, in Italy

Region	*Product*	*Year*	*No. positive/ total samples*	*Fumonisin B_1 content (mg/kg)*
n.d	Maize (a)	1992-93	7/7	0.1 - 6.8 (b)
North	Maize lines (c)	1990-91	26/26	0.01 - 2.9 (b)
n.d	Maize meal (d)	1992-93	8/8	0.50 - 4.7 (b)
n.d	Foods (e)	1992-93	6/6	0.90 - 6.6 (b)
n.d	Foods (f)	1992-93	5/5	0.06 - 0.3
n.d	Foods (g)	1992-93	6/10	up to 0.08 (b)
Apulia/retail	Polenta	1994	20/20	0.2 - 3.8 (h)
n.d	Maize	1991-92	4/4	0.08 - 8.4 (i)
n.d	Cow feed	1991-92	3/3	0.1 - 3.0 (i)
n.d	Poultry feed	1991-92	5/5	0.1 - 1.2 (i)
n.d	Pig feed	1991-92	10/12	up to 1.4 (i)
n.d	Rabbit feed	1991-92	1/1	0.1
North	Maize-based feed	1995-99	503/503	62 - 51690 (i)
Retail	Cereals, Flours	2002	35/111	0.01 - 2.9 (i)
Retail	Bread	2002	8/24	0.03 - 0.2 (i)
Retail	Durum wheat pasta	2002	17/17	0.08 - 0.8 (l)
Retail	Breakfast cereals	2002	7/14	0.08 - 0.2 (i)
Retail	Biscuits	2002	2/24	0.02 - 0.2 (i)
Retail	Baby/infant foods	2002	0/12	-

(a)-Commercial maize intended for human and animal consumption. (b)-Total FB_1 and FB_2. (c)-Maize genotypes (inbred lines). (d) - Maize flour or polenta (6) and maize grits (2). (e) - Puffed (extruded) maize. (f) - Sweet maize. (g)-Miscellaneous food commodities (maize flakes, popcorn kernels or products, tortillas chips). (h)-Associated with FB_2 (0.06-0.91), and traces of FB_3. (i)-In several samples FB_2 was also present. (l) FB_2.

5. OCCURRENCE OF MYCOTOXINS IN MILK AND MILK PRODUCTS.

Investigations on the occurrence of mycotoxins in foods of animal origin, regarded mostly the presence of AFM_1 in milk and milk products. The results obtained from several surveys carried out by different Italian Institutions, showed evidence of a diffuse micro-contamination of AFM_1 in milk, cheese and other milk products. Some representative studies are here briefly reported, whereas the extensive data are illustrated in table 9.

Pietri *et al.* (1997) analysed for AFM_1 a total of 223 samples of "Grana padano" cheese manufactured in 4 years (1991-94) by dairies in 11 provinces of the Po valley. Aflatoxin M_1 was extracted with chloroform, purified by an immunoaffinity column, and determined by HPLC using a fluorescence detector. The results obtained showed that only one sample exceeded the maximum tolerated level in cheese in some European countries (250 ng/kg). Most samples (91%) were in the range 5-100 ng/kg and only 15 (6.7%) in the range 100-250 ng/kg. Mean contamination levels of 1992 and 1994 were significantly higher ($P<0.05$) than those of 1993 and 1991. No significant differences were observed among provinces of dairies of origin.

Finoli & Vecchio (1994), using an immunoaffinity column for clean-up and HPLC with fluorescence detector for quantitation, analyzed samples of goat dairy milk and derived milk products for AFM_1. The toxin was detected: in 25 of 80 (4.0 to 37.0 ng/kg) milk samples from farms; in 8 of 23 (19.0 to 160.0 ng/kg) cheese samples from farms; and in 11 out 20 (18.0 to 200.0 ng/kg) cheese samples collected from retail markets.

Minervini *et al.* (1998a) screened for AFM_1 samples of cow (118) and buffalo (45) milk, and samples of cow (98) and buffalo (83) dairy products. Aflatoxin M_1 was found in 55% and 44% of cow and buffalo milk samples at levels ranged from 5.5 to 77 ppt, respectively. Low levels of AFM_1 (except one sample of ripe cheese with 540 ppt) were detected in 29.5 and 9.6 of cow and buffalo dairy products, respectively.

The above mentioned extraordinary level of AFB_1 contamination of maize-based feeds in the Northern Italy during the 2003 season (see 4.1) caused also the highest incidence of milk contamination by AFM_1 recorded during the last decade. Data kindly provided by Pietri (personal communication) showed that AFM_1 occurred in the milk samples analysed over the law limit (0.05 ng/g) for around 50-60% of samples investigated in October 2003. Although this percentage decreased in further analysis performed, the percentage of highly contaminated samples by AFM_1, still has been constantly high during the following five months.

6. OCCURRENCE OF MYCOTOXINS IN BIOLOGICAL FLUIDS AND TISSUES.

Investigations on this subject concerned almost exclusively the presence of ochratoxins in human serum, milk and urine, as well as in maternal serum, placenta and funiculum. The obtained data indicated a high exposure to the toxins not only for adults but also for children during breast-feeding, whereas the established placental transport of high levels of toxin pointed out the risk for the foetus during crucial prenatal development. Some indicatives studies are briefly reported here, whereas the extensive data are illustrated in Table 10.

Breitholtz-Emanuelsson *et al.* (1994) determined OTA in human serum samples, collected in the south of Italy in November 1992, using ion-pair HPLC and fluorescence detector. Samples were collected from healthy people (65 subjects), and from patients with five different kidney disorders: transplanted (13), chronic glomerulonephritis (8), renal calculus or cyst (6), chronic renal failure (13), and dialysed (28). The mean concentration of OTA in the health group was 0.53 ng/ml serum. The highest mean concentration (1.4 ng/ml serum) was found in the group of dialysed patients. Moreover, the highest incidence of samples containing more than 0.44 ng OTA/ml serum was found in the same dialysis group, compared to the other groups.

Micco *et al.* (1995) in order to assess the incidence of OTA in human biological fluids, investigated the concentration of toxin in breast milk of donor mothers in Italy. Of 111 samples, 22 were contaminated in the range 0.1-12 ng/g.

Miraglia *et al.* (1998b) collected forty samples of maternal blood serum and their related fluids from placenta and funiculum from 40 donors, to assess the possibility of active OTA placental transport, and the consequent foetal exposure to the toxin. The OTA concentration was 0.11 to 0.98 ng/ml in 39% of maternal serum. Ochratoxin A was also detected in placenta (0.38 to 10.60 ng/kg) and in funiculum (0.27 to 9.50 ng/kg). These results indicated an active placental transport of OTA and the possibilities for the toxin to reach the foetus during crucial prenatal development.

Table 9. Occurrence of aflatoxin M_1 in milk and milk products, in Italy

Region	*Year*	*No. positive/ total samples*	*Aflatoxin M_1 content (ng/kg)*
Commercial fresh and heat treated whole milk (a)			
North	1994	20/20	31.0 - 41.0
North	1996	25/80	4.0 - 37.0 (d,g)
South	1997	20/47	5.4 - 44.0 (b)
South	1997	48/94	5.3 - 77.0 (g)
South	1997	17/24	5. - 72.6
North/Center	1989-95	398/997	1.0 - 50.0
North	1999	99/99	5.0 - 70.0
Cheese			
North	1996	8/23	19.0 - 160.0 (d,g)
North/Import	1996	11/20	18.0 - 200.0 (d)
North	1991-94	219/223	5 - >250
South	1997	9/18	60 - 140
South	1997	4/21	60 - 70 (e)
South	1997	7/53	50 - 160
South	1997	4/8	62 - 101 (f)
South	1997	4/60	50 - 230 (f)
Apulia	1997	2/40	140 and 160 (c)
North/Center	1984-95	323/1593	50 - 500

(a)-Cow milk, if not oterwise stated. (b)-Buffalo milk. (c)-In two samples of hard cheese (caciocavallo and provolone, respectively). (d)-Goat products. (e)-Soft cheese. (f)-Buffalo products. (g)-Farm milk.

Data not mentioned in the text are retrieved from Caggioni & Pietri 1999; Finoli & Vecchio 1997; Gelosa & Buzzetti 1994; Minervini *et al.*, 1998b.

Pascale *et al.* (2001) applied immuno-affinity columns to the analysis of OTA in human serum and urine of three individuals affected by kariomegalic interstitial nephritis (KIN) and six healthy individuals from the same family. The levels of OTA in serum were higher in KIN patients (range 1.107-2.206 ng/ml) than in healthy individuals (range 0.207-0.934 ng/ml). The levels of OTA in urine ranged from 0.012 to 0.046 ng/ml, except for one sample from a patient affected by KIN (0.140) ng/ml).

Table 10. Occurrence of ochratoxin A in biological fluids, in Italy

Region	*Origin*	*Year*	*No. positive/ total samples*	*Ochratoxin A content (ng/g)*
		Blood		
Imported	Pig	1994	64/88	0.05 - 1.1*
South	Human (a)	1992	65/65	0.1 - 2.0
South	Human (b)	1992	13/13	0.20 - 1.0
South	Human (c)	1992	8/8	0.05 - 1.4
South	Human (d)	1992	6/6	0.2 - 1.0
South	Human (e)	1992	13/13	0.2 - 1.4
South	Human (f)	1992	28/28	0.2 - 14.0
South	Human (g)	1997-98	3/3	1.1 - 2.2
South	Human (a)	1997-98	6/6	0.2 - 0.9
Center	Human	1997-98	15/40	0.1 - 1
		Milk		
Center	Human (h)	--	9/30	0.1 - 1.0
Center	Human (i)	--	13/81	0.7 - 12.0
		Urine		
South	Human (g)		3/3	0.01 - 0.1
South	Human (a)		6/6	0.01 - 0.1

(a)-Health individuals. (b)-Transplanted patientis. (c)-Patients affected by chronic glomerulo-nephritis. (d)-Patients affected by renal calculus or cyst. (e)-Patients affected by chronic renal failure. (f)-Patients treated with dialysis. (g)-Patients affected by kariomegalic interstitial nephritis (h)-Hospitalized donor mothers. (i)-No-hospitalized donor mother.
*= Minervini *et al.*, 1994.

7. CONCLUSIONS

The data presented in this paper, referring mainly to the last decade literature, provide an overview of the Italian situation on the occurrence of toxigenic fungi and mycotoxins, although most of these data were obtained without standard sampling procedures and/or analytical methods. More fitting comprehensive strategies are required for monitoring programmes intended to estimate the real incidence of mycotoxins in foods and feeds in Italy, or to provide data for assessment of dietary exposure to mycotoxins, and eventually to assess the human and animal risk exposure at national level. To this purpose, the application of suitable sampling procedures plays a basic role together with the adoption at national level of protocols of standard analytical methods recommended by International Institutions.

Finally, the present study does not report the occurrence of several mycotoxins considered toxicologically less important than those focused in this paper (e.g. other *Fusarium* toxins, *Alternaria* toxins etc). Nevertheless, attention should be paid to the possible co-occurrence of these mycotoxins in foods and feeds with the most studied and investigated ones. Their possible additive toxic effects (if not synergistic) should be evaluated to obtain a complete picture of the risk that food and feed contamination by mycotoxins pose to animal and humans. Indeed, co-occurrence of the most toxic mycotoxins in particular food and feed commodities (e.g. maize) is frequently observed and reported. The accurate evaluation of the effects of multiple mycotoxin contamination on humans will be the future challenge for scientists working in this field.

8. AFFILIATIONS

Institute of Sciences of Food Production, Research National Council (CNR) Via Amendola 122, 70126, Bari, Italy. antonio.moretti@ispa.cnr.it

9. REFERENCES

Avantaggiato G., Quaranta F., Desiderio E., & Visconti A. (2002). Fumonisin contamination of maize hybrids visibly damaged by Sesemia. Journal of Science of Food and Agriculture. 83, 13-18.

Balmas V., & Corazza L. (1994). *Fusarium graminearum* Gruppo 2 agente causale della scabbia del frumento duro nell'Italia centrale. L'Informatore agrario, 11, 59-62

Balmas V., Maccaroni M., Santori A,. & Corazza L. (1999). Fusariosi della spiga di cereali autunno-vernini: indagine epidemiologica e valutazione della resistenza varietale di frumento duro nell'Italia centrale. L'Informatore agrario, 12, 37-42.

Barbagallo R.N., & C. Russo (1999). HPLC Determination of aflatoxins in pistachios of different geographical origin. Industrie Alimentari, 38, 533-545.

Battilani P., Giorni P., & Pietri A. (2003). Epidemiology of toxin-producing fungi and ochratoxin A occurrence in grape. European Journal of Plant Pathology, 109, 715-722.

Beretta B., Gaiaschi A., Galli C.L., & Restani P. (2000). Patulin in apple-based foods: occurrence and safety evaluation. Food Additives and Contaminants, 17, 399-406.

Bottalico A. (1977). Presenza di zearalenone nelle spighe di Granturco (Zea mays L.) infette da Gibberella zeae (Schw.) Petch, nel Metapontino. Phytopathologia Mediterranea, 16, 14-17.

Bottalico A., Logrieco A., A. Ritieni, Moretti A., Randazzo G. & Corda P. (1995). Beauvericin and fumonisin B_1 in preharvest *Fusarium moniliforme* maize ear rot in Sardinia. Food Additives and Contaminants, 12, 599-607.

Breitholtz-Emanuelsson A., Minervini F., Hult K. & Visconti A. (1994): Ochratoxin A in human serum samples collected in Southern Italy from healthy individuals and individuals suffering from different kidney disorders. Natural Toxins, 2, 366-370.

Brera C., Miraglia M., & Colatosti M. (1998). Evaluation of ochratoxin A in roasted coffee and derived beverage samples. Rev. Méd. Vét., June 1998, 698.

Caggioni C., and Pietri A. (1999). Le aflatossine nel latte: dove nasce il problema e come prevenirlo. Informatore agrario, 45, 45-50.

Casulli F., Pancaldi D., De Lillo E., & Alberti I. (1995). Observations on wheat crown rot and head blight caused by *Fusarium* spp. in Italy. In: Book of Abstracts of International Seminar on *Fusarium* - mycotoxins, taxonomy and pathogenicity, Martina Franca, Italy, May 9-13, 1995, 139-140.

Cirillo T., Ritieni A., Galvano F., & Amodio Cocchieri R. (2003). Natural co-occurennce od deoxynivalenol and fumonisin B_1 in Italian marketed foodstuffs. Food Additives and Contaminants, 20, 566-571.

Doko M.B., & Visconti A. (1994). Occurrence of fumonisins B_1 and B_2 in corn and corn-based human foodstuffs in Italy. Food Additives and Contaminants, 11, 433-439.

Finoli C., & Ferrari M. (1994). Aflatossine in spezie ed erbe aromatiche, 1994. Industrie Alimentari, 33, 732-736.

Finoli C., & Vecchio A. (1997). Aflatoxin M_1 in goat dairy products. Microbiol. Alim. Nutr., 15, 47-52.

Gelosa M. and I. Buzzetti (1994): Effetto del trattamento termico del latte sul contenuto di aflatossina M_1. Rivista della Società Italiana di Scienze Alimentari, 23, 115-117.

Logrieco A., Moretti A., Ritieni A., Bottalico A., & Corda P. (1995): Occurrence and toxigenicity of *Fusarium proliferatum* from preharvest maize ear rot, and associated mycotoxins, in Italy. Plant Disease, 79, 727-731.

Logrieco A., Doko B., Moretti A., Frisullo S., & Visconti A. (1998). Occurrence of fumonisin B_1 and B_2 in *Fusarium proliferatum* infected *Asparagus* plants. Journal of Agricultural and Food Chemistry, 46, 5201-5204.

Lops R., Pascale M., Pancaldi D & Visconti A. Infezioni fungine e presenza di deossinivalenolo in cariossidi di frumento prodotte in diverse regioni italiane. L'Informatore Fitopatologico, 4, 60-66.

Micco C., Miraglia M., Brera C., Corneli S., & Ambruzzi M.A. (1995). Evaluation of ochratoxin A level in human milk in Italy. Food Additives and Contaminants, 12, 351-354.

Minervini F., M. Pascale, M. Caramelli and A. Visconti (1994): Presenza di ocratossina A nel sangue umano e suino. In: Atti della Società Italiana delle Scienze Veterinarie, 48, 1225-1229.

Minervini F., G. Stea, & Visconti A. (1998a). Indagine sulla presenza di aflatossine nei mangimi, latte e prodotti caseari bovini e bufalini. In: Atti del Progetto Speciale <<Miglioramento della qualità dei prodotti alimentari >>, Firenze, 9 luglio 1998, 97-104.

Minervini F., G. Stea & Visconti A. (1998b). Aflatossine in mangimi, latte bovino e prodotti caseari in Puglia. In: Atti della Società Italiana delle Scienze Veterinarie, 52, 421-422.

Miraglia M., Brera C., Onori R., Corneli S., Colatosti M., Cava E., Ippoliti D., & Quaglia M. (1998a). Mycotoxin contamination in Italy over ther last decade. In: Mycotoxins and Phycotoxins - Development in Chemistry, Toxicology and food Safety (Miraglia M., H. Van Egmond, C. Brera and J. Gilbert, eds.), Alaken Inc., Fort Collins, Colorado, 601-608.

Miraglia M., Brera C., Corneli S., Cava E., Montanino G., & Miraglia E. (1998b). Occurrence of ochratoxin A (OA) in maternal serum, placenta and funiculum. In: Mycotoxins and Phycotoxins - Development in Chemistry, Toxicology and food Safety (M. Miraglia, H. Van Egmond, C. Brera and J. Gilbert, eds.), Alaken Inc., Fort Collins, Colorado, 165-179.

Moretti A., Logrieco A., Doko B., Frisullo S., Visconti A., & Bottalico A. (1995) *Fusarium proliferatum* from asparagus, in Italy: occurrence, fertilità and toxigenicity. Cereal Research Communication, 25, 785-786.

Pancaldi D., & Torricelli R. (1999). Presenza di specie di *Fusarium* su cariossidi di frumento duro in Emilia-Romagna nel triennio '94-'96. L'Informatore agrario 3, 42-46.

Pascale M., Doko M.B., & Visconti A. (1995). Determinazione di fumonisine nella polenta mediante cromatografia HPLC. In: Atti 2° Congresso Naz. "Chimica degli Alimenti, La Grafica Ed., Messina, 1967-1069.

Pascale M., & Visconti A. (2001). Rapid method for the determination of ochratoxin A in urine by immunoaffinity column clean-up and high-performance liquid chromatography. Mycopathologia, 152, 91-95.

Pascale, M., Bottalico A., Pancaldi D., Perrone G., & Visconti A. (2002). Occurrence of deoxynivalenol in cereals from experimental fields in various Italian regions. Petria, 12, 123-129.

Pascale, M., Bottalico A., Pancaldi D., Perrone G., & Visconti A. (2002). Occurrence of deoxynivalenol in cereals from experimental fields in various Italian regions. Petria, 12, 123-129.

Pasquini M., Cardone A.M., Casini F., Casulli F., Chierico M., Contoli S., Converso R., Corazza L., Del Vecchio A., Iori A., Lombardi T., Lombardo G.L., Mazzocchetti A., Natalizi G., Notario T., Pancaldi D., Raimondo I., Ratti C., Riccardi M., Rubies-Autonell C., Sereni L., Vallega V., Virzì N., &Vitale S. (2001). Monitoraggio delle malattie sui cereali autunno-vernini: le principali malattie del frumento. L'Informatore agrario, 1, 33-39.

Pietri A., Bertuzzi T., Bertuzzi P., & Piva G. (1997). Aflatoxin M_1 occurrence in samples of Grana Padano cheese. Food Additives and Contaminants, 14, 341-344.

Pietri A., Bertuzzi T., Pallaroni L., & Piva G. (2001). Occurrence of ochratoxin A in Italian wines. Food Additives and Contaminants, 18, 647-654.

Pietri A., 03..31.2004, personal communication.

Pietri A., Bertuzzi T., Pallaroni L., & Piva G. (in press). Occurrence of mycotoxins and ergosterol in maize harvested during five years in northern Italy. Food Additives and Contaminants (in press).

Ritieni A., A. Moretti, A. Logrieco, A. Bottalico, G. Randazzo, M. Monti, R. Ferracane and V. Fogliano. (1997). Occurrence of Fusaproliferin, Fumonisin B_1, and Beauvericin in Maize from Italy. Journal of Agricultural and Food Chemistry, 45, 4011-4016.

Ritieni A. Patulin in Italian commercial apple products. (in press). Journal of Agricultural and Food Chemistry, (in press).

Romani S., Sacchetti G., Chaves Lopez C., Pinnavia G.G., & Dalla Rosa M. (2000) Screening on the occurrence of Ochratoxin A in green coffee beans of different origins and types. Journal of Agricultural and Food Chemistry, 48, 3616-3619.

Solfrizzo M., Avantaggiato G., & Visconti A. (1998). Use of various clean-up procedures for the analysis of ochratoxin A in cereals. Journal of Chromatography A, 815, 67-73.

Tafuri A., Ferracane R., & Ritieni A. (in press). Ochratoxin in Italian marketed cocoa products. Food Chemistry, in press

Visconti A., Pascale M., & Sarzi Amadè P.L. (1997). Impiego di minicolonne ad immunoaffinità nell'analisi di deossinivalenolo (vomitossina) in campioni di frumento. Bollettino Chimica dell' Igiene, 48, 127-131.

Visconti A., & Pascale M. (1998). Determination of zearalenone in corn by means of immunoaffinity clean-up and high-performance liquid chromatography with fluorescence detection. Journal of Chromatography A, 815, 133-140.

Visconti A., Pascale M. & Centonze G. (1999). Determination of ochratoxin A in wine by means of immunoaffinity column clean-up and high-performance liquid chromatography. Journal of Chromatography A, 864, 89-101.

Visconti A., Pascale M. & Centonze G. (2000). Determination of ochratoxin A in domestic and imported beers in Italy by immunoaffinity column clean-up and liquid chromatography. Journal of Chromatography A, 888, 321-326.

J. PERKOWSKI, J. CHEŁKOWSKI & P. GOLIŃSKI

OCCURRENCE OF MYCOTOXINS IN CEREALS, PLANTS, FOODS AND FEEDS IN POLAND

1. INTRODUCTION

Toxigenic fungi occurrence and mycotoxins production in Poland have been studied since 1969. We have examined mostly cereal grain and feeds for toxigenic fungi, mycotoxins potential of isolated strains and natural contamination of food and feed components with mycotoxins. *Fusarium* toxins and ochratoxin A were found to be significant contaminations of agricultural products, present in high percentages of cereal grain samples (Chełkowski 1998, Goliński *et al.,* 1996, Perkowski & Wiwart 2002).

Small grain cereals are produced in Poland in harvest in over 20 million tons per year and maize grain production significantly increased as well during the last years.

In this report we focus mainly on the above cereals discussing their mycotoxins contamination. Our other interests were mycotoxin contamination in apple juice, sward and dust in cereal industry. Nowadays research on mycotoxins is becoming more and more comprehensive and should help in the public health and sanitary condition improvement.

2. TOXIGENIC FUNGI

A significant number of toxigenic fungal species has been isolated in Poland during the last decade and in previous years.

Most of them belong to four genera: *Alternaria, Fusarium, Aspergillus* and *Penicillium.*

Species of above mentioned genera were isolated mostly from cereal grains, cereal products, and feeds, as well as from soil.

These subjects were already described and presented by Chełkowski *et al.,* 2001.

3. AFLATOXINS

Due to climate conditions in Poland, the presence of aflatoxins in cereals and cereals products is related to the metabolites analysis in imported products only. A good example is reported by Czerwiecki 2001.

Investigations on aflatoxins content of several commodities available in Poland, such as cereal products, bread, cereal - milk formulas for babies and children, milk an powder red milk conducted in 1990 - 92, showed incidental contamination in one out of 351 samples. Only aflatoxin G_1 was detected at the level 0.6 μg/kg (detection

A. Logrieco and A. Visconti (eds.),
An Overview on Toxigenic Fungi and Mycotoxins in Europe, 161-172.

limit of TLC method). There was no evidence of aflatoxin M_1 present in 270 samples of milk and special milk formula products for children. The above results indicate both, that there is no aflatoxins problem in diary production in Poland and that for production of milk formula products for babies and children the best quality raw materials were used. The detection limit for TLC technique was 0.025 and 0.5 μg/L or kg of sample. In further investigations during period 1995 - 2000 out of 105 different imported products, e.g. peanuts, aflatoxin B_1 was determined in 10 % of peanut samples and in 30 % of pistachio nuts at levels 0.5 - 3.0 μg/kg. In some samples (e.g. wine samples), trace levels of aflatoxin G_1 were determined. The detectability of the improved method was 0.1 - 0.3 μg/kg.

4. *FUSARIUM* TOXINS

Presence of 18 *Fusarium* metabolites: DON, 3-AcDON, 15-AcDON, NIV, FUS, T-2 toxin, HT-2 toxin scirpentriol, neosolaniol, T-2 tetraol, DAS, 4,7 dideoxynivalenol, zearalenone, moniliformin, beauvericin and fumonisins B_1, B_2 and B_3 has been demonstrated in Poland in cereals grain samples. This report is focused on the occurrence of the most important toxins such as DON, NIV, T-2, HT-2 and MON. In Table 1 published results since 1997 are summarized.

In addition to the toxins above, the occurrence of scirpentriol (STO) in small grain cereals is worth to underline (Perkowski *et al.*, 2003b). Scirpentriol was detected in 1%, 37% and 8 % of 248 wheat, 32 barley and 99 oat grain samples, respectively, and the maximum concentration was 83 μg/kg. Samples of wheat and oat grain with visible scab symptoms were also analyzed, and STO (mean level 255 μg/kg) was detected only in oat samples infected with *F. sporotrichioides* and *F. poae* as the dominating species. In 15 barley samples - subdivided based on seed size into fractions of $\leq$ 2.5 and > 2.5 mm - smaller kernels exhibited in average 94 % of the STO content in total samples (in kernel fraction > 2.5 mm 28 μg/kg, $\leq$ 2.5 mm 297 μg/kg). In oats, accumulation of STO was highest in the chaff, lower in the stem part adjacent to wips and lowest in the grain.

In 15 studied, naturally contaminated cultivars significant correlations between concentrations of most analysed *Fusarium* toxins are shown in Table 1 and results were related to phytopathological observation (Perkowski *et al.*, 2003a).

In this work co-occurrence of several *Fusarium* toxins was observed. Occurrence of several toxins simultaneously has also been indicated by Tomczak *et al.*, 2002 (Table 3). Incidence of *Fusarium* spp. and some mycotoxins on grain and ears of winter wheat, depending on applied cultivar and fungicides, were examined (Sadowski *et al.*, 2002). It was noted that the symptoms of *Fusarium* blight (FHB) were most numerous in monoculture, but there were slight differences between the analysed systems. Symptoms of FHB on head did not represent the incidence of Fusarium spp. on grain. In every year of investigations a significantly lower infestation of grains with these fungi was observed in the organic system and higher in monoculture. On grains collected from the organic system *F. avenaceum* and *F.*

poae were dominant, while *F. avenaceum* and *F. graminearum* were dominant in grains collected from monoculture.

The above studies are only a part of publications touching FHB in Poland. Work on the mechanism of fusariosis and selection of more resistant species may be considered as very advanced and include inoculation of heads with selected *Fusarium* spp.

Table 1. Natural contamination with Fusarium *toxins (average) and percentage of positive samples (%) in Polish cereals (based on data from Perkowski* et al., *1997, Grabarkiewicz Szczęsna* et al., *2001, Tomczak* et al., *2002, Perkowski & Kaczmarek 2002, Perkowski* et al., *2003a, Basiński & Perkowski 2001, Perkowski & Basiński 2002).*

Cereals	*Number of samples*	*Fusarium toxins content (µg/kg)*									
		DON		*NIV*		T-2		*HT-2*		*MON*	
		%	µg/kg	%	µg/kg	%	µg/kg	%	µg/kg	%	µg/kg
Wheat	126	63	51	43	55	7	17	4	16		
Barley	32	91	120	72	53	62	38	50	49		
Oat	24	79	69	93	56	50	110	42	192		
Wheat	25	96	104	76	97					28	63
Wheat*											
1998 Żuławy	61	65	190	85	180					66	120
Wielkopolskie		60	3710	80	1080					20	80
1999 Żuławy	68	95	2270	57	150					100	340
Wielkopolskie		89	2610	30	80					66	190
Triticale	8	100	60								
Barley	15	100	85	100	61	100	47	93	55		
Oat	191	20	14	24	55						
Oat	99					15	60	24	21		

* results of kernels from heads with visible FHB

5. OCHRATOXIN A

Ochratoxin A (OTA), the toxic secondary metabolite of storage fungi *Aspergillus* and Penicillium species – is not only an important contaminant of cereals but also plays a major role in the alimentary chain of animals and human beings. The metabolite was frequently found in cereals, feed components, animal tissues and human blood. The presence of indicates thehygienic status of food and feed as a social problem of prime concern. Ochratoxin A has been identified in Poland for the first time in cereal grain samples in 1975 and since then surveys on OTA presence in cereal grain and feed samples are continued.

Table 2. Correlation coefficients between Fusarium *toxins for kernel fractions # I (> 2.5 mm) # II (< 2.5 mm), according to Perkowski* et al., *2003a.*

Fraction # II (< 2.5 mm)	*Fraction # I (> 2.5)* DON	*NIV*	*15-AcDON*	*ZEA*	*T-2 tetraol*	*DAS*	*HT-2 toxin*	*T-2 toxin*
DON	0.41**	0.77**	0.68**	0.39**	0.24	0.20	0.17	0.37*
NIV	0.43**	0.13	0.62**	0.39**	0.37**	0.25	0.10	0.37*
15-AcDON	0.70**	0.47**	0.55**	0.45**	0.122	0.30*	0.22	0.36*
ZEA	0.18	0.41**	0.37*	0.30*	0.42**	0.15	0.46**	0.38*
T-2 tetraol	0.46**	0.37**	0.56**	0.32**	0.46**	0.16	0.70**	0.31*
DAS	0.09	0.33*	0.24	0.16	0.28	0.32*	0.31*	0.44**
HT-toxin	-0.07	0.45**	0.37*	0.45**	0.55**	0.28	0.41**	0.37*
T-2 toxin	0.06	0.18	0.15	0.60**	0.32*	0.41**	0.41**	0.54**

* significant at P < 0.05; ** significant at P < 0.01

Table 3. Percentage of samples contaminated with three mycotoxins (DON + NIV + MON), two mycotoxins (DON + NIV or DON + MON) and one Fusarium *mycotoxin in wheat kernels samples collected in 1998 and 1999.*

Year	*Toxin detected* DON, NIV, MON	DON, NIV	DON, MON	NIV, MON	DON	NIV	MON	ND
1998								
Number of samples	20	17	0	8	2	6	3	5
Percentage (%)	32.8	27.9	0	13	3.3	10	5	8
1999								
Number of samples	24	1	25	1	12	0	2	3
Percentage (%)	35.3	1.5	36.8	1.5	17.6	0	2.9	4.4

The mycotoxin has been detected additionally in animal tissues – mainly in porcine kidneys and blood samples, as well as in human blood samples (Szebiotko *et al.*, 1981).

In Table 4 results on different cereals contamination with toxin are summarized. In samples collected during 1975 – 1990, OTA was present in 12 % of all analyzed samples (1255 total) of cereals in range of concentration 5 - 2400 μg/kg. Rye was the most frequently contaminated, while the highest concentration of the metabolite was found in wheat.

Table 4. Contamination of cereals with ochratoxin A in Poland in years 1975 - 1991 according to references (adapted from Goliński et al., *1991, Jarczyk* et al., *1998.)*

Grain	*Number of samples*			*OTA amounts (µg/kg)*
	Analyzed	*Contaminated Number*	*%*	*Range*
Wheat	239	28	11.7	5-2400
Rye	228	62	27.2	5-800
Barley	616	54	8.8	5-1200
Oats	49	2	4.1	80-800
Maize	123	2	1.6	25-400
All cereals (total)	1255	148	11.8	5-2400

Percentage of positive samples was higher in 1996 (about 46 %), however contamination level was lower, up to 13 µg/kg. Similar results for feed survey were obtained (Table 5). Before 1990, out of 1240 tested samples 1.5 % was contaminated with OTA at concentration range 10 – 200 µg/kg, while during the period 1990 – 1999 the percentage of positive samples was about 21 %.

In other summarized results (Czerwiecki 2001) out of 351 cereal products samples, particularly special formulas for babies and children, investigated in 1990 – 1992, OTA was detected only in one out of 204 grain/milk and grain and rice groats. The level of contamination was 3.5 µg/kg. There was no evidence of OTA presence in other samples of tested kinds of products for babies or children. However, in case of different bread sorts, this mycotoxin was present in 10 out of 110 examined samples (about 10 % of samples). Nevertheless, it should be underlined, that 70 % of contaminated bread were rye bread samples. The contamination level of bread was 3 – 16 µg/kg (mean 6.7 µg/kg). Other cereal products, such as oats flakes and barley flakes, were contaminated by OTA too, but the highest concentration observed was 140 – 160 µg/kg in barley flakes.

In 1999 samples were collected in the period from February to May in the southern Wielkopolska region. Similarly in years 2000 and 2001. Among contaminated feed samples 15 % (1999) and 46,9 % (2000 – 2002) contained OTA levels higher than 5 ng/g. More than 400 samples of Polish cereal grain of harvest years 1997 and 1998, obtained from conventional and ecological farms, were investigated on the ochratoxin A presence (Table 6).

The results of study in Poland raise the following conclusions.

The average percentage of ochratoxin A occurrence in all types of cereal grains from ecological and conventional farms was almost equal, however, particularly high contamination of wheat from conventional farms was observed. This was the reverse to the 1977 results.

Higher concentrations of ochratoxin A in wheat samples from conventional farms were associated with a higher frequency of fungal contamination.

In 2000 and 2001 the Ministry of Agriculture and Village Development prepared a report on soil, plant, agricultural and food product quality monitoring investigations (Michna & Szteke 2001, 2002). In this report grain contamination with ochratoxin was summarized (Table 7).

Czerwiecki (Czerwiecki 2001) reported that, in 2000 and 2001, in some Polish cereal and cereal products out of 330 samples 26.4% was ochratoxin A positive in range 0.1 - 667 µg/kg.

Presence of ochratoxin A in feeds for pigs can be a reason of mycotoxic porcine nephropathy and the metabolite occurrence in porcine tissues and blood. Table 8 summarizes results of OTA residues in examined porcine blood and kidneys collected in slaughterhouses. The metabolite was detected with similar percentages of positive samples during all years of screening. Kidneys samples exhibited lower levels of OTA concentrations comparing to blood samples.

Occurrence of ochratoxin A in the blood serum of slaughtered pigs and the contamination of feed with OTA indicate the threat of mycotoxins to the health of humans and animals. It is therefore essential to continue studies and screening on a larger scale and to use the obtained results in consumers' health protection, as well as to introduce the necessary regulations and limits.

Table 5. Contamination with ochratoxin A of feeds, cereal products and feed concentrates (own results and the literature data) (adapted from Goliński et al., *1991, Jarczyk* et al., *1998, Czerwiecki 2001, Waśkiewicz* et al., *2003).*

N_o	*Years*	*Source*	*Number of samples*			*Concentration (µg/kg)*	
			Analyzed	*OTA Contaminated*		*Average*	*Range*
				Number	*%*		
1	up to 1990	mixed feeds	1240	18	1.5		10-2400
2	1990-1991	mixed feeds	208	20	9.6	44.6	2.5-120
		feeds concentrates	36	5	13.8	198	120-280
3	1996	mixed feeds	32	27	84.4	2.8	0.02-23
4	1999 (own results)	mixed feeds	40	15	37.5	2.3	0.3-14
5	1990-1992	cereal products	351	14	3.9		3-160
6	1999	fFeed	40	15	37.5	2.3	0.3-13.5
7	2000-2002	feed	180	118	65.6		

Table 6. Occurrence of ochratoxin A in cereal samples both types of farms (1997, 1998) according to Czerwiecki et al., *2002a, 2002b.*

		Conventional farms			*Ecological farms*		
Year	*Cereal*	*No.of samples*	*Positive samples (no./%)*	*OTA (μg/kg)*	*No.of samples*	*Positive samples (no./%)*	*OTA (μg/kg)*
1997	Rye	52	3/5.8	0.82-2.5 (mean 1.38)	48	18/37.5	0.21-10.0 (mean 3.17)
	Wheat	32	-	-	39	3/7.7	0.48-1.20 (mean 0.83)
	Barley	26	1/3.9	0.30	40	3/7.9	6.7-57.0 (mean 25.73)
	Total	110	4/3.6	0.30-2.50 (mean 1.11)	127	24/19.9	0.21-57.0 (mean 5.70)
1998	Rye	37	4/10.8	4.73-8.80 (mean 6.75)	46	5/10.9	2.0-35.3 (mean 14.5)
	Wheat	37	18/48.6	0.60-1024 (mean 267)	34	8/23.5	0.8-1.60 (mean 1.17)
	Barley	36	2/5.5	1.20-9.70 (mean 5.45)	17	2/11.8	1.43-35.3 (mean 18.4)
	Total	110	23/20.9	0.60-1024 (mean 202)	97	15/15.5	0.80-35.3 (mean 7.92)

Table 7. Ochratoxin A in Polish cereals sample in years 1999 and 2000 (according to Michna & Szteke 2001, 2002).

Commodity		*No. of samples*	*Range μg/kg*	*nd*[1)]		*OTA ≤5 μg/kg*		*OTA >5 μg/kg*	
				N	*%*	*n*	*%*	*n*	*%*
1999	Wheat	517	0-850	236	45.6	178	34.4	103	20.0
	Rye	376	0-1070	98	26.1	131	34.8	147	39.0
	Oat	80	0-220	-	-	10	12.5	70	87.5
2000	Wheat	289	nd-675.0	239	82.7	25	8.65	25	8.65
	Rye	277	nd-1100	187	67.5	46	16.6	44	15.9

[1)] nd = not detected

6. PATULIN

Patuline is a biologically highly active and toxic mycotoxin found in fruit, vegetables and their products, produced by *Penicillium expansum,* causing wet rot of apple fruits. Patulin content is a good measure of quality of fruits juices.

In 2000 different apple juices of 29 different producers were monitored on patulin presence (Michna & Szteke 2001, 2002). In none of the analyzed samples the toxin concentration limit (30 μg/L) for juice (10°Bx extract) introduced by Polish

Health Ministry was exceeded. Detectable levels of this mycotoxine (above 5 µg/L) were observed in 33 samples (18 % of all tested). In most cases patuline concentration was below 10 µg/L and higher contamination levels were determined in 6 samples (up to 26 µg/L).

Table 8. Residues of ochratoxin A (OTA) in porcine samples (adapted from Goliński et al., *1985, 1991, Kotowski* et al., *1993, 2000).*

Years	*Porcine samples*	*No. of samples*			*OTA (ng/kg)*	
		Analyzed	*Contaminated*			
			No.	*%*	*Average*	*Range*
Up to 1990	Serum	894	334	37.4	5.5	4-520
	Kidney	235	79	33.6	0.9	1-9
1991-1992	Serum	105	63	60.0	16.6	0-122
	Kidney	85	35	41.2	0.4	0-3.1
1999	Serum	45	26	57.8	5.8	0.3–70

In 2001 apple juice (150 samples) was also tested for patulin presence. Detectable amounts (above 5 µg/L) were found in 46 samples, which is almost 30 % of the examined. In most cases patuline concentration was lower than 10 µg/L (over a half of the positive results). Concentration above 20 µg/L was found in 4 samples and in one of them only 70 µg/L was observed, which is above the acceptable maximum level by regulation.

Based on the obtained results it is clear that the patulin concentration level depends on the producer and production date of the juice. Presence of patulin was mostly found in juices with 'Best before' date between December and August. It is highly possible that for their production concentrated juice, produced from apples (especially of low quality) with a long-term period of storage in cold storage plants were used, where the possibility of bacterial and fungal infections development is high.

7. OCHRATOXIN AND ZEARALENONE IN SWARD

The concentration of mycotoxins in sward depending on time of pre-utilization in summer and date of harvest in winter was analysed during the 2000 – 2002 vegetation seasons. Higher concentrations of ochratoxin A – respectively: 0.55 and 0.50 ng/g of dry matter (DM) – were found in the pasture sward harvested in December and January when compared to the sward collected in November (0.36 ng/g DM). The average concentration of zearalenone varied from 2.74 ng/g DM in December, to 6.00 ng/g DM in November and 9.77 ng/g DM in January. The highest concentration of ochratoxin A was 1.82 ng/g DM and zearalenone 47.89 ng/g DM. The highest number of samples in which ochratoxin A exceeded the level of 0.3 ng/g DM was noted in sward harvested in January (61.1 %), while in the case of zearalenone, the number of samples containing this mycotoxin at a level exceeding

3.0 ng/g DM varied from 55.6 % in December to 66.7 % in November and January. It indicates that the percentage of ochratoxin A positive samples and the toxin concentration level in the sward of winter pastures increased during the time of sward re-growth.

Table 9. Concentration of ochratoxin A and zearalenone in winter pasture sward (ng/g DM) according to Goliński et al., *2003.*

Date of harvest	*Pre-utilization*	*Investigation years*			
		2000/01	*2001/02*	*Mean*	*Mean*
			OTA (ng/g)		*ZEA*
November	June	0.10	0.74	0.42	5.90
	July	0.00	0.51	0.26	4.42
	August	0.00	0.79	0.40	7.69
December	June	0.37	0.48	0.43	3.42
	July	0.13	0.62	0.38	2.63
	August	0.83	0.83	0.83	2.16
January	June	0.14	0.72	0.43	4.05
	July	0.62	0.68	0.65	3.95
	August	0.17	0.67	0.42	21.30

8. FUNGI AND MYCOTOXINS IN SAMPLES OF GRAIN AND GRAIN DUST COLLECTED ON FARM IN EASTERN POLAND

Agricultural workers may be exposed to dust borne fungi and their toxic metabolites, which could be spread in large quantities into air of breathing zone at harvesting, threshing, loading and storage of grain, flax threshing, cleaning of herbs and other works. The exposure may be a reason of allergic and immunotoxic diseases such as: bronchial, asthma, allergic alveolitis, allergic rhinitis, atopic conjunctivitis, organic dust toxic syndrome, chronic fatigue-like syndrome. Mould fungi of *Aspergillus, Penicillium, Alternaria, Cladosporium* and *Mucor* genera were identified as the etiologic factors of the above-mentioned diseases. The diseases called mycotoxicoses may be caused by the harmful effects of mycotoxins, which are secondary metabolites of mould fungi often present in dust mentioned above.

Mycotoxins present in grain (aflatoxins, ochratoxin A, moniliformin and trichothecenes) produced by fungi belonging to genus *Fusarium*: deoxynivalenol, nivalenol, zearalenone, T-2 toxin and others, exhibit serious potential risk for agricultural workers. Special attention is needed in case of toxins produced by *Fusarium* (fusariotoxins), which have been scarcely studied with respect to occupational risk in plants with high level of dust generated during technological processes. Recent data on this subject are presented in Table 10.

In samples of wheat dust, thyme herb and hay dust bedding for farm animals *Alternaria alternata, Aspergillus candidus, Geotrichum candidum* and *Fusarium* and *Penicillium* species fungi were found. The above mentioned are toxigenic allergens and may cause potential threat to the health of farmers exposed to them by inhalation during work.

In samples of bedding ochratoxins (OTA), deoxynivalenol (DON), 3-AcDON and NIV, in wheat dust samples DON, 3-AcDON and NIV and in thyme samples OTA, MON, DON, 3-AcDON and NIV were found.

Table 10. Mean concentration of total fungi, Fusarium *spp and selected mycotoxins determined in samples of grain (G) and grain dust (D) collected in farm (F) in eastern Poland and during combine threshing (T) according to Krysińska* et al., *2001, 2003.*

Sample type	*Total fungi Malt agar cfu/g·10³*	*Total* Fusarium *spp (SNA medium) cfu/g·10³*	*Mycotoxin µg/kg (range mean)*				
			MON	*DON*	*NIV*	*Total* Fusarium *toxin*	*OTA*
G	0	0	25-200 0	110-950 0	15-1280 7.5	15-1480 127.5	0.5-2.5 0.5
D	977.5	1150.0	25-780 145	15-46 0	15-880 0	27-1090 350	0.5-52.4 0.5
F	5.0	0	25-88 25	15-68 0	0	25-103 33	0.4-0.8 0.2
T	615.0	527.5	25-149 70	15-215 22	15-360 15	15-468 123	0.4-1.2 0.5

9. CONCLUSION

The results of this report and described studies indicate that OTA should be perceived as real contamination of Polish cereals and cereal products. The frequency of occurrence and the level of OTA contamination seems to be similar to most of the European countries. On the other hand, aflatoxins in cereals and their products do not seem to be a problem in Poland, however attention should be given to some imported products such as peanuts. In addition, important studies on aflatoxin M_1 in milk samples in the future are needed.

Fusarium toxin presence in small grain cereals is well documented in Poland, but their mean concentration is lower in comparison to other European countries.

It is becoming common to find mycotoxins in plants other than cereals e.g. grasses tissues. Cereal grains and aerogenic dusts may be a threat to workers employed in cereals technological plants.

In view of the study findings described above, it is apparent that permanent mycotoxin monitoring becomes vital.

10. AFFILIATIONS

Juliusz Perkowski, August Cieszkowski Agricultural University of Poznań, Department of Chemistry, ul. Wojska Polskiego 75, 60-625 Poznań, Poland; julperk@au.poznan.pl

Jerzy Chełkowski, Institute of Plant Genetics Polish Academy of Sciences, ul. Strzeszyńska 34, 60-479 Poznań, Poland; jche@igr.poznan.pl

Piotr Goliński, August Cieszkowski Agricultural University of Poznań, Department of Chemistry, ul. Wojska Polskiego 75, 60-625 Poznań, Poland, piotrg@au.poznan.pl

11. REFERENCES

Basiński, T., & Perkowski, J. (2001). Occurrence of *Fusarium* toxins in grain of oat in Poland in years 1997-1998. Biuletyn Naukowy UWM, 65-70 (in Polish).

Chełkowski, J. (1998). Distribution of *Fusarium* Species and Their Mycotoxins in Cereal Grains. In: Sinha, K.K., & Bathnagar, D. (editors). Mycotoxins in Agriculature and Food Safety. Marcel Dekker Inc, New York, 45-64.

Chełkowski, J., Perkowski, J., Grabarkiewicz Szczęsna, J., Kostecki, M., & Goliński, P. (2001). Toxigenic fungi and mycotoxims in cereal grain and feeds in Poland. In: Logrieco A. (editor) Cost Action 835 European Commission, Occurrence of toxigenic fungi and mycotoxins in plant, food and feed in Europe. Directorate-General for Research, EUR 19695, 111-130.

Czerwiecki, L. (2001). Ochratoxin A and other mycotoxins in Polish cereals and foods. Mycotoxin Research, 17A, 2, 125-128.

Czerwiecki, L., Czajkowska, D.,& Witkowska–Gwiazdowska, A. (2002a). On ochratoxin A and fungal flora in Polish cereals from conventional and ecological farms. Part 1: Occurrence of ochratoxin A and fungi in cereals in 1997. Food Additives and Contaminants, 19, 5, 470-477.

Czerwiecki, L., Czajkowska, D.,& Witkowska–Gwiazdowska, A. (2002b). On ochratoxin A and fungal flora in Polish cereals from conventional and ecological farms. Part 2: Occurrence of ochratoxin A and fungi in cereals in 1998. Food Additives and Contaminants, 19, 11, 1051-1057.

Goliński, P., Hult, K., Grabarkiewicz-Szczęsna, J., Chełkowski, J., & Szebiotko, K. (1985). Spontaneous occurrence of ochratoxin A residues in porcine kidney and serum samples in Poland. Applied Environmental Microbiology, 49, 4, 1014-1015.

Goliński, P., Grabarkiewicz-Szczęsna, J., Chełkowski, J., Hult, K., & Kostecki, K. (1991). Possible sources of ochratoxin A in human blood in Poland. IARC Scientific Publications, 115, 153-158, Lyon, France.

Goliński, P., Perkowski, J., Kostecki, M., Grabarkiewicz-Szczęsna, & J. Chełkowski, J. (1996). *Fusarium* Species and *Fusarium* toxins in wheat in Poland - comparison with neighbour countries. Sydovia, 48, 12-22.

Goliński, P., Kostecki, M., Golińska, B., Golińska, B.T., & Goliński, P.K. (2003). Accumulation of mycotoxins in forage of winter pasture during prolonged utilization of sward. Polish Journal of Veterinary Science, 6, 2, 81-86.

Grabarkiewicz–Szczęsna, J., Kostecki, M., Goliński, P., & Kiecana, I. (2001). Fusariotoxins in kernels of winter wheat cultivars field samples collected during 1993 in Poland. Nahrung/Food, 45, 28-30.

Jarczyk, A., Jędrychowski, L., Wróblewska, B., & Jędryczko, R. (1998). The estimation of contamination levels with fungi and ochratoxin A in grains and mixed meals for swine in North-East Poland. Polish Journal Food and Nutrition Science, 7/48, 4, 631-644.

Kotowski, K., Kostecki, M., Grabarkiewicz-Szczęsna, J., & Goliński, P. (1993). Ochratoxin A residue in kidneys and blood of pigs. Medycyna Weterynaryjna, 49, 12, 554 - 556.

Kotowski, K., Grabarkiewicz-Szczęsna, J., Waśkiewicz, A., Kostecki, M., & Goliński, P. (2000). Ochratoxin A in porcine blood and in consumed feed samples. Mycotoxin Research, 6, 66-71.

Krysińska–Traczyk, E., Kiecana, I., Perkowski, J., & Dutkiewicz, J. (2001). Levels of fungi and mycotoxins in samples of grain and grain dust collected on farms in eastern Poland. Annals of Agricultural and Environmental Medicine, 8, 269-274.

Krysińska–Traczyk, E., Perkowski, J., Kostecki, M., Dutkiewicz, J., & Kiecana, I. (2003). Flamentous fungi and mycotoxins as potential occupational risk factor among farmers engaged in combine harvesting of cereals. Medycyna Pracy, 54, 2, 133-138

Michna, W., & Szteke, B. (2001). Report of Monitoring survey ort quality of soil, plants and agricultural and food products in 2000. Ministerstwo Rolnictwa i Rozwoju Wsi. Rada Monitoringu Jakości Gleb, Roślin, Produktów Rolniczych i Spożywczych. Warsaw (in Polish).

Michna, W., & Szteke, B. (2002). Report of Monitoring survey ort quality of soil, plants and agricultural and food products in 2000. Ministerstwo Rolnictwa i Rozwoju Wsi. Rada Monitoringu Jakości Gleb, Roślin, Produktów Rolniczych i Spożywczych. Warsaw (in Polish).

Perkowski, J., Stachowiak, J., Kiecana, I., Goliński, P., & Chełkowski, J. (1997). Natural occurrence of *Fusarium* mycotoxins in Polish cereals. Cereal Research Communication, 25, 379-380.

Perkowski, J., & Basiński, T. (2002). Natural contamination of oat with group A trichothecene mycotoxins in Poland. Food Additives and Contaminants, 19, 478-482.

Perkowski, J., & Kaczmarek, Z. (2002). Distribution of deoxynivalenol and 3-acetyldeoxynivalenol in naturally contaminated and infected by *Fusarium culmorum* triticale samples. Nahrung/Food, 46, 415-419.

Perkowski, J., & Wiwart, M. (2002). Quality hazards of cereals caused by fungi from genus *Fusarium* sp. Polish Journal of Food and Nutrition Sciences, 11/52 SI2, 23-28.

Perkowski, J., Kiecana, I., & Kaczmarek, Z. (2003a). Natural occurrence and distribution of *Fusarium* toxins in naturally contaminated 15 barley cultivars. European Journal of Plant Pathology, 109, 4, 331-339.

Perkowski, J., Kiecana, I., Stachowiak, J., & Basiński, T. (2003b). Natural occurrence of scirpentriol in cereals infected by *Fusarium* species. Food Additives and Contaminants, 20, 572-578.

Sadowski, Cz., Wyczling, D., Łukonowski, A., Lenc, L., & Baturo, A. (2002). Occurrence of *Fusarium* and some mycotoxins on grain and ears of winter wheat depending on culltivar and fungicides applied. Journal of Applied Genetic, 43A, 103-110.

Szebiotko, K., Chełkowski, J., Dopierała, G., Godlewska, B., & Radomyska, W. (1981). Mycotoxins in cereal grain. Part I. Ochratoxin, citrinin, sterigmatocystin, penicillic acid and toxigenic fungi in cereal grain. Nahrung/Food, 25, 415-421.

Tomczak, M., Wiśniewska, H., Stępień, L., Kostecki, M., Chełkowski, J., & Goliński, P. (2002). Deoxynivalenol, nivalenol and moniliformin in wheat samples with head blight (scab) symtoms in Poland (1998-2000). European Journal of Plant Pathology, 108, 625-630.

Waśkiewicz, A., Kotowski, K., Kostecki, M., & Goliński, P. (2003). Ochratoxin A in feed samples. Biuletyn Naukowy UWM., 22, 45-49 (in Polish).

A. PEITO & A. VENÂNCIO

AN OVERVIEW OF MYCOTOXINS AND TOXIGENIC FUNGI IN PORTUGAL

1. INTRODUCTION

Fungi are cosmopolitan and ubiquitous and have a strong ecological link with vegetable products. The natural mycobiota existing in conjunction with food and feed are of interest to society, since many of these fungi can produce compounds of interest (*e.g.*, antibiotics, enzymes or organic acids). On the other hand, they may also be detrimental by promoting food and feed deterioration. Often, the mycobiota of food commodities is dominated by toxigenic species, such as *Aspergillus*, *Fusarium* and *Penicillium*, which may produce mycotoxins with effects on human and animal health. Aflatoxins, ochratoxins, patulin and the *Fusarium* toxins are considered to be the most important of the mycotoxins. Losses arising from the presence of mycotoxins have a large impact on the economy of countries.

In this report, we summarised the results obtained in Portuguese institutions with responsibility for food and feed inspection, and in mycological food and feed research. The aim is to show the natural occurrence of some mycotoxins produced by species from *Aspergillus*, *Fusarium* and *Penicillium* in products of the food and feed chains. Some of these products are imported and others are produced in Portugal. Data on the occurrence of aflatoxins, ochratoxin A, patulin and some *Fusarium* toxins are reported.

2. AFLATOXINS IN FOOD

2.1. Aflatoxins B_1, B_2, G_1, G_2, and M_1

The *Direcção Geral de Fiscalização e Controlo de Qualidade Alimentar* (DGFCQA, Directorate General of Food Inspection and Quality Control) is the responsible authority for the sampling and analysis of food commodities on sale in retail stores. It is also the responsible organisation for mycotoxin control at import points. In both cases, all products exceeding the legal limits in the EU (Commission Regulation n. 466/2001) are withdraw from shelves or rejected, and not allowed to enter the food chain, as imposed by law.

From 1999 to July 2003, several samples of dried fruits, spices and milk were analysed. Samples of dried fruits and nuts (1242), 83 samples from difference spices, and 69 samples of milk were collected in stores and at import control points for aflatoxins analysis (Tables 1, 2 and 3, respectively).

A. Logrieco and A. Visconti (eds.),
An Overview on Toxigenic Fungi and Mycotoxins in Europe, 173-184.

Table 1. Occurrence of aflatoxins in dried fruits and nuts from 1999 to July 2003 (number of contaminated samples in brackets)

Product	*Number of samples*	*Positive samples (%)*	*Aflatoxins content Min. – Max. (µg/kg)*			
			B_1	B_2	G_1	G_2
Almonds	56	14	0.3 – 147 (8)	11.5– 20.8 (4)	0.9 – 15.5 (3)	1.6 – 2.7 (3)
Peanuts	745	18	0.4– 777.3 (132)	0.3– 125.1 (71)	0.2 – 19.2 (28)	0.2– 28.3 (11)
Hazelnuts	22	14	< 0.6 – 0.9 (3)	nd	< 0.9– 4.1 (3)	< 0.8 (2)
Cashew-nuts	23	22	< 0.3 – 0.8 (5)	0.1– 0.3 (2)	0.2 – 0.8 (2)	nd
Dry figs	303	82	0.6– 141.5 (250)	0.3 – 17.9 (127)	0.8– 172.6 (88)	0.8 – 23 (17)
Nuts	15	6	nd	nd	2.5 (1)	nd
Pine seeds	3	0	nd	nd	nd	nd
Pistacchio nuts	58	64	0.6– 323.5 (37)	0.3 – 37.2 (21)	0.3 – 3.9 (2)	< 0.4 (1)
Raisins	17	0	nd	nd	nd	nd

nd = not detected

Samples were analysed by HPLC, following the methodology recommended by CEN (European Committee for Standardization). For those commodities where specific methodology was not available, adaptations were made to existing methods. The limit of detection (LOD) for aflatoxin B_1 (AFB_1), aflatoxin B_2, aflatoxin G_1 and aflatoxin G_2 varied with the product in analyses (Table 1). For aflatoxin M_1 the LOD was 0.003µg/L.

The results indicate high levels in some dried fruits and nuts, such as almonds, peanuts, dry figs and pistachio nuts (Table 1). Aflatoxins were also observed in curry, garlic, nutmeg, and white pepper, but only in curry and nutmeg were the recommended limits for AFB_1 exceeded (5 µg/kg) (Table 2). The levels of aflatoxin M_1 in milk samples were well below the EU limit of 0.05 µg/L (Table 3).

Table 2. Occurrence of aflatoxins in spices from 1999 to July 2003 (number of contaminated samples in brackets)

Product	*Number of samples*	*Positive samples (%)*	*Aflatoxins content Min. – Max. (µg/kg)*			
			B_1	B_2	G_1	G_2
Chilli	1	100	11.9 (1)	0.7 (1)	nd	nd
Barbecue mixture	6	66	0.2 – 2.1 (4)	0.1– 0.17 (2)	0.2 – 2.3 (2)	0.6 – 3.1 (2)
Cinnamon	1	0	nd	nd	nd	nd
Curry	11	90	0.5 –10 (10)	0.1 – 1.1 (4)	0.1 – 4.4 (3)	0.7 – 4.6 (7)
Coriander	2	50	1.9 (1)	0.17 (1)	< 0.2 (1)	nd
Cumin	1	0	nd	nd	nd	nd
Nutmeg	9	77	0.6 –10 (7)	0.4 – 1.6 (4)	0.8 – 1.5 (3)	nd
Origum	1	0	nd	nd	nd	nd
Paprika	30	90	0.1 – 3.6 (27)	0.1 – 2.4 (26)	0.2 – 4.7 (25)	2.3 (1)
Piri - piri	1	100	6.2	0.4	nd	nd
White pepper	9	100	0.4 – 0.8 (8)	0.3 (9)	1.0 – 1.2 (6)	0.5 – 1.4 (9)
Garlic	9	100	0.6 (1)	nd	nd	0.5 – 1.1 (9)
Saffron	2	50	1.5 (1)	0.26 (1)	nd	nd

nd = not detected

Table 3. Occurrence of aflatoxin M_1 in milk samples

Year	*Number of samples*	*Positive samples*	*Aflatoxin M_1 content (µg/L) min.-max.*
1999	54	51	0.010 – 0.022
2000	2	2	0.01 – 0.01
2001	12	7	<0.018 – 0.024

2.2. Aflatoxin B_1 and mycotoxigenic fungi in feeds

The National Laboratory for Veterinary Research (*Laboratório Nacional de Investigação Veterinária* - LNIV) is the driving force of mycotoxin control in feeds. Mainly bovine, poultry and swine feeds are analysed (Martins, 2003). All bovine feeds were for dairy cattle, which are responsible for the production of aflatoxin M_1 in milk for human consumption. From 1999 to 2002, 558 feed samples were analysed (Table 4) The LOD for AFB_1 detection method was 1 µg/kg. Besides the mycotoxin determination, these samples were also screened for the presence of fungi (Figures 1, 2, and 3).

Table 4. Occurrence of aflatoxin B_1 and filamentous fungi in feedstuffs

Product	*number of samples*	*positive samples (%)*		*min / max (µg/kg)*
		AFB_1	*fungi*	
Bovine feed	399	34	4	5 / 15
Poultry feed	85	16	93	1 / 20
Swine feed	74	7	71	1 / 2

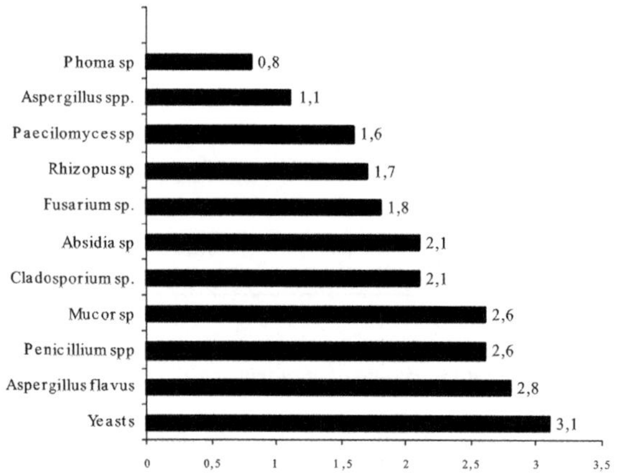

Figure 1. Means of log colony-forming units of fungal species in positive bovine feed samples (dairy cattle)

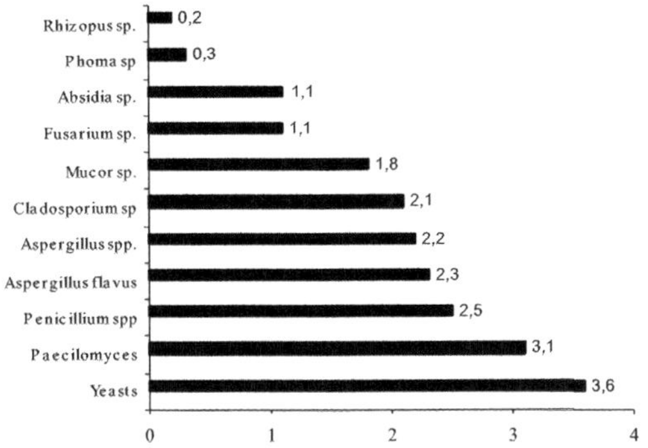

Figure 2. Means of log colony-forming units of fungal species in positive poultry feed samples

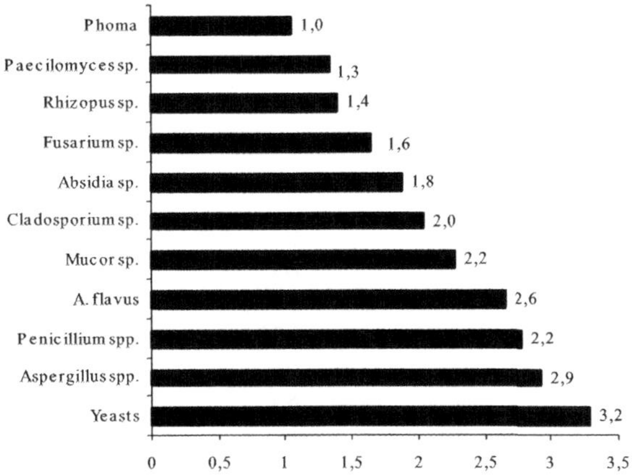

Figure 3. Means of log colony-forming units of fungal species in positive swine feed samples

When comparing these data with those previously reported (Martins & Martins, 2001), it is concluded that there was a decrease in AFB_1 in terms of percentage positive samples and concentration. These may be the result of the new technology employed during harvest and storage (rapid drying, correct storage of the harvested crops and use of effective anti-mould preservatives). The high percentage of bovine samples without mycological contamination may be explained by the fact that all samples were granulated. The maximum limit of aflatoxin B_1 recommended is 5μg/kg for bovine feed (dairy cattle), and 20 μg/kg for poultry and swine feed (Decreto-Lei n. 182/99). Among the 34 % of positive samples of bovine feed, just 9.5 % (13 samples) exceeded the recommended maximum limit. In the case of poultry feeds, only 2.4 % of samples (2) attained this limit.

The occurrence of AFB_1 in these feeds was correlated with the presence of *Aspergillus flavus*. As it can be seen from Figures 1 to 3, the number of strains from this species is among the predominant ones.

3. OCHRATOXIN A

From 1999 to July 2003, 410 samples were analysed for OTA in different foods and drinks (Table 5). Some of them were collected by DGFCQA in retail stores, mainly in the area of Lisbon; others arrived at the laboratory of *Departamento de Tecnologia das Indústrias Alimentares* for quality control analysis. They were analysed by HPLC according to the methodology proposed by CEN, with a LOD depending on the type of sample. Sixty-four samples of coffee were analysed by immunochemical method (RidaScreen Fast Ochratoxin A assay), being the LOD of this method 5 ppb (Lourenço *et al.*, 2002).

Table 5. Occurrence of ochratoxin A in different food commodities from 1999–July 2003 (adapted from Vicente et al.*, 2001; Lourenço* et al.*, 2002;DGFCQA, personal communication)*

Product		*Number of samples*	*Positive samples (%)*	*OTA level (µg/kg)*
Coffee	Green	257	89	<0.3 – 30.1
	Roasted	66	77	<1.0– 12.1
Raisins		9	89	<0.3 – 13.9
Bonnet – pepper		3	66	1.1 – 1.1
Paprika		3	100	1.1 – 4.3
Nutmeg		3	100	0.8 – 8.5
Pepper		1	0	nd
Barbecue mixture		15	86	<0.3 – 52.8
Breakfast cereals		15	60	<0.3 – 0.7
Beer	National	13	13	0.003 – 0.01
	Imported	15	60	0.002 – 0.064
Grits		5	100	0.202 – 0.204
Malt		5	100	0.121 – 0.201

Current EU legislation for OTA, includes its presence in (a) cereal and cereal products, and (b) dried vine fruits or raisins (Commission Regulation n. 472/2002). Limits for other food commodities are being considered. According to Table 5, just some samples of raisins may exceed the EU limit, although one sample – barbecue mixture – presented a very high value of OTA – 52.8 μg/kg. Coffee samples, either green or roasted, have a high incidence of OTA. Limits for coffee may be desirable, especially because of its widespread consumption (Paterson *et al.*, 2001).

The presence of OTA in grapes and grape products – namely wine, is being assessed by different research groups in Portugal (Abrunhosa *et al.*, 2003; Alves, 2003). In the literature data was found for Port Wine, for table wine, and for grapes for wine making (Table 6). The LOD varies between 0.02 and 0.084 μg/L, and 0.03 and 0.10 μg/kg for wine and grapes, respectively. Although there are yet no limits for the presence of OTA in wine, the International Office of Vine and Wine recommends a value of 2 μg/L, from the 2005 vintage. This value was reached in only 1 from 401 wine samples (Alves, 2003).

Table 6. Occurrence of OTA in grapes, Port wine, and table wine (from Alves, 2003; Festas et al.*, 2000; Serra* et al.*, in press)*

Product	*Number of samples*	*Positive samples (%)*	*LOD μg/L*	*Mean / max. μg/L*
Port Wine	189	17	0.084	0.137 / 0.474
	31	0	0.02	--
Table wine	151	24	0.084	0.179 / 2.08
	30	0	0.02	--
Grapes	58	1	0.105	0.110
	11	3	0.03	0.014 / 0.061

3.1. Ochratoxin A producing fungi in grapes

The incidence of filamentous fungi producing OTA from Portuguese wine grapes is being evaluated, by conducting surveys in vineyards, from four winemaking regions. From setting to the harvesting period, berries were sampled by plating methods for filamentous fungi isolation and identification. From the *Aspergillus* and *Penicillium* strains isolated, 14 % of the *Aspergillus* strains and none of the *Penicillium* strains were OTA producing ones (Serra *et al.*, 2002; Serra *et al.*, 2003). All *Aspergillus* strains were tested *in vitro* for OTA production, with *A. carbonarius* (97 %) and *A. niger* aggregate (4 %) strains being OTA producers. Almost all ochratoxigenic strains were isolated at harvest time (Table 7), mainly in regions with a Mediterranean climate. The vineyards from regions with Atlantic influences - Vinhos Verdes region, with high rainfall, exhibited the lowest occurrence of *Aspergillus* and ochratoxigenic strains.

Table 7. Ochratoxigenic fungi isolated from grapes at different maturation stages of grapes

Region	*Vineyard*	*% of berries with ochratoxigenic fungi*		
		Green berry	*Veraisson*	*Ripe berry (harvest)*
Vinhos Verdes	1	nd	nd	nd
	2	nd	nd	2 % *A. carbonarius*
	3	nd	nd	nd
Douro	4	nd	nd	8 % *A. carbonarius*
	5	nd	nd	20 % *A. niger* aggregate
	6	nd	nd	2 % *A. niger* aggregate
Ribatejo	7	nd	2 % *A. carbonarius*	12 % *A. carbonarius*
	8	nd	nd	2 % *A. carbonarius*
	9	nd	nd	nd
Alentejo	10	nd	nd	38 % *A. carbonarius* 2 % *A. niger* aggregate
	11	nd	nd	2 % *A. carbonarius*

nd = not detected

4. PATULIN

Data on the occurrence of patulin in 36 apple and pear samples collected in Portuguese retail stores from August to October of 2001 and analysed by HPLC are presented (Table 8) (Majerus & Kapp, 2002; Novo & Felgueiras, 2002). The method of analysis follows the Portuguese recommendation, with the LOD calculated as 5μg/L.

Recent EU legislation (Commission Regulation 1425/2003) for patulin defines a limit value of 50μg/kg in fruit juices and nectar, and in based spirit drinks; a value of 25μg/kg in solid apple products; and a value of 10 μg/kg in products intended for children and young infants. Values reported in Table 7 are well below these limit values, as long as they are not intended for children and young infants. Juices with milk are mainly consumed by children; however, none of these samples contained detectable amounts of patulin.

Martins *et al.* (2002) reported the occurrence of patulin in 351 Portuguese apples with small rotten spots. Even after having used a TLC technique with a LOD of 120μg/kg, these authors were able to detect patulin in 68.6 % of those apples. In some of these samples the level of patulin was higher than 80mg/kg.

5. *FUSARIUM* TOXINS

The occurrence of fumonisin B_1 (FB_1) in corn and poultry feed, and in oat and horse feed was assessed due the high sensitivity of horses to this mycotoxin, and possible effects in humans. In Table 9, data on 46 samples are presented, the determination being made by HPLC with immunoaffinity clean up, and the LOD calculated as 10 ng/kg. The recommended level of FB_1 in total feed fixed by US

Food and Drug Administration (FDA) is 1 mg/kg for horse feed and 50 mg/kg for poultry feed. From 46 samples, 36 were positive for FB_1 (78.3 %). The highest FB_1 content was detected on a corn sample (32.2 mg/kg). All horse feed samples had contents lower than the recommended values for FB_1. The control of mycotoxins in animal feeding stuffs is very important as they may cause reduced productivity or other symptoms. Hence, they may be transmitted to humans.

Table 8. Occurrence of Patulin in Apple and Pear juices (adapted from Majerus & Kapp, 2002)

Product	*Number of samples*	*Positive samples %*	*µg/L-Patulin min -max.*	*mean*
Clear apple juice	7	-----	nd	
Cloudy apple juice (nectar)	8	88	4.0 – 25.2	10.5
Apple and other fruit juices	5	20	5.6	
Apple juice and milk	6	0	nd	
Apple puree	3	0	nd	
Cloudy pear juice (nectar)	7	29	12.6 – 23.4	18.0
Pear and other fruit juices	2	100	7.2 – 9.6	8.4
Pear juice and milk	2	0	nd	
Pear puree	2	0	nd	

nd = not detected

Table 9. Fumonisin B_1 content in feedstuffs (from Novo et al., *2000)*

Product	*Number of samples*	*Positive samples (%)*	*mg FB_1/Kg min. - max.*	*mean*
Corn	12	67	0.025 – 32.20	11.911
Oat	5	40	0.132 – 0.421	0.277
Poultry feed	22	90	0.031 – 7.437	1.177
Horse feed	7	86	0.06 – 0.500	0.307

The occurrence in Portugal of three *Fusarium* toxins in foods – deoxynivalenol (DON), T2-toxin and zearalenone (ZEA) – is mentioned in the EU report, SCOOP task 3.2.10 on the occurrence of these mycotoxins (Gareis, 2003), although just a few determinations were made on these toxins (Table 10).

Table 10. Occurrence of some Fusarium *toxins in different food commodities (adapted from Gareis, 2003)*

toxin	*Product*	*Number of samples*	*Positive samples (%)*	*LOD*	*mean / max. µg/Kg*
ZEA	Wheat	1	0	2.5	nd
	White wheat flour	1	0		nd
	Wheat bran	2	100		13 / 15
	Cereals breakfast	11	64		5.1 / 11
DON	Wheat	3	33	25	256 / 744
	White wheat flour	3	33		119 / 333
	Wheat bran	4	50		761 / 1821
	Cereals breakfast	10	100		161 / 426
T2-Toxin	Wheat	9	0	37	nd
	Corn	10	0		nd

nd = not detected

6. CONCLUSION

Toxigenic fungi are cosmopolitan and ubiquitous, in the agricultural environment, such as soil, cereal grain, others grains, oil seed, fruits, spices, etc.. Filamentous fungi are responsible for crop losses, because of their presence, and the possible production and accumulation of mycotoxins in these food commodities.

Current legislation in Portugal comprises the presence of aflatoxins, OTA, and patulin in some food commodities, and of aflatoxins in feedstuff. Data herewith reported concern the period from January 1999 to July 2003. Probably because of this legislation, larger amounts of data are available for aflatoxins and ochratoxin A, and few data were found for patulin (legislation for patulin is very recent), and some *Fusarium* toxins.

It is worthwhile mention that some data supplied by DGFCQA concern samples collected at import control points, and not samples from food commodities on sale in local stores. This explains the high values of total aflatoxin exhibit by several samples of dried fruits and nuts (Table 1). Samples collected at retail stores had much lower levels.

OTA is by far the second most studied mycotoxin in Portugal. The incidence of OTA in food commodities seems to be high, but at low levels. Just in a few cases (roasted coffee, raisins and a barbecue mixture), OTA reaches levels of some concern.

Research efforts are ongoing on the evaluation of the presence of OTA in Portuguese wines. In this case the incidence of OTA seems to be low, and at levels far from the limits currently in discussion. In only one situation, it was possible to find a wine sample with an OTA level close to the proposed limit.

Concerning other mycotoxins (patulin and *Fusarium* toxins), the current amount of information is not enough to define trends, and more research is required.

The work of Martins *et al.* (2002) with rotten apples emphasizes the risk of human exposure to patulin through the consumption of fruits or fruit based foods and drinks manufactured from apples with small rotten areas.

7. ACKNOWLEDGEMENTS

We wish to thank DEng's Elvira Esteves, Alexandra Salgado and Lurdes Camilo from DGFCQA for kindly supplying their data on the occurrence of mycotoxins in food commodities. We wish to thank also Dr[a] Marina Martins, from LNIV, for kindly supplying her data on the occurrence of aflatoxins and fungi in feeds.

8. AFFILIATIONS

Amália Peito, DTIA - Departamento de Tecnologia das Indústrias Alimentares, Instituto Nacional de Engenharia e Tecnologia Industrial. Estrada Paço do Lumiar, 1649-038, Lisboa, Portugal; Amalia.Peito@mail2.ineti.pt

Armando Venâncio, CEB - Centro de Engenharia Biológica, Universidade do Minho, Campus de Gualtar, 4710-057 Braga, Portugal; avenan@deb.uminho.pt

9. REFERENCES

Abrunhosa, L., Serra, R., Lima, N., & Venâncio, A. (2003). Ochratoxin A risk assessment in Portuguese wines: a one-year case study, Bulletin de l'OIV, 76, 618-634.

Alves, A. (2003) Rastreio de ocratoxina A em uvas e vinhos Portugueses. In: Cerdeira, A., Venâncio, A., Alves, A., Barros, P. & Tomás Simões (Editors) Impacto da Contaminação Fúngica sobre a Competitividade de Vinhos – Ocratoxina A. Micoteca da Universidade do Minho, Braga, 53-70.

Commission Regulation (EC) No 466/2001 of 8 March. Aflatoxins in foodstuffs. Official Journal of the European Communities. 16/03/2001.L77/1.

Commission Regulation (EC) No 472/2002 of 16 March. Ochratoxin in foodstuffs.

Commission Regulation (EC) No 1425/2003 of 11 August. Amending Regulation (EC) n° 466/2001 as regards patulin.

Decreto Lei 182/99, 22 de Maio. Diário da República Portuguesa n°119/99, I série A. Aflatoxins in feedstuffs.

Festas, I., Herbert, P., Santos, L., Cabral, M., & Alves, A. (2000). Ochratoxin A in some Portuguese wines: method validation and screening in Port wine and Vinho Verde. American Journal of Enology and Viticulture, 51, 150-154

Gareis, M. (co-ordinator) (2003). Collection of occurrence data for *Fusarium* toxins in food and assessment of dietary intake by the population of EU Member States. Report on Task 3.2.10 for Scientific Cooperation, Directorate-General Health and Consumer Protection.

Lourenço, A., Távora, C., & Peito, A. (2002). Determination of ochratoxin A in coffee by immunochemical method. Proceedings Book of Congress Food Safety. Seminário de Vilar, 24-25 May, Porto, 64

Majerus, P., & Kapp, K. (2002). Assessment of dietary intake of Patulin by the population of EU Member States. Report on Task 3.2.8 for Scientific Cooperation, Directorate-General Health and Consumer Protection.

Martins, M. (2003). Occurrence of Aflatoxins in Raw Material and in Feedstuffs in Portugal 2000-2002. Proceeding Book of Congress Food Protection, Monte da Caparica, Portugal, 12-13 Maio, 30

Martins, M.L., & Martins, H.M. (2001). Moulds and aflatoxins contamination in swine, bovine and poultry feeds in Portugal. In: Logrieco, A. (editor). Occurrence of toxigenic fungi and mycotoxins in plants, food and feed in Europe. Cost Action 835, 131-146.

Martins, M.L., Gimeno, A., Martins, H.M., & Bernardo, F. (2002). Co-occurrence of patulin and citrinin in Portuguese apples with rotten spots. Food Additives and Contaminants, 19, 568-574.

Novo, R., & Felgueiras, I. (2002). Apple and Pear Juices. Proceedings Book of Congress Qualidade e Segurança Alimentar, ISA, Maio, Lisboa, 22

Novo, R., Fernandes, A., & Felgueiras, I. (2000). Fumonisin B_1 in Animal Feedingstuffs. Proceedings Book of Congress Food Safety. Fundação Eng. António de Almeida, 10-11 Novembro, Porto, 45-46.

Paterson, R.R.M., Baker, P.S., & van der Stegen, G.H.D. (2001) Ochratoxin A in coffee. In: Baker, P.S. (Editor) Coffee Futures: A source book of some critical issues confronting the coffee industry, The Commodities Press, 16-25.

Serra, R., Abrunhosa, L. & Venâncio, A. (2002) Micoflora das Uvas e Micotoxinas. In: Santos, I.M., Venâncio, A., & Lima, N. (Editors) Ecologia dos Fungos, Micoteca da Universidade do Minho, Braga, 47-54.

Serra, R., Abrunhosa, L., Kozakiewicz, Z., & Venâncio, A. (2003) Black *Aspergillus* species as ochratoxin A producers in Portuguese wine grapes. International Journal of Food Microbiology, 88, 1, 63-68.

Serra, R., Mendonça, C., Abrunhosa, L., Pietri, A., & Venâncio, A. (in press). Determination of Ochratoxin A in wine grapes: comparison of extraction procedures and method validation, Analytica Chimica Acta, in press

Vicente, D., Felgueiras, I., Peito, A., & Catarino, M. (2001). Determinação de Ocratoxina A em Cerveja. Proceedings Book of Congress 5º Encontro de Química dos Alimentos. 8-11 Maio, Porto, 276-278.

M. ITTU, A. TRIF & N. BELC

TOXIGENIC FUNGI AND MYCOTOXINS IN ROMANIA: CHALLENGES AND APPROACHES

1. INTRODUCTION

Pre-harvest and post-harvest occurrence of agriculturally important toxigenic species of *Fusarium, Aspergillus* and *Penicillium*, is a worldwide phenomenon. Unprecedented damaging *Fusarium* epidemics registered after 1990 on wheat in Europe, the USA and Canada have re-emerged scientific and social concerns on these pathogens, since many of them cause not only severe reduction in crop yield, but are often associated with contamination of related mycotoxins. From *field to fork,* many mycotoxins with harmful effect on humans and livestock health, such as aflatoxins, ochratoxins, patulin and the *Fusarium* toxins are hazardous during storage, food processing and transport. The major sources of mycotoxins studied up to now are products made from cereals, in particular wheat and corn, but with regard to toxigenic strains of *Aspergillus* and *Penicillium* grapes must be also carefully considered.

The very dynamic food and feed processing emphasized in our days the tremendous needs for „free contaminant" products for both, domestic consumption and for export. The higher exigencies of consumers concerning the diversification of food process and safety increasing standards are more challenging as before.

Therefore, specialised international organisations, programs and agencies (FAO/WHO, IPCS, IARC, and JECFA), and the European Commission, paid considerable efforts aimed to setting up statutory maximum tolerable daily intake values (TDI-values) of individual mycotoxins, in order to protect consumers and minimizing restrictions on trade.

Within the EU, legal limits or guidelines for different mycotoxins in food products, as well as those aimed to the improvement of assessment methodology have been established in several Commission Regulations (EC). In many countries, the regulations for sampling and analysis of food commodities, on sale and imported, have been already imposed by law. In Romania, this process is in real progress. Some limits were provided by the Directive O.G 84/91/2002, issued by the agricultural and health resorts, for the content of aflatoxins (B1, B1+B2+G1+G2 and M1) in cereals, milling products, peanuts and dried fruits, patulin in juices, zearalenone (ZEA), T-2 toxin (T2) and ochratoxin A (OTA) in cereals and milling products, foodstuffs and meat.

In this review, the most relevant aspects on results, challenges and approaches with regard to the toxigenic fungi and related mycotoxins in cereals and fodders cultivated in Romania are reviewed, with emphasis on the prevalent occurrence in

A. Logrieco and A. Visconti (eds.),
An Overview on Toxigenic Fungi and Mycotoxins in Europe, 185-194.

different commodities. Moreover, the quantification of DON in several combinations wheat genotypes/ Romanian *Fusarium* isolates under field artificial inoculation is also stressed, as a pre-requisite for improvement of breeding methodology and hence for preventing and/or reducing the levels of primary contamination with both, pathogen and mycotoxin.

2. FACTORS OF CONTAMINATION

Production of field crops is a significant component of the Romanian economy. More than 60% from the total of arable (about 11000 thousand hectares), in various environments, is cultivated with cereals: maize (31.6), wheat and rye (27. 2 %), barley (5.6) (Table 1). In favourable conditions, most of them but also fodders and grapes could be targeted of attacks by toxigenic fungi.

*Table 1. Agricultural crops susceptible to toxigenic fungi and mycotoxin contamination *)*

Crop	*Acreage*		*Production*	
	Hectares x1000	%	Tonne x1000	%
	9401.5	63.3		
Cereals for grain:	6294.9	67.0	18870.9	100
Wheat & rye	2558.6	27.2	7763.8	41.1
Maize	2974.0	31.6	9119.2	48.3
Barley	528.8	5.6	1580.0	8.4
Pastures	3421.4	23.0		
Fodder crops	1011.4	29.6	15261.3	100
Old and new perennials	668.1	19.5	11535.7	75.6
Annuals for hay and green fodder	269.8	7.9	3146.2	20.6
Plants used for silage	37.7	1.1	579.4	3.8
Vineyards + orchards	267.4	3.5		
Bearing vineyards	244.4	91.4	1121.7	100

*) Romanian statistical yearbook 2002. NIS 2003 Bucharest, Romania.

In different field crops, years and locations significant losses have been registered. This suggest a consequently contamination with mycotoxins.

Most reports are related to *Fusarium* diseases of wheat and maize cultivated in the West, North and center of Romania, which are usually characterised by wetter conditions at anthesis (81-108mm on average). However, extended areas with high attacks are not quite unusual in some years (e.g. 1990, 1998,1999).

Additionally, the rotation wheat/maize that became very common in the last years and the preference of farmers for low costs crop practices, contribute to the increasing risks of occurrence.

This emphasizes the concern to develop new Romanian cultivars with improved resistance to toxigenic fungi, in order to prevent and/or to minimize the potential contamination. Also, due to the different patterns over time and region as determined by variation in climate, cultivars and crop practices, the continuous monitoring of the occurrence of toxigenic fungi and mycotoxin status is therefore recommended, especially regarding the crop areas with potential risks of epidemics.

3. NATURAL OCCURRENCE OF TOXIGENIC FUNGI AND MYCOTOXINS

3.1.1. Toxigenic fungi in the field

The natural occurrence of *Fusarium* head blight (FHB, scab) on wheat is the most documented survey. Data on incidence of this pathogen over Romania are provided by the Central Laboratory for Phytosanitary Quarantine (CLPQ), Ministry of Agriculture, forests, waters and environment. A survey of FHB performed from 1998 to 2002, on 1464–2067 thousand hectares/yr, revealed means values of incidence ranging from 19 to 37 %, but only 1-5 % of these attacks could be considered as very damaging. In five counties of the 41 existing in Romania, FHB was endemic with high intensities, ranging from 10 to 33% (Bistriţa Nasaud, Cluj, Alba, Giurgiu and Caraş) (Fig. 1).

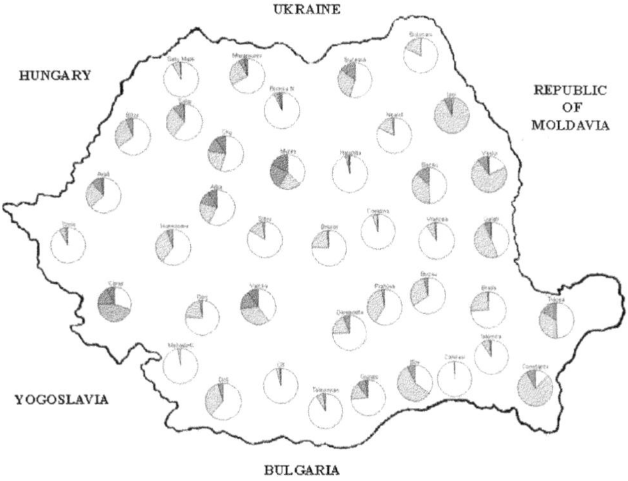

Figure 1. Natural occurrence of FHB in wheat (CLPQ-1998-2002)

3.1.2 Toxigenic fungi in agricultural commodities

Toxigenic fungi *Fusarium, Penicillium* and *Aspergillus* were commonly detected at variable levels in different feedstuffs and fodder sorts investigated in western Romania (Nichita and Sarandan, 1980, Nichita and al., 1996, Trif and al., 2003). However, the level of contamination is considered rather low. Intense level of contamination was found on average in less than 9% of samples collected in 1993 and 1994 (Nichita and al., 1996).

The highest incidence in fodder samples was reported for *Penicillium*, respectively 52% in 1993 and 1994 (Nichita and al., 1996) and between 93-97 % in the samples of each sort investigated at the Laboratory of Mycology, Directorate for Veterinary Medicine of Timis, Romania, over 1998-2002, respectively rough forage (N=262), silage and pastures (N=100), raw stuffs (N=501) and mixed feed (N=442) (Trif and al., 2003).

The frequency of *Fusarium* varied according to the year of investigation and the sort of fodder. Studies performed for two years (1993 and 1994), revealed the presence of *Fusarium* on average, in 6% of samples (Nichita *et al.*, 1996), while the maximum value for this trait was 37% in the period 1998-2002 (Trif *et al.*, 2003). When various sorts of fodder were analysed, the range of variation was comprised between 21 (raw stuffs) and 37% (rough forage). It was concluded that this fungus could have a significant impact mainly when high rainfalls associated with lower temperatures occur during vegetation period. Data are shown in Table 2.

Table 2. Natural occurrence of toxigenic fungi in some fodder sorts (western Romania, 1993-2002)

Sort/Period	*Toxigenic fungi (%)*			*Reference*
	Fusarium	*Penicillium*	*Aspergillus*	
Fodder (1993-1994)	6	52	26	Nichita, Curtui, Miclea, Trif, 1996
Fodder sorts (1998-2002)				
Rough forage	37	97	87	Trif, Sircuta, Muselin, 2003
Silage and pastures	28	93	89	
Raw stuffs	21	94	90	
Mixed feed	23	97	94	

3.2. Mycotoxin contamination

This trait was less investigated in Romania, as far as the analytical opportunities were not very available in the past. Unfortunately, consistent data on the natural occurrence of mycotoxins are still missing. The existing information is restricted to a few host plants and commodities/mycotoxins combinations investigated mainly in western (Timis district) and central Romania (Transylvania), under weather conditions that were very conducive to toxigenic infections. Moreover, the mycotoxin analyses were performed mostly in laboratories from Germany, within framework of German-Romanian cooperation in agricultural research.

3.2.1. Fusarium *toxins*

Analyses performed by ELISA and HPLC methods at the Institute for Hygiene and Technology of Food of Animal Origin, Veterinary Faculty, University of Munich, Munich, Germany, revealed in wheat samples (n=25) and maize samples (n=30) harvested in 1997 from western Romania (Arad, Bihor and Timis), the prevalence of *Fusarium* toxins, as compared with toxins produced by *Penicillium* and *Aspergillus* (Curtui *et al.*, 1998).

Deoxynivalenol (DON) and acetyl analogues were found with frequencies of 46% (maize) and 100% (wheat), the maximum concentration being relatively high in wheat samples (5,600 µg kg-¹) and in maize samples (160,000µg kg-¹), respectively. Incidence of zearalenone (ZEA) was 100% in wheat samples and only 13.0 in maize samples, but the concentrations were on the contrary, low in wheat (maximum concentration 170 µg kg-¹) and 1,200 µg kg-¹ (one sample of maize) (Table 3). Different data were obtained by HPLC analysis of maize samples, harvested in 2001 and 2002 from the center of Romania (Transylvania) where the concentration of ZEA raised to 2500 µg kg-¹ (Macri *et al.*, 2003). (*Data not shown*).

Diacetoxyscirpenol (DAS) and fumonisin B1 (FB1) were sporadically found only in maize, while T-2 toxin (T2) seems to play a minor role in wheat and maize samples harvested in 1997 (Curtui *et al.*, 1998) (Table 3). Similar data for T2 were obtained in combined feed samples investigated during four years (1998-2002) (Trif *et al.*, 2003) (Table 4).

3.2.2. Aspergillus *and* Penicillium *toxins*

From this category of toxins, in samples harvested in 1997 in western Romania only ochratoxin A (OTA) in wheat and citrinin (CT), in maize were found. (Curtui *et al.*, 1998) (Table 3). This finding suggests that an increasing concentration of the both toxins, having adverse effects in human and animal pathology, after prolonged storage could be expected. Further HPLC and enzyme immunoassays investigations of blood serum, kidney, liver and muscle samples per animal, collected from slaughtered pigs (n=52) in three counties, located in central and western Romania, have confirmed the previous presumption concerning the potential adverse effect of OTA. Data have shown the presence of OTA in Romania. The values registered for both incidence and mean level of contamination were comparable with other European countries, but concentration of OTA has not exceeded the maximum tolerated level in Romania (5ng/g) (Curtui *et al.*, 2001).

Analyses performed by Thin-layer chromatography (TLC), from 1998 to 2002, at the Laboratory of Mycology, Directorate for Veterinary Medicine, Timis, Romania, indicate on average the positive toxigenic contamination in 7.1% of samples in some fodder sorts from the 437 samples totally investigated (Trif et *al.*, 2003). OTA was the most prevalent (36-50 %), in combined feed and raw stuffs, respectively. Aflatoxin B1 (AFB1) occurred in raw stuffs (17%) and in combined feed samples (20%). In the same sorts of fodders, citrinin and aflatoxin G1 (AFG1) were also identified at lower frequencies (16 % and 2.4, respectively) (Table 4).

Table 3. Incidence and levels of mycotoxin contamination in wheat and maize in western Romania, 1997(Curtui et al, *1998)*

Mycotoxin	*Incidence/levels*	*Wheat (N=25)*	*Maize (N=30)*
DON	Frequency (%)	100	46,6
	Maximum (μg kg-¹)	5,600	160,000
	Minimum (μg kg-¹)	97	50
	Mean (μg kg-¹)	1,500	13,113
3-AcDON	Frequency (%)	48	27
	Maximum (μg kg-¹)	250	7,900
	Minimum (μg kg-¹)	41	65
	Mean (μg kg-¹)	93	1,200
15-AcDON	Frequency (%)	32	37
	Maximum (μg kg-¹)	380	42,00
	Minimum (μg kg-¹)	70	69
	Mean (μg/kg-¹)	150	4,500
ZEA	Frequency (%)	100	13
	Maximum (μg kg-¹)	170	1,200
	Mean (μg/kg-¹)	23	420
T2	Frequency (%)	24	3
	Maximum (μg kg-¹)	63	-
	Mean (μg kg-¹)	26	63
DAS	Frequency (%)	0	3
	Mean (μg kg-¹)	-	2,6
FB1	Frequency (%)	0	3
	Mean (μg kg-¹)	-	140
OTA	Frequency (%)	4	0
	Mean (μg kg-¹)	37	-
CT	Frequency (%)	0	3
	Mean (μg kg-¹)	-	580

4. DON CONTENT IN WHEAT ARTIFICIALLY INOCULATED WITH *FUSARIUM*

Breeding of host resistance, particularly the resistance to DON, is considered the most reliable, low cost and friendly to environment method to prevent and/or reduce the losses produced by FHB on wheat under the limit of risk. For more consistent results of assessment under artificial inoculation, *Fusarium* isolates, highly DON producers and good methods of inoculation are pre-requisite conditions. Since the

Table 4. Occurrence of toxins in some fodder sorts from western Romania, 1998-2002 (Trif et al.*, 2003)*

Fodder sort	*Samples*		*Occurrence of toxins in positive samples (%)*				
	Total	*Positive (%)*	*T2*	*CT*	*AF B1*	*AF G1*	*OTA*
Rough forrage	2	-					
Silage & Pastures	2	-	-	-	-	-	-
Raw stuffs for mixed feed	173	3.5	-	-	17	-	50
Mixed feed	260	9.6	4	16	20	2.4	36
TOTAL	437	7.09					

DON content, ranging in accordance to environment, cultivar and crop practice could significantly deteriorate the grain grade until the rejection from processing, it is highly recommended to use this parameter even at early stages of breeding.

In the wheat-breeding program from Romania, development of new cultivars with increased resistance to contamination with FHB and related toxins became in the past years a goal of outmost importance. In this respect, much effort is paid for improvement of assessment methods with regard to artificial inoculation and criteria of selection. It is generally supposed that artificial inoculation performed with highly toxigenic isolates will allow the selection of wheat genotypes more resistant to bothpathogen and related mycotoxin.

In order to evaluate the variation of DON into the collection of *Fusarium* isolates that is in use for artificial field inoculation, quantification of DON in several combinations wheat genotypes/ Romanian *Fusarium* isolates has been preliminary investigated. The wheat genotypes and the *Fusarium* isolates have been previously characterised, based on several components of resistance and aggressiveness, other than resistance to DON and DON content, respectively.

Samples of six wheat genotypes artificially inoculated in experimental plots from ARDI-Fundulea, Romania with two Romanian *Fusarium graminearum* isolates (1996), have been analysed by HPLC at the Department of chemistry, Rolnicza Academy *Augusta Cieszkowskiego*, Poznan-Poland (Ittu *et al.*, 2000). Other sets of samples wheat/*Fusarium* from 1998-1999 (30) and 2002 (4) were analysed by the enzyme immunoassay method (ELISA kit-Ridascreen ®FAST-DON), at the University of Hohenheim, State Plant Breeding Institute (720), Stuttgart, Germany within the frame of a collaborative research project with dr. Habil. Thomas Miedaner (Ittu et Miedaner, 2001) and the Institute for Animal Biology and Nutrition, Bucharest, Romania within the research national project ORIZONT A 19/2000 (*not published)*.

The minimum, maximum and average content of DON in combinations wheat/*Fusarium,* varied with the wheat genotype, *Fusarium* isolate and environment. The preliminary results indicate high content of DON (325.95 mg/kg) in one sample (Table 5). However, for selection of the most suited types allowing the development of assessment method based on this trait, further studies with the same combinations wheat genotypes/*Fusarium* isolates, investigated in more environments are necessary.

Table 5. DON production in wheat genotypes artificially inoculated in field with Romanian Fusarium graminearum isolates in 1996, 1998, 1999 and 2002

Number of isolates/ samples	*DON content (Range of variation)*		*Reference*
2/12 [1]	Minimum	0.22 µg/g	Ittu, et al., 2000
	Average	1.21 µg/g	
	Maximum	2.53 µg/g	
15/30 [2]	Minimum	1.25 mg/kg	Ittu & Miedaner, 2001
	Average	59.64 mg/kg	
	Maximum	325.95 mg/kg	
1/ 4 [3]	Minimum	1.7 mg/kg	Not published
	Average	1.9 mg/kg	
	Maximum	2.4 mg/kg	

1-HPLC analyses performed at the Department of Chemistry, Rolnicza Academy Augusta Cieszkowskiego), Poznan-Poland, 2-3. Enzyme immunoassay method (ELISA kit-ridascreen ®fast-DON at University of Hohenheim, State Plant Breeding Institute (720), Stuttgart, Germany (2) and 3- Animal Biology And Nutrition, Bucharest, Romania (3).

5. CONCLUSIONS

Production of field crops is a significant component of the Romanian economy. More than 60% from the total of arable (about 11000 thousand hectares) is cultivated with crops, such as cereals (maize, wheat, barley, rye, etc), fodders and grapes, potentially susceptible to the toxigenic fungi (Table 1). Therefore, occurrence of *Fusarium, Penicillium* and *Aspergillus* on the crop plants and fodder sorts from Romania has been reported. The most documented is the occurrence of *Fusarium* head blight (FHB) in wheat. A survey of FHB performed from 1998 to 2002, on 1464–2067 thousand hectares/yr, revealed means values of incidence ranging from 19 to 37 %, but only 1-5 % of these attacks could be considered as very damaging (Figure 1). In fodders, the frequency of *Fusarium* strains varied according to the year of investigation, between 6-37% of samples and additionally, according to the sort of fodder. In fact, as shown by investigation, the incidence was higher in rough forage (37%), as compared with raw stuffs (21%)(Table2).

Concerning the other toxigenic fungi, high level of contamination was found, on average at less than 9% of samples investigated in 1993 and 1994. *Penicillium* strains were the most prevalent toxic pathogens detected in fodder samples analysed in 1993 and 1994, (52%) and much higher from 1998 to 2002, respectively 93-97%.

Knowledge on the natural occurrence of mycotoxins in commodities from Romania is not very consistent. Current legislation in Romania comprises differentiated limits for consumption of babies and adults related to aflatoxins (B1, B1+B2+ G1+G2 and M1) in cereals, milling products, peanuts and dried fruits, patulin in juices, zearalenone (ZEA), T-2 toxin (T2) and ochratoxin A (OTA) in cereals and milling products, foodstuffs and meat.

Available data show that *Fusarium* toxins were prevalent in western Romania (1997) on wheat and maize. DON and acetyl analogues were found with frequencies of 46% in maize and of 100% in wheat. The maximum concentration was relatively high, 5,600 μg kg^{-1} in wheat samples and 160,000μg kg^{-1} in maize samples, respectively. Incidence of ZEA was 100% in wheat samples and only 13,0 in maize samples, but the concentrations were on the contrary, low in wheat (maximum concentration 170 μg kg^{-1}) and 1,200 μg kg^{-1} (one sample of maize). However, a higher contamination of maize with ZEA (up to 2,500 μg kg^{-1}) was also reported. DAS and AFB1 have been sporadically found only in maize (Table 3). Contamination with T-2 seems to play, according with level of detection and occurrence, a minor role in samples of wheat and maize harvested in 1997 (Table 3), as in combined feed samples (1998-2002) (Table 4).

Aspergillus and *Penicillium* toxins were found either in wheat (OTA) or maize samples (CT) harvested in 1997 in Western Romania. These findings suggest that an increasing concentration of the both toxins, having adverse effects in human and animal pathology, could be expected after prolonged storage. Analyses performed during four years (1998-2002), indicate the positive toxigenic contamination of 7.1% of the samples of different fodder sorts. The most prevalent was OTA with an incidence of 36-50 %, in samples of combined feed and raw stuffs, respectively. AFB1 occurred in raw stuffs (17%) and in combined feed samples (20%). In the same sorts of fodders, CT and AFG1 were also identified but at lower rates (16 and 2.4%, respectively).

Quantification of DON in several combinations wheat genotypes/ *Fusarium* isolates under field artificial inoculation, demonstrated the possibility to identify high DON producers types (up to 325.95 mg/kg).

For the increasing agricultural production, food industry and trade, the current knowledge on the occurrence of toxigenic fungi and related mycotoxins is not enough to define the dimension of this phenomenon in Romania. More research and legislation efforts are required. In addition, due to the variation in climate, cultivars and agricultural practices from Romania, the continuous monitoring of the occurrence of toxigenic fungi and mycotoxin status is therefore recommended, especially regarding the crops and areas with potential risks of epidemics.

6. AFFILIATIONS

Mariana Ittu-Agricultural Research-Development Institute (ARDI)-Fundulea, Romania; ittum@ricic.ro/gittu@pcnet.ro.

Alexandra Trif-Faculty of Veterinary Medicine, University of Agricultural Sciences and Veterinary Medicine of Banat–Timisoara, Romania; al_trif@yahoo.com.

Nastasia Belc-Institute for Food Bioresources, Bucharest, Romania; inraa@bioresurse.ro.

7. REFERENCES

Curtui, V. G., M. Gareis, E. Usleber and E., Martlbauer. (2001). Survey of Romanian slaughtered pigs for the occurrence of mycotoxins ochratoxins A and B, and zearalenone, Food Additives and Contaminants, 18: 730-738.

Curtui, V.G., E. Usleber, R. (1998). A survey on the occurrence of mycotoxins in wheat and maize from western Romania, Mycopathologia, 143: 97-103.

Ittu, M. and Miedaner T. (2001). Genetic and environmental variation of DON production and aggressiveness in the wheat/*Fusarium graminearum*, pathosystem, International Symposium Bioactive fungal metabolites-impact and exploitation. Swansea, UK, 22-27 April, 2001: 68.

Ittu, M., Grabarkiewicz-Szczesna, J., Kostecki, M. and Golinski, P. (2000). Deoxynivalenol accumulation and the scab symptoms in six Romanian wheat genotypes inoculated with *Fusarium graminearum*, Mycotoxin Research, 16:15-22, ISSN 0178-788.

Macri A., M. Schollenberger, Z. Dancea, W. Drochner. (2003). Zearalenone and ergosterol contents in corn samples of Romania. Abstracts of the 25th Mycotoxin Workshop, Giessen, 19-21 May, 2003: 45.

Nichita G., V. G. Curtui, V. Miclea, A.Trif. (1996). Mycotic and mycotoxic pollution of fodders-a risk factor for animal health. Proceedings of the Scientific Communication Meeting of „ Aurel Vlaicu University", Third Edition, 16-17 May 1996, Arad, Romania, 11: 65-68 (In Romanian).

Ordinul de guvern nr. 84/91 13-28 februarie 2002 pentru aprobarea Normelor privind contaminantii din alimente, publicat in M. Of. Nr. 255 din 16 aprilie 2002.

Romanian statistical yearbook 2002. (2003) NIS, Bucharest, Romania,869 p.

Trif A., G. Sircuta, F. Muselin.(2003). Mycotic and mycotoxic contamination of some sorts of fodders in Timis district during 1998-2002. Proceedings of the First Romanian Seminar „Mycotoxin contamination of cereals-risks and challenges for farmers, food processors and consumers", 30 January 2003, Bucharest. (In press)

M. LEVITIN

TOXIGENIC FUNGI AND MYCOTOXINS IN CEREALS GRAIN AND FOOD IN RUSSIA

1. INTRODUCTION

The study of toxigenic fungi started at the end of 19th – at the beginning of 20th century after the appearance of *Fusarium* head blight epidemics in the Far East and after the registration of man and animal poisoning caused by infected grain (Voronin, 1890, Palchevsky, 1891, Jaczewski 1904, Naumov, 1916). More active researches of diseases caused by toxigenic fungi was developed in 30-40 years after the identification of stachybotryotoxicosis in horses and alimentary toxic aleukia in man (Sarkisov, 1947). In these years Sarkisov and his colleagues initiated the new direction of mycological researches, named mycotoxicology (Sarkisov, 1954). Presently in Russia mycotoxicological researches are carried out in different medical and agricultural laboratories. In this short review, information is provided about the phytosanitary state in Russia for the last 10-15 years in connection with the widespread of toxigenic fungi on cereals and contamination of food by mycotoxins.

2. TOXIGENIC FUNGI

During the last years *Fusarium* head blight has become the most dangerous and widespread disease of cereals in Russia. This is one hundred years after the first epidemics in the Far East of Russia. One of the worst regions is North Caucasus, which is also is one of the main regions producing grain. The warm and humid climate of this district is very favorable for the development of fusariosis.

Only in the Krasnodar district (North Caucasus) there were three large epidemics of head blight in the period of 1988 – 1992. The loss of wheat crop reached 25-50% and the contamination of cereal grain by mycotoxins increased more than 25 times (Levitin *et al.*, 1994). During 1989-1992 on average about 23% samples of cereals (wheat, barley, rye) in Russia were contaminated by deoxynivalenol. Amongst them, 9% of samples contained DON in concentrations exceeding the permissible level. In 0.4% of samples of bread and groats products concentrations of mycotoxins exceeded hygienic standards (Tuteljan, 1995).

In our laboratory the study of *Fusarium* species on cereals of Russia has allowed to identify 19 species (Ivaschenko *et al.*, 1997; Levitin *et al.*, 1998; Ivaschenko *et al.*, 2000, Levitin, 2001). The dominant species causing *Fusarium* head blight on cereals in the warm and humid climate in the North Caucasus region are *F. graminearum, F. verticillioides, F. proliferatum, F. poae* and *F. sporotrichioides.* However, *F. graminearum* is the most serious and dangerous pathogen of these. The

A. Logrieco and A. Visconti (eds.),
An Overview on Toxigenic Fungi and Mycotoxins in Europe, 195-199.

Central region of Russia is characterized by a moderately continental cool climate. Amongst the *Fusarium* species, *F. avenaceum, F. culmorum, F. poae, F. sporotrichioides* are dominant. In moderately continental humid climate of the Northwest region of Russia – *F. avenaceum, F. culmorum, F. poae* and *F. sporotrichioides* are dominant. In West Siberia - *F. verticillioides* and *F. sporotrichioides* are dominant. In East Siberia - *F. acuminatum* and *F. avenaceum* are dominant. The biggest diversity of *Fusarium* species on cereals is observed in the Far East. The South of the Far East is characterized by a warm and very humid climate. Such an environment as well as that in North Caucasus is very favorable for the development of fusariosis. In this zone the dominant species causing an epidemics is *F. graminearum*.

The researches of Kononenko (1999) are devoted to the monitoring for *Fusarium* fungi on cereals in the central part of Russia. In the Moscow district 83 samples of seed infected by *Fusarium* were analyzed. Nine *Fusarium* species were isolated, among them *F. culmorum, F. avenaceum, F. sporotrichioides* were dominant. Nearly all isolates produced DON in average concentration 24.5 mg|kg.

In Bashkiria the analysis of species composition of *Fusarium* fungi on rye were carried out by Garifullina (1994). Everywhere the infection of rye ears was observed. The number of plants with the visible characters of disease reached 24%, but microbiological investigations showed that 94% of ears were infected by *Fusarium* fungi. *F. sambucinum* and *F. sporotrichioides* were dominant species.

The monitoring for *Fusarium* species composition was also carried out in the Ural (Pirjazeva, 2001). The infection of cereal seeds was 3.5%; the barley and oats seeds were infected very strongly. From infected seed samples 11 *Fusarium* species were isolated. *F. poae* and *F. avenaceum* were dominant species.

Toxigenic fungi *Penicillium* and *Aspergillus* often were isolated from seeds during their storage. Microbiological analysis of seeds, used in bread making industry, showed that 87-96% of seed samples contained *Penicillium* (L'vova *et al.*, 1992). About 79% of the isolates produced mycotoxins. Two wheat, five maize samples and one rice sample contained ochratoxin A in concentration 36-450 μ/g.

In connection with receipts of imported grain in Russia, particular attention has been given to the carrying out of mycological analysis. In particular, from grain imported to Russia in 1986 from Europe and USA, 908 fungal isolates were selected and identified (Dmitrieva, 1994). From corn imported by USA 72.7% isolates of *Aspergillus*, 6.5 % of *Penicillium*, 14.3% of *Alternaria* and 6.5% *Fusarium* were selected. Analysis of wheat grain imported by Germany showed that 54.3% isolates selected from seed surface were contaminated by *Aspergillus*. Isolates of *Penicillium* were selected from internal layers of seed. The barley samples from Great Britain were infected mainly by *Penicillium*, although in some samples *Aspergillus* was detected. The wheat and barley samples from Sweden were infected mainly by *Aspergillus* and *Fusarium*. Among *Fusarium* species *Fusarium avenaceum* was dominant. The number of infected samples of grain imported from different countries was more 37%.

The high frequency of DON contamination was detected in grain imported from mainly Canada, USA and France (Zacharova *et al.*, 1994). The DON concentration in 29 wheat samples was 0.05 – 0.95 mg/kg.

3. MYCOTOXINS

The problem of cereal grain contamination by mycotoxins is very important and actual for Russia. It is known that *Fusarium* mycotoxins are the most widespread, and among them DON, Zearalenone and T-2 toxin are the most dangerous. According to the data, obtained from researches of the Institute of Nutrition of the Russian Academy of Medical Sciences, the frequency of the DON detection in harvest of wheat (1986-1988) was 60-100% in Russia (Kravchenko *et al.*, 1998). Constantly high levels of grain contamination by DON were observed in the North Caucasus region during the period 1989-1995. Among 1695 samples of cereals 82% contained DON, of which 48% samples contained DON in concentrations exceeding the permissible level. A very high level of grain contamination by DON was observed during the *Fusarium* head blight epidemics in the North Caucasus. Therefore, practically 100% of cereal samples contained DON during epidemics in 1989, 1992 and 1993; among them 57-70% of samples contained DON in concentrations exceeding the permissible level. Zearalenone is less widespread. This toxin was observed in 68% of wheat samples in harvest of 1992, its concentration was 0.01-1.48 mg/kg (L'vova *et al.*, 1997). In the following years the frequency of ZEA detection sharply decreased.

The DON detection frequency in barley was lower than in wheat (Kravchenko *et al.*, 1998). Among 400 barley samples DON was observed on average in 4% of samples with maximum concentration 0.91 mg/kg. More rarely DON was detected in rye grain. Among 311 grain samples of rye (harvest of 1989-1995) DON was observed only in 2% of samples in concentration 0.09 – 1.11 mg/kg. Based on the data of consumption of groat products in 1989-1995 the authors calculated the daily dose of DON per person. This dose was 0.07 – 1.40 μ/kg of body mass (the permissible daily dose of DON is 3 μ/kg).

In other regions of Russia the quantity of cereal samples contaminated by mycotoxins was less than in the North Caucasus. The frequency of DON detection in wheat samples was 27% in West Siberia and 28% in Volga's region. The maximum level of DON did not exceed 0.32 mg/kg (Zacharova *et al.*, 1994).

The mycotoxin contamination of cereal grain in Asian part of Russia was studied in the laboratory of mycotoxicology at the All-Russian Institute for Veterinary Sanitation, Hygiene and Ecology during 1995 - 2002. Among 1545 samples 537 (34.8%) were contaminated by *Fusarium* toxins (Kononenko, Burkin, 2003). The frequency of T-2 toxin detection was 69.4% in the Ural district, 86.9% in West Siberia and 29.6% in the Far East. The average concentration of T-2 toxin in grain was 10 – 100μ/kg in the Asian part of Russia. In 24 samples (5.4%), mainly from the Far East, concentration of T-2 toxin exceeded the permissible level (110.0 – 625.5 μ/kg). Besides T-2 toxin in wheat samples from the Far East, 8-oxotrichotecene (4 deoxynivalenol and its analogy) and ZEA were detected. In 2001 the quantity of grain samples contaminated by 8-oxotrichotecene and ZEA was 52.9%; in 2002, 90.7%. 23.5% of the contaminated samples contained 8-oxotrichotecene in a concentration of 1000 μ/kg, indicating a precarious mycotoxicological situation in this region of Russia.

In this laboratory also monitoring of contamination of food and feed by ochratoxin A was carried out (Bukin *et al.*, 2002). Among 21 investigated regions of the Asian part of Russia, contaminated grain samples were revealed in 9 districts of Ural, West and East Siberia and Far East. The range of mycotoxin concentration was 3.0 – 292.0 μ/kg. In 1995 – 1996, severe intoxication of animals was observed in some farms in Kursk and Orlov districts (central part of Russia). The analysis of barley revealed very high levels (440 – 3250 μ/kg) of contamination by ochratoxin A.

The results of monitoring for the safety of food (1988-1993) in Russia showed that in 0.95 – 3.82 % food samples the concentration of aflatoxin B1 and in 0.72 – 3.04% of samples the concentration of patulin exceeded the permissible level. Aflatoxins were detected mainly in imported corn and barley (Monisov *et al.*, 1994).

State Sanitary-Epidemiological Service of Russian Federation controls the health of the Russian people. It carries out sanitary-hygienic monitoring for safety of food and quality of feed. In June 200, the new rules concerning the permissible level of mycotoxins in grain began to act in Russia. According these regulations the following levels of mycotoxins in grain are permitted (mg/kg): DON – 0.7-in wheat grain and 1.0 in barley grain; ZEA - 1.0 – in wheat grain, barley grain and corn; T-2 toxin - 0.1 – in different cereals grain, aflatoxin B1 – no more than 0.005 mg/kg. The use of wheat and barley grain as food is permitted only if *Fusarium* infected grain does not consist more than 1% and the concentration of mycotoxins does not exceed the permissible level.

4. ACKNOWLEDGEMENTS

I thank Tapani Yli-Mattila for critical reading of this manuscript and for valuable comments.

5. AFFILIATIONS

Mark Levitin, Laboratory of Mycology and Phytopathology, All-Russian Institute of Plant Protection, 196608 St.-Petersburg, Pushkin, Russia; levitin@IF3960.spb.edu

6. REFERENCES

Burkin, A.A., Soboleva, N.A., Kononenko, G..P. (2002). Researches of corn contamination by mycotoxins. In: Modern Mycology in Russia. Moscow, 262- 263 (in Russian).

Dmitrieva, I.V. (1994). Mycoflora of imported grains and researches of its toxigenecity. Ph.D. Moscow (in Russian).

Garifullina, G.F. (1994). Species composition and structure of population *Fusarium* fungi on winter rye in Bashkiria. Ph.D. Moscow (in Russian).

Ivaschenko, V.G., Shipilova, N.P., Kirtsideli, I.Yu. (1997). Ecological monitoring of grain cereals fusariosis causing agents in the North-Western Russia. Mycology and Phytopatholology, .31, 64-69 (in Russian).

Ivaschenko, V.G, Shipilova, N.P., Levitin, M.M. (2000). The diversity of *Fusarium* species on cereals in the Asian part of Russia. Mycolology and Phytopathology, 34, 54-58 (in Russian).

Jaczewski, A.A. (1904). About tempulent corn. Sheet of the information about disease control. Spb., Russia, 89-92 (in Russian).

Kononenko, G.P., Burkin, A.A. (2003). Fusariotoxins in cereals grain: regional peculiarity. In: Success in Medical Mycology. Moscow; 1, 141-144 (in Russian).

Kononenko, G.P., Malinovskaya, L.S., Soboleva, N.A., Pirjazeva, E.A., Zotova, E., Shalyganova, O.N. (1999). Distribution and production of toxins by fungi from infected cereals causing fusariosis on cereal grain in Moscow region. Mycology and Phytopathology, 33, 118-124 (in Russian).

Kravchenko, L.V., Zacharova, L.P., Obolski, O.L. (1998). Fusariotoxins in grain: the results of monitoring. Hygiene of nutrition, 1, 3-5 (in Russian).

Levitin, M.M. (2001). Distribution and toxicology of *Fusarium* species on cereals in Russia. In: Logrieco, A. (editor). Occurrence of toxigenic fungi and mycotoxins in plants, food and feed in Europe. Cost Action 835, 151-157.

Levitin, M.M., Ivaschenko, V.G., Shipilova, N.P., Gagkaeva, T.Yu. (1994). *Fusarium* head blight of cereals and problems of breeding for resistance. Plant Sci., 31, 158-160.

Levitin, M.M., Ivaschenko, V.G., Shipilova, N.P., Gagkaeva, T.Yu. (1998). Species and intraspecies variety of *Fusarium* fungi on cereals. In: Proc. Intern. Conf. devoted to 80-th anniversary of Mycology and Algology department of Moscow University. Moscow, 64-66 (in Russian).

L'vova, L.S., Kizlinko, O.L, Shulgina, A.P., Omelchenko, M.D., Zakharova, L.P., Piminova, V.V., Gagkaeva, T.Yu. (1997). Peculiarities of dezoxinivalenol and zearalenon synthesis in wheat grain infected by fusariosis of spike. Mycology and Phytopathology, .31, 52- 58 (in Russian).

L'vova, L.S., Orlova, N.Yu., Omel'chenko, V.D. (1992). The Penicillium fungi are producers of ochratoxin A in grain. Applied Biochemistry and Microbiology, 28, 889-893 (in Russian).

Monisov, A.A., Tuteljan, V.A., Hotimchenko, S.A., Tereshkova, L.P. (1994). Problems of food security in Russia. Questions of nutrition, 3, 33- 39 (in Russian).

Naumov, N.A. (1916). Tempulent corn. Observations under some species of genus *Fusarium*. Spb., Russia (in Russian).

Palchevsky, N.A. (1891). Disease of cereal grain in the South-Ussuriiskij region. Spb., Russia (in Russian).

Pirjazeva, E.A. (2001). Sanitary-mycological characteristics of fodder grain from Ural and West Siberia regions of Russia. Ph.D. Moscow.

Sarkisov, A. Ch. (1947). Veterinary Mycology. Veterinarija, 12, 25-27 (in Russian).

Sarkisov, A.Ch. (1954). Mycotoxicosis. Moscow, Sel'chozgiz (in Russian).

Tuteljan, V. (1995). Medical-biological requirements to quality and safety of grain products and bread-loaf goods. J. The bread baking in Russia, 2, 17-19 (in Russian).

Voronin, M.C. (1890). About tempulent corn in the South-Ussuriiskij region. Spb, Russia (in Russian).

Zacharova, L.P., Obolski, O.L., L'vova, L.S., Kravchenko, L.V. (1994). Problem of cereal contamination by deoxynivalenol in Russia. Questions of nutrition, 3, 40 – 44 (in Russian).

J. LEVIĆ, S. STANKOVIĆ, A. BOČAROV-STANČIĆ,
M. ŠKRINJAR, & Z. MAŠIĆ,

THE OVERVIEW ON TOXIGENIC FUNGI AND MYCOTOXINS IN SERBIA AND MONTENEGRO

1. INTRODUCTION

The available literature data indicate that the importance of toxigenic fungi and mycotoxins is much greater for the area of Serbia than for the area Montenegro. The knowledge on the occurrence of moulds (*Gibberella saubinetii*), as well as on animal and human diseases caused by mouldy food, especially by maize that has been traditionally grown as food and feed in the area of Serbia, dates as far back as the 19th and the beginning of the 20th century (Radić, 1872; Lozanić, 1904; Ranojević, 1914). However, the largest outbreaks of epidemic moulds, mostly caused by *Fusarium* species, were registered in maize in 1955, 1968, 1972, 1974 and 1984 and in wheat in the early 1970s (Lević, 2004). The greatest outbreaks of animal diseases, especially diseases of pigs such as oestrogenism, vomiting and feed refusal, dermatoxicity, haemorrhages and others were recorded during the same and/or subsequent years. Alimentary intoxication of horses, described in some cases in 1955, indicates, to a great extent, clinical signs of eleucoencephalomalacia. This type of horse illness could be a result of the use of mouldy feedstuffs, mostly husked maize ears and shelled kernels (Rajić, 1963), or of the occurrence of *F. moniliforme* that prevailed in epidemic ear rot in 1955 (Marić, 1981). Nevertheless, it has never been completely observed nor identified.

The large scale diseases of animals were not recorded after the late 1980s, except for some problems in animal feeding in certain farms in Vojvodina caused by a very high level of zearalenone (ZEA) in samples of maize, soya bean and sunflower meal that were imported in Serbia in the period of 1996-2001 (Jajić *et al.*, 2001). Diseases in both, animals and humans, occurring as a result of mycotoxins synthesised by species of the genera *Aspergillus* and *Penicillium*, have not been recorded in Serbia yet, although they are significant contaminants of stored products, feed and food. Moreover, climatic conditions in Serbia favour the large-scale occurrence of *Fusarium* species and biosyntheses of some fusariotoxins.

The present study is an attempt to overview the occurrence of toxigenic fungi and their second metabolites, as well as, microbiotical and mycotoxicological safety of feed and food in Serbia.

A. Logrieco and A. Visconti (eds.),
An Overview on Toxigenic Fungi and Mycotoxins in Europe, 201-218.

2. OCCURRENCE OF POTENTIALY TOXIGENIC FUNGI

During a long period of time in Serbia, 29, 28 and 30 different species of the genera *Fusarium*, *Penicillium* and *Aspergillus*, respectively, isolated mainly from cereals (Bočarov-Stančić, 1996), and afterwards from sunflower (Noory, 1983), feedstuffs (Mrđen *et al.*, 1990; Škrinjar *et al.*, 1992a, 1995a), wheat flour (Škrinjar *et al.*, 1997), oilseed products (Bočarov-Stančić *et al.*, 2001c), vegetable and fruit products (Ljubojević & Šutić, 1989; Žakula *et al.*, 1985), sugar confectionery (Šutić *et al.*, 1989), raw milk (Škrinjar *et al.*, 1995b), meat products (Škrinjar & Horvat-Skenderović, 1989), and other feed and food, were determined.

In general, *A. flavus*, *A. versicolor*, *F. oxysporum*, *F. solani*, *F. graminearum*, *P. aurantiogriseum* (=*P. verrucosum* var. *cyclopium*) were most frequently isolated from either feed or food. However, there are many indeterminate species or strains of *Fusarium* isolated from field crops or stored products, as wall as, *Aspergillus* and *Penicillium* isolated from feed and food.

During the last 20 years, there were no significant changes in the distribution of genera *Aspergillus* (5%), *Fusarium* (41%) and *Penicillium* (46%) in maize grain (Noory, 1983; Lević et al., 2003), except in the species composition and frequency. Noory (1983) reported the dominance of *F. oxysporum* and a two-fold lower occurrence of *F. verticillioides (F. moniliforme)* and *F. graminearum*, while Lević et al. (2003) detected the dominance of *F. subglutinans* and *F. verticillioides*, the unchanged distribution of *F. graminearum* and the tendency of *F. proliferatum* to spread (Table 1). In general, these four species can be considered the most important pathogens of maize under agricultural conditions of Serbia, but with

Table 1. Frequencies (%) of Aspergillus, Fusarium *and* Penicillium *species in maize grain during the period of 1994-1999 and in 2002 (adapted from Lević et al., 2003)*

Species	*No. of year of fungal occurrence*	*Fungal frequency (%)*	
		Interval	*Average*
Aspergillus spp.	2	1.7-4.1	2.9
F. avenaceum	1	1.7	1.7
F. chlamydosporum	2	0.3-1.7	1.0
F. equiseti	3	0.5-9.2	3.5
F. graminearum	7	1.7-29.9	15.0
F. oxysporum	3	1.9-8.8	5.8
	7	2.2-21.5	9.9
F. proliferatum			
F. semitectum	2	0.2-5.0	2.6
F. solani	5	0.3-13.3	4.8
F. sporotrichioides	5	0.3-8.3	4.3
F. subglutinans	7	13.3-0.3	36.1
F. verticillioides	7	9.2-77.8	40.8
Penicillium spp.	3	3.3-45.9	20.0

different frequencies over years (Bočarov-Stančić *et al.*, 1997; Lević *et al.*, 1997).

The occurrence of species *Aspergillus* and *Penicillium* in post-harvest maize grain was determined in two and three years, at the frequency of 1.7-4.1% and 3.3-45.9% or on the average in 2.9% and 20.03% samples, respectively (Table 1).

F. graminearum was a principal causal agent of head blight of wheat for the last three decades (Lević, 2004), while the occurrence of *F. culmorum* was frequent sometimes (Milošević *et al.*, 1995). The dominance of the genus *Alternaria* (69.3-86.8%) was determined in 2002 by Balaž *et al.* (2003). The same authors determined a less dominance of *Fusarium* species (11.3-20.6%), while the distribution of *Aspergillus* (0-1.2%) and *Penicillium* (0-8.9%) species was respectively very low and somewhat greater. According to studies of kernels of five wheat varieties carried out in four locations over two years, Lević and Dopuđa (unpublished data) determined that species *F. graminearum* and *F. verticillioides* were present up to 61% and 10%, respectively, in 2002, while species *F. poae* (up to 12%) and *F. prolifearatum* (5%) were more abundant in 2003. The same authors determined other *Fusarium* species in a frequency of up to 4-5% (*F. sporotrichioides*) or up to 1-3% (*F. acuminatum, F. avenaceum, F. culmorum, F. equiseti, F. oxysporum, F. tricinctum*). In these studies, the frequency of *Alternaria* species was the highest and varied from 24 to 56% and from 40 to 78% in 2002 and 2003, respectively, while *Aspergillus* (1%) and *Penicillium* (7-11%) species were at the level established by Balaž *et al.* (2003).

Furthermore, the dominance of *Alternaria* spp. (up to 72%) was determined in kernels of brewing barley harvested in 1995, as well as, the distribution of *Fusarium* species, especially *F. poae* (53%) was significant. Unlike wheat, *Aspergillus* (*A. flavus* and *A. versicolor*) and *Penicillium* (*P. aurantiogriseum*) species were evenly distributed in barley kernels (Bočarov-Stančić *et al.*, 2001a).

Aspergillus species, notably species *A. versicolor*, *A. flavus* and *A. ochraceus*, were more important for sunflower than for maize (Noory, 1983) or wheat (Bočarov-Stančić, 1996). Noory (1983) established that the total distribution of *Aspergillus*, *Fusarium* (mainly *F. semitectum* and *F. solani*) and *Penicillium* species in sunflower seeds was 15, 5 and 10%, respectively, and the frequency of *P. aurantiogriseum* in sunflower seeds was approximately equal in maize kernels.

Results gained by Stojšin *et al.* (2000) indicate that ZEA producers (*F. graminearum*, *F. oxysporum* and *F. equiseti*) are often isolated from the root of sugar beet. This is a very important factor, because pelleted sugar beet pulp is used as dairy cattle feed.

Škrinjar and Horvat-Skenderović (1989) found that 87% of samples of the hard sausage were contaminated by moulds including *Aspergillus* and *Penicilium* species. These two authors also determined the highest number of species on the surface of the sausage casings during the storage, and especially during long-term storage and after curing. Out of totally isolated species, 38% belonged to the genus *Penicillium*, while the species *P. aurantiogriseum* prevailed.

Toxigenicity of fungi originating from Serbia was proven for 7, 14 and 4 species of genera *Aspergillus*, *Fusarium* and *Penicillium*, respectively (Table 2).

3. MYCOTOXINS AND THEIR PRODUCERS

Table 2. Mycotoxins produced by Aspergillus, Fusarium *and* Penicillium *species originating from Serbia*

Mycotoxin	*Fungal species (reference)*
Aflatoxins	*A. flavus*, *A. tamari* (Noory, 1983; Lazić-Savić & Šutić, 1989)
Aflatoxin B_1 (AB_1)	*A. sclerotiorum*, *P. aurantiogriseum* (Bočarov-Stančić *et al.*, 1996)
Beauvericin (BEA)	*F. proliferatum*, *F. subglutinans* (Stanković *et al.*, 2004)
Fumonisin B_1 (FB_1)	*F. proliferatum*, *F. subglutinans*, *F. verticillioides* (Stanković *et al.*, 2004)
Fusaroproliferin (FUS)	*F. proliferatum*, *F. verticillioides* (Stanković *et al.*, 2004)
Ochratoxin A (OA)	*A. ochraceus*, *A. sclerotiorum*, *A. sulphureus*, *P. aurantiogriseum*, *P. variabile* (Škrinjar & Dimić, 1992; Bočarov-Stančić *et al.*, 1996)
Trichothecenes type A:	
Diacetoxyscirpenol (DAS)	*F. acuminatum*, *F. avenaceum*, *F. culmorum*, *F. equiseti*, *F. oxysporum*, *F. semitectum*, *F. sporotrichioides*, *F. sulphureum* (Bočarov-Stančić *et al.*, 1986a, 1986b, 2003, unpublished)
T-2 toxin	*F. avenaceum*, *F. culmorum*, *F. equiseti*, *F. graminearum*, *F. oxysporum*, *F. proliferatum*, *F. solani*, *F. sporotrichioides*, *F. subglutinans*, *F. tricinctum*, *F. verticillioides* (Bočarov-Stančić *et al.*, 1986c, Noory, 1983; Mašić *et al.*, 1997)
Zearalenonone (ZEA)	*F. graminearum*, *F. oxysporum*, *F. proliferatum*, *F. verticillioides* (Noory, 1983; Oberan, 1989; Bočarov-Stančić *et al.*, 2003)

Some strains of *Aspergillus* and *Penicillium* species produced both, aflatoxina and ochratoxins, while other strains of these species produced either aflatoxina or ochratoxina (see Table 2).

Under agricultural conditions of Serbia, species and strains of the genus *Fusarium*, particularly *F. graminearum*, have a relatively high potential for biosynthesis of ZEA (Bočarov-Stančić, 1996). If the scale of Mirocha and Christensen (1974) is applied to the results obtained by Noory (1983) it can be stated that the populations of *F. graminearum* in Serbia vary from very productive (3,740 mg kg^{-1}) to very weak (2.3 mg kg^{-1}) producers of ZEA.

Although the majority of populations of species of the genus *Fusarium* are weak producers of trichothecenes type A (Bočarov-Stančić *et al.*, 1986a), Muntañola Cvetković *et al.* (1991) considered that favourable ecological conditions permitted the presence and distribution of some populations with a higher potential for the synthesis of trichothecenes type A.

Results obtained by Stanković *et al.,* (2004) showed that isolates of *F. proliferatum* had the highest potential for the biosynthesis of fumonisin B_1 and beauvericin Strains of *F. verticillioides* possessed the highest potential for the synthesis of fusaproliferin, while isolates of *F. proliferatum* and *F. subglutinans* displayed approximately the same potential for the synthesis of beauvericin (Table 3). All strains observed in this study were previously identified based on morphological and biological properties described by Leslie (1995).

Table 3. Mycotoxin production potential of F. verticillioides, F. proliferatum *and* F. subglutinans *originating from Serbia (adapted from Stanković* et al., *2004)*

Fusarium species	*Interval ($mg\ kg^{-1}$)*	*Frequency of isolates (%)*		
		FB1	*BEA*	*FUS*
F. verticillioides	0		99.2	3.6
(28 strains)	in trace			21.4
	< 200	7.1		
	300-400	7.1		67.8
	500-700	25.0	3.8	7.2
	1000-1500	42.9		
	2000-3000	17.9		
F. proliferatum	0	5.3	14.0	15.7
(57 strains)	in trace	3.5	5.3	19.2
	< 200	12.3		21.1
	300-400	10.6	66.6	38.7
	500-700	7.0	14.1	5.3
	1000-1500	52.5		
	2000-3000	7.0		
	>3000	1.8		
F. subglutinans	0	64.4		92.9
(14 strains)	in trace		28.6	7.1
	< 200	21.4		
	300-400		57.1	
	500-700	7.1	14.3	
	1000-1500	7.1		

Few data confirm the link between particularly toxigenic fungal species and mycotoxins in infected plants or contaminated feed. One of the first such studies was the *in vivo* investigation of the potential of a greater number of isolates of different *Fusarium* species for biosynthesis of T-2 toxin (Bočarov-Stančić, 1996). According to these results, *F. graminearum* strains were exclusive producers of the T-2 toxin in both pure culture (3,000-15,000 $\mu g\ l^{-1}$) and wheat kernels infected by *F. culmorum* and *F. graminearum* and contaminated by T-2 toxin (320-750 $\mu g\ kg^{-1}$). Bočarov-Stančić *et al.* (2003) established the link between the presence of ZEA in wheat and a high potential of *F. oxysporum*, originating from these kernels, to biosynthesise this mycotoxin (10,400 μg kg^{-1}). Bagi *et al.* (2000) and Protić (2000) observed the composition of fungi and mycotoxins in freshly harvested wheat under conditions of artificially infected wheat spikes by *F. graminearum*. Furthermore, Protić (2000) also presented results on mycotoxins after the technological production process of bread made from infected wheat. The ability of *F. verticillioides* to produce ZEA was tested by Škrinjar *et al.* (1995a), since this fungus commonly occurred in feed mixture samples not contaminated by ZEA. It was established that *F. verticillioides* was also ZEA-negative under laboratory conditions.

4. NATURAL OCCURRENCE OF MYCOTOXINS IN FEED AND FOOD

The natural occurrence of mycotoxins in Serbia is usually analysed in samples that are submitted for testing mainly on the basis of monitoring of sanitary conditions of feed or when it is suspected that mycotoxins are a possible cause of health disorders in animals, while food is most often tested if it has been imported.

4.1. Aflatoxins

Aflatoxins are rarely found in Serbia, but there is a possibility for a larger presence due to the import of soya bean meal, earthnut cakes, tapioca meal and others (Bočarov-Stančić *et al.*, 2001b). According to studies by Bočarov-Stančić, (1996) kernels of maize imported from the USA for the needs of food industry contained a much higher concentration of aflatoxins (200 μg kg^{-1}) than local maize (table 4). Furthermore, Bočarov-Stančić *et al.* (2001c) detected aflatoxin B_1 (AB_1) in 12.1% of samples of imported food in the concentration of 2 μg kg^{-1} (cacao), 4 μg kg^{-1} (rice, coffee, American earthnut) and 69 μg kg^{-1} (chocolate tree). Moreover, these authors registered the same mycotoxins in maize grits and flour (2 μg kg^{-1}), wheat grits and sunflower (4 μg kg^{-1}) originating from Banat, the region of Vojvodina. Generally AB_1, ranging from 1 to 50 μg kg^{-1} (Table 4), was the most frequently detected aflatoxin in Serbia.

Škrinjar *et al.* (1992a) reported that some samples of dairy cattle feed were contaminated by AB_1 and AM_1 (5-16 μg kg^{-1}), but none of the row milk samples tested was contaminated with these aflatoxins.

4.2. Ochratoxins

The natural occurrence of ochratoxin (OTA) is generally more important than the natural occurrence of aflatoxins under Serbian environmental conditions. The OTA content in feedstuffs varied from 10 to 400 μg kg^{-1} (Table 5).

Table 4. Natural occurrence of AB_1 in feeds and foods (adapted from Lazić-Savić & Šutić, 1989; Ljubojević & Šutić, 1989; Škrinjar et al., *1989a, 1992a; Šutić* et al., *1989; Bočarov-Stančić* et al., *1995, 2001a, 2001c, 2004; Mašić* et al., *1994)*

Sample	*Positive samples (%)*	*Content ($\mu g\ kg^{-1}$)*
Bread	4.2	5
Dairy cattle feed	3.3-26	1.0-30.0
Juice, compote, marmalade & jam	25-42.0	<30 >50
Livestock feed & their components	3.7-35.2	3-16
Maize grain & pig feed mixture		10-40
Malt spent grain		50
Malting barley	16.6	3-6
Mash concentrates	3.53	10-50
Milk		10-50
Soya bean, sunflower & their processed products	40.0-45.4	3-16
Tomato and paprika products	3.5-10.0	20-30

The OTA presence was determined in 5.71% samples of pork and 11.9% samples of offal in the concentration of 31 do 89 $\mu g\ kg^{-1}$, respectively (Mašić *et al.*, 1989). In addition, OTA has been investigated in some meat products (Škrinjar *et al.*, 1999a, 2001), alimentary products, such as confectionary (Škrinjar *et al.*, 1999b, 2000), the so-called organic/healthy food (Škrinjar *et al.*, 2003a), spices (Škrinjar *et al.*, 2003b), etc. It is interesting to mention that OTA was also detected in red wine (Škrinjar, 2004).

The contamination of samples of hard sausages (31) by moulds, aflaoxins B1, B2, G1, G2, OTA and ZEA was investigated at the retail level by Škrinjar and Horvat-Skenderović (1989). These authors found that five hard sausage samples (16%) were contaminated by OTA in the concentration of 40 $\mu g\ kg^{-1}$ on the fifth and the tenth day of storing, although raw materials used to manufacture hard sausages were not contaminated by the investigated mycotoxins.

Škrinjar and Horvat-Skenderović (1992) also examined the possibility of ochratoxin A reproduction in hard sausages. Obtained results show that *P. aurantiogriseum* producing OTA in wheat kernels (42.5-65.0 $\mu g\ kg^{-1}$) also produced OTA in hard sausages during drying and storing.

The OTA presence was detected in 9.5% samples of imported food in the concentrations of 6 $\mu g\ kg^{-1}$ (cacao and earthnut), 13 $\mu g\ kg^{-1}$ (pepper, cacao powder) and 18 $\mu g\ kg^{-1}$ (earthnut), as well as, in grits, flour and germs of maize of local origin (6 $\mu g\ kg^{-1}$) (Bočarov-Stančić *et al.*, 2001b).

4.3. Fusarium *toxins*

There are many data on fusariotoxins ZEA and T-2 toxin very often occurring as natural contaminants in Serbia. However, less data on DAS, very little on DON and none on either its derivates or moniliformin (MON) are available. This disagrees with the occurrence of producers and favourable conditions for biosynthesis of these mycotoxins in Serbia.

Table 5. Natural occurrence of OTA in feeds and foods (adapted from Škrinjar et al.*, 1989b, 1992b, 1997; Šutić* et al.*, 1989; Mašić* et al.*, 1994; Radišić et al., 1994; Bočarov-Stančić* et al.*, 1995, 1996, 2001c, 2004; Dimić & Škrinjar, 1995)*

Sample	*Positive samples (%)*	*Content (μg kg⁻¹)*
Brewing barley	33.3	100-150
Confectionery products	14.0	10
Feedstuffs	1.4-32.8	48-96
Forage & grain feed	7-83.0	14.5-400
Formula feed for pigs		10-60
Maize dregs & germs		10-20
Maize grain		10-30
Peanut		up to 10
Mash concentrates	17.7	10-70
Soya bean meal		10-20
Soya bean, sunflower & their processed products	50-54.5	126-160
Spice mixture for Frankfurter hard sausage		32.0
Wheat flour		6-40

4.3.1. Zearalenone

The biosynthesis of zearalenones and their natural occurrence in Serbia is mainly associated to *F. graminearum*, and it is the highest in years with abundant precipitation and lower temperatures at the end of summer and the beginning of autumn. Under such conditions the frequency of the occurrence and the concentration of zearalenones can be 20 fold-higher in maize grain than in dry years (Stojanović *et al.*, 1978). Effects of such conditions reflect on mycotoxicological properties of maize-based formula feed for piglets and other pig classes, in which maize is a basic component. Bočarov-Stančić *et al.* (2004) established that 77.9% of these formulas contained ZEA in 2001 when a higher precipitation sum caused both, a greater infection of maize by *Fusarium* species and ZEA biosynthesis.

Chronologically, the ZEA level in maize grain varied from 200 to 7,000 $\mu g\ kg^{-1}$ in Serbia in the period 1981-1983, and 61% of samples contained these mycotoxins in a concentration of 200-20,000 $\mu g\ kg^{-1}$ (Kordić *et al.*, 1986). Such high ZEA concentrations at such a high frequency were not determined later (Table 6). The highest number of maize samples contained mycotoxins in 1986 (62.26%), and 20.75% of these samples were contaminated by ZEA in concentrations of up to 1,200 $\mu g\ kg^{-1}$ (Jurić *et al.*, 1994). A total of 30.96% of maize grain samples were contaminated by mycotoxins in Vojvodina in the period of 1986-1992 and 11.61% out of these 30.96% contained ZEA without causing any greater problem (Jurić *et al.*, 1994).

Table 6. Natural occurrence of ZEA in feeds and foods

Sample (%)	*Content ($\mu g\ kg^{-1}$)*	*Reference*
Barley (31%)	440-800	Bočarov-Stančić *et al.*, 1998
Bread	21-50	Protić, 2000
Black pepper grain	192-288	Dimić & Škrinjar, 1995
Feed (12.45%)	1,660-47,500	Jurić & Pupavac, 1996
Feedstuffs (72.6-77.9%)	260-2,560	Bočarov-Stančić *et al.*, 2004
Formula feed for pigs	100-800	Bočarov-Stančić *et al.*, 1995
Maize (6.3-100%)	50-10,000	Lević *et al.*, 1993a; Bočarov-Stančić *et al.*, 1996, 1997; Jajić, *et al.*, 2001
Maize dregs & germs	100-150	Bočarov-Stančić *et al.*, 1995
Mash concentrates	100-1,600	Bočarov-Stančić *et al.*, 1995
Pelleted sugar been pulp	140-150	Škrinjar *et al.*, 1990
Soya bean & sunflower meal (50-100%)	>3,000	Jajić *et al.*, 2001
Soya bean meal	150-200	Bočarov-Stančić *et al.*, 1995
Soya bean, sunflower & their products (80-3.6%)	260-10,240	Bočarov-Stančić *et al.*, 2001c
Sunflower meal	100	Bočarov-Stančić *et al.*, 1995
Wheat (87.5%)	up to 1,600	Stojanović, 1999
Wheat flour	up to 200	Škrinjar *et al.*, 1997; Stojanović, 1999

A relatively high ZEA content in maize can be explained by a common practice in Serbia to dry maize naturally. Kordić *et al.* (1986) previously established that ZEA occurs in naturally dried maize (61.2%) twice as much as in artificially dried maize (33.7%). More recently, Bočarov-Stančić *et al.* (1998) observed a higher ZEA level in naturally dried maize with initial moisture of 25-29.5% (390-1,120 $\mu g\ kg^{-1}$ than in maize stored with moisture of 11.9-15.45% (270-960 $\mu g\ kg^{-1}$).

4.3.2. Trichothecenes

Trichothecenes became a subject of serious studies after the 1980s (Duletić & Muntañola-Cvetković, 1989), although there had been indications of their existence since the 1960s (Milić *et al.*, 1969).

A natural occurrence of trichothecene type A, T-2 toxin (Table 7) and DAS (Table 8) is very important for the region of Serbia, while it is very difficult to comment on the importance of trichothecenes type B as there are very little literature data. DON was only detected in bread (1-7 $\mu g\ kg^{-1}$) with 25% of bran of wheat kernels that were 100% infected by the species of the genus *Fusarium*, while it was not found in bread with 10% of bran (Protić, 2000).

Beside precipitation (Bočarov-Stančić *et al.*, 2001b), the mode of maize drying was a second factor affecting the natural occurrence of T-2 toxin and DAS (Bočarov-Stančić *et al.*, 1997).

Table 7. Natural occurrence of T-2 toxin in feeds and foods (adapted from Bočarov-Stančić et al., *1993, 1995, 1997, 1998, 2001c, 2004; Lević* et al., *1993a)*

Sample	*Positive samples (%)*	*Content ($\mu g\ kg^{-1}$)*
Feedstuffs	45.2-53.3	100-1,000
Formula feed & concentrate for pigs		100-3,750
Maize	38.4-52.0	100-2,500
Maize dregs & germs		300
Mash concentrate		200-1,600
Soya bean, sunflower & their products	40.0-45.4	250-3,000
Wheat	33.0	500-750

Naturally dried maize contained 250-1,750 μg kg^{-1} and 500-2,000 μg kg^{-1} of T-2 toxin and DAS, respectively, while the corresponding values for artificially dried maize amounted to 250-750 μg kg^{-1} and 250 μg kg^{-1}, respectively, (Bočarov-Stančić *et al.*, 1997).

Table 8. Natural occurrence of T-2 toxin in feeds (adapted from Bočarov-Stančić et al., *1997, 2001c, 2004;)*

Sample	*Positive samples (%)*	*Content ($\mu g\ kg^{-1}$)*
Feeds & its components	20.5-75.4	100-2,000
Maize	9.6-25,0	500-1,000
Soya bean, sunflower & their products	31.8	100-2,000

4.3.3. Fumonisins

There are little data on the natural occurrence of fumonisins in Serbia, which is in disagreement with the dominance of this toxin producer on maize ear/kernel (Lević *et al.*, 1997) and with confirmed capacities of strains of these species to biosynthesise fumonisin (Table 3). Furthermore, European corn borer, widely distributed in Serbia, can increase the ear infection by *F. verticillioides* (by 63.1%) (Lević *et al.*, 1993b).

The presence of fumonisins (FB1+FB2) in the concentration of 250-640 μg kg^{-1} was detected in wheat (Protić, 2000). Furthermore, the presence of just FB1 in the concentration of 45-70 μg kg^{-1}, i.e. 450 μg kg^{-1} was determined in two out of 15 maize samples harvested in 2002 and in one out of five samples of imported soya bean meal, respectively (Đilas *et al.*, 2003).

5. MICROBIOLOGICAL AND MYCOTOXICOLOGICAL SAFETY OF FEEDS AND FOODS

The presence of a particular number of fungal spores in feedstuffs and mycotoxicological safety of feed and food in Serbia and Montenegro are regulated by the law (Tables 9 and 10).

Table 9. The largest number of fungi permitted in feeds in Serbia and Montenegro (Official Gazette of FRY 2/92, article 9)

Feeds	*No. of fungal spores*
Feeds of plant origin	300,000
Feeds of animal origin	10,000
Formula feed for young animals	50,000
Formula feed for adult animals	300,000

Based on criteria imposed by law (table 10), Mrđen *et al.* (1990) determined that due to the presence of spores of moulds (*Aspergillus*, *Penicillium*, *Fusarium*, etc.) 22.67% of commercial mixed feed were not safe, 10.76% of components of the plant origin (maize, barley, meal), collected mainly in Vojvodina, were not safe, and 1.82% of components of animal origin, mostly imported, were not safe. Due to the presence of *Aspergillus* (78.3%), *Penicillium* (58.7%) and *Fusarium* (43.5%) species, Mašić *et al.* (2002) established in 2001 that 43.5-48.1% of formula feed for pigs (pre-starter, starter and grower) and 14.0-17.2% of mixed feed for initial and final fattening of pigs, in-pig sows and nursing sows were not safe.

However, the presence of these fungi in feed is not always the indication of the presence of mycotoxins. For instance in 1999, Đilas *et al.* (2001) reported that, although 43, 46 and 58% of commercial mixed feed for pigs, poultry and cattle, were not microbiologically safe due to the presence of *Aspergillus*, *Penicillium* and *Fusarium*, respectively, these mixtures did not contain any mycotoxins analysed (aflatoxins B1, B2, G1 and G2, OTA, ZEA and trichothecenes).

Mycotoxicological studies preformed in 1992 and 1993 showed that all components of formula feed and mash concentrates were sanitary safe according to the regulation (Bočarov-Stančić *et al.*, 1995). Maize was the only exception as unsafety of 55.5% and 33.3% was established in 1992 and 1993, respectively. Similar differences were determined for mash concentrates (36.4% and 33.3%) and formula feed for pigs (17.4% and 10%). Mycotoxicological unsafety of samples was a result of the presence of ZEA (up to 1,500 $\mu g\ kg^{-1}$) and T-2 toxin (up to 3,750 $\mu g\ kg^{-1}$).

According to regulations, Mašić *et al.* (2003) analysed 339 samples of cereals and industrial by-products and determined, on the basis of the presence of aflatoxins, OTA, ZEA and trichothecenes (T-2 toxin + DAS), that 5.6%, 18.29%, 19.7% and 18.52% of samples, respectively, exceeding the legal limits, were not mycotoxicologically safe (Table 11). Furthermore, the same authors analysed 246 samples of formula feed for pigs, poultry and cattle and determined the unsafety of 13.01%, 11.79%, 9.76% and 9.09% of samples.

Table 10. Reference values for maximum tolerable concentrations of mycotoxins in feeds and foods in Serbia and Montenegro (Official Gazette of FRY 2/92, article 11)

Sample	*μg kg⁻¹*				
	AB_1 $+AG_1$	AM_1	*OA*	*ZEA*	*T-2 DAS*
Feedstuffs	50		-	-	-
Mash concentrates for chicks and turkey poults	10		100	100,000	300
Mash concentrates for laying hens	20		250	100,000	600
Mash concentrates for adult fowl	20		1,000	100,000	600
Formula feed for piglets	10		100	500	300
Formula feed for pig growers (to 25 kg)	20		100	500	300
Formula feed for fattening pigs	20		200	1,000	600
Formula feed for nursing sows and in-pig sows, gilts and boars	20		200	500	600
Formula feed for calves	10		-	3,000	300
Formula feed for beef cattle and all classes of adults	50		-	5,000	600
Formula feed for milking cows	10		-	3.000	600
Formula feed for adult sheep	50		-	3,000	600
Formula feed for adult goats	50		-	3,000	600
Spices	30		10		
Milk and milk products	0.5	0.5	10		
Foodstuffs	5		10		
Cereal products			10		
Other foodstuffs	0.5-5		10		

In the maize samples, out of which some originated from Hungary, Mašić *et al.* (2003) detected at the end of 2001 the presence of ZEA, trichothecenes, OTA and aflatoxin B1 in average concentrations of 750 μg kg-1, 240 μg kg-1, 69 μg kg-1 and 36 μg kg-1, respectively. On the average, sunflower meal, i.e. formula feed for gilts and piglets, were contaminated by ochratoxin, i.e. ZEA and aflatoxin B1 at concentrations of 78, i.e. 440 and 37 μg kg-1, respectively. Mash concentrates for fattening poultry, pullets, cockerels and laying hens were contaminated by OA and trichothecenes (T-2 toxin and DAS) at concentrations of 140-165 and 230-300 μg kg-1, respectively.

The natural occurrence of trichothecenes above maximum tolerable concentrations for young domestic animals (<300 μg kg^{-1}) in the period 2001-2003 was the essential cause of mycotoxicological unsafety of the greatest number of feed samples in Vojvodina (Bočarov-Stančić *et al.*, 1995, 2004). In the case of T-2 toxin, concentrations exceeding legal limits were found in 18%. The maximal concentration observed was 2,000 μg kg^{-1} for DAS and 1,000 μg kg^{-1} for T-2 toxin.

Table 11. Mycotoxicological unsafety (%) of feeds in 2002 (adapted from Mašić et al.*, 2003)*

Sample	*Total samples*	*Mycotoxins*			
		Aflatoxins	*OTA*	*ZEA*	*T-2 DAS*
Cereals & industrial by-products:					
Maize kernels	122	10.65	27.05	36.88	31.82
Barley kernels	32	9.37	21.05	15.62	15.38
Maize silage	11	0.00	0.00	9.90	0.00
Soya bean meal	115	1.74	7.83	5.22	0.00
Sunflower meal	36	2.78	27.78	13.89	9.09
Feeding yeast	23	0.00	13.40	13.04	0.00
Total feed	339	5.60	18.29	19.17	18.52
Formula feed for pigs, poultry & cattle:					
Formula feed for piglets & growers (to kg)	61	11.47	21.31	31.15	0.00
Formula feed for fattening & breeding pigs	51	7.84	5.88	15.62	0.00
Mash concentrates for breeding & fattening fowls	46	13.04	10.87	0.00	19.23
Mash concentrates for laying hens	78	17.95	10.26	0.00	11.63
Formula feed for milking cows	10	10.00	0,00	0.00	0.00
Total formula feed & mash concentrates	246	13.01	11.79	9.76	9.09

6. CONCLUSION

Although there are numerous data on the occurrence of toxigenic fungi in Serbia not only in maize and small grains, but also in other plant species, data on the presence of certain mycotoxins in post-harvest and stored products and feed and food, on effects of particular mycotoxins on animals and on measures that alleviate their adverse effects on animals, are not a result of systemic studies at the national level. These data mainly vary from case to case pointing to the following situation within this field in Serbia: 1. Studies are not related to the whole territory of Serbia, but are mostly aimed at the region of Vojvodina, the granary of the country; 2. Genera *Fusarium*, *Penicillium* and *Aspergillus* are predominately noted, but there are little data on species of these genera, probably due to insufficient experience of researches in identification of species; 3. The potential of the mycotoxins biosynthesis was observed for a small number of identified species, as well as, for a small number of strains of a certain species; 4. Few data on the origin of samples and the sampling methods, especially related to imported products of the plant and animal origin; are available 5. Analyses are mainly performed on submitted samples, while there are no analyses on systematically collected samples; 6. Mycotoxins are analysed less in pre-harvest plants and more in stored products; 7. Feed, particularly maize as its main component, was mostly analysed, while food was not; 8. Only mycotoxins that are defined by regulations (aflatoxins, ochratoxins and trichothecenes) were analysed either in fungal cultures or feed and food, although there are assumptions that other mycotoxins are also present; 9. The TLC technique is mainly used in analyses of mycotoxins; 10. The link between a fungal species and mycotoxins is insufficiently documented; 11. Absorbents are preventively applied with the aim to control myoctoxicoses in animals; 12. The scope of mycotoxicological studies is much wider than the scope of gained results published and analysed in the present study.

7. AFFILIATIONS

Jelena Lević, Maize Research Institute, Zemun Polje, Slobodana Bajića 1, 11185 Belgrade-Zemun, Serbia and Montenegro, levkez@yahoo.com

Slavica Stanković, Maize Research Institute, Zemun Polje, Slobodana Bajića 1, 11185 Belgrade-Zemun 1, 11081 Zemun, Serbia and Montenegro, sstojkov@mrizp.co.yu

Aleksandra Bočarov-Stančić, Centre for Biology, Petra Drapšina 15, 23000 Zrenjanjn, Serbia and Montenegro, tamaras@ptt.yu

Marija Škrinjar, Faculty of Technology, University of Novi Sad, Bulevar Cara Lazara 1, 21000 Novi Sad, Serbia and Montenegro, lea2@neobee.net

Zoran Mašić, Veterinary Scientific Institute, Novi Sad, Rumenački put 20, 21000 Novi Sad, Serbia and Montenegro, zoran@niv.ns.ac.yu

8. REFERENCES

Bagi, F., Balaž, F., & Škrinjar, M. (2000). Pathogenicity and zearalenone production by different *Fusarium graminearum* isolates in artificially infected wheat grain. Cereal Research Communication, 28, 477-484.

Balaž, F., Bagi, F., Škrinjar, M., & Stojšin, V. (2003). Cereal seed mycopopulations in 2002 (in Serbian). Žito i hleb, 4-5, 149-220.

Bočarov-Stančić, A. (1996). The influence of ecological and other factors on distribution of molds and mycotoxins in cereals and possibility of their decontamination (in Serbian with the summary in English). Ph.D. Thesis, Faculty of Agriculture, University of Novi Sad, Novi Sad, Yugoslavia.

Bočarov-Stančić, A., Adamović, M., Milovac, M., Tomašević-Čanović, M., & Daković, A. (2004). Review of the presence of mycotoxins in Vojvodina. Agroznanje (in press).

Bočarov-Stančić, A., Gojkov, M., & Laco, D. (2001b). Occurrence of mycotoxins in food and food additives sampled in middle Banat region in the period of 1993-1998 year (in Serbian). Book of Proceedings of the 4th Yugoslav Symposium Chemistry and Protection of Environment, Zrenjanin, Yugoslavia, 193-195.

Bočarov-Stančić, A., Jovanović, Đ., & Muntañola-Cveković, M. (1986a). DAS and T-2 toxin biosynthesis by Fusarium isolates from agricultural cultures in Yugoslavia (in Serbian with the summary in English). ANU B&H, Special Publications, LXXX, Department of Medical Sciences, 12, 129-146.

Bočarov-Stančić, A., Kpučin, Ž., & Nešić S. (2001c). Natural occurrence of mycotoxins in oilseed and oilseed products (in Serbian). Book of Proceedings of the 42nd Meeting Production and Oil Plant Processing, Herceg Novi, Yugoslavia, 365-370.

Bočarov-Stančić, A., Laco, D., Tomašević-Čanović, M., Adamović, M., & Daković A. (2003). Toxicity of *Fusarium* isolates from wheat grain contaminated with zearalenone (in Serbian with the summary in English). Book of Proceedings of the 10th Symposium Food Technology for Animal Safety and Quality, Vrnjačka Banja, Serbia and Montenegro, 299-304.

Bočarov-Stančić, A., Mašić, Z., Škrinjar, M., & Zurovac-Kuzman, O. (1996). Toxygenicity of *Aspergillus* and *Penicillium* strains isolated from Yugoslav cereals. Book of Abstracts of the 1st Congress of Biologists of Macedonia, 106.

Bočarov-Stančić, A., Milovac, M., & Bokorov, M. (2001a). Mycological and mycotoxicological investigations of barley vareites from Vojvodina. Book of Proceedings of the 14th International Congress Cereal bread 2000, Novi Sad, 70-73.

Bočarov-Stančić, A., Milovac, M., & Mašić, Z. (1998). A survey of natural occurrence of mycotoxins in Yugoslav cereals in 1990-995. Book of Proceedings of the 3rd International Symposium Interdisciplinary Regional Research (Hungary, Romania, Yugoslavia), Novi Sad, Yugoslavia, 2, 665-668.

Bočarov-Stančić, A., Muntañola-Cveković, M., & Oberan, Lj. (1986b). DAS and T-2 toxin biosynthesis by *Fusarium* cultures isolated from wheat (in Serbian with the summary in English). ANU B&H, Special Publications, LXXX, Department of Medical Sciences, 12, 147-160.

Bočarov-Stančić, A., Muntañola-Cveković, M., & Oberan, Lj. (1986c). Zearalenone biosynthesis by *Fusarium graminearum* isolates at different laboratory conditions (in Serbian with the summary in English). ANU B&H, Special Publications, LXXX, Department of Medical Sciences, 12, 105-118.

Bočarov-Stančić, A., Protić, N., & Živković, B. (1995). Review of microbiological and myxotoxicological investigations of feeds during 1992 and 1993 in the region of Vojvodina (in Serbian with the summary in English). Biotehnologija u stočarstvu, 11, 63-68.

Bočarov-Stančić, A., Škrinjar, M., 7 Ač, M. (1993). Distribution of toxigenic fungi and mycotoxins in maize grain (harvest 1991) (in Serbian). Book of Proceedings of 2nd Symposium Chemistry and Environmental Protection, Vrnjačka Banja, Serbia and Montenegro, 399-400.

Bočarov-Stančić, A., Škrinjar, M., Mašić, Z., Pavkov, S., & Gološin, B. (1997). Natural occurrence of *Fusarium* spp fusariotoxins in Yugoslav corn kernels. Cereal Research Communications, 25(3/2), 581-582.

Đilas, S., Abramović, B., & Mašić, Z. (2003). Occurrence of fumonisins and methods of determination (in Serbian with the summary in English). Book of Abstracts and Brief Communications of the 15th Meeting of Serbian Veterinarians, Zlatibor, Serbia and Montenegro, 110-118.

Đilas, S., Živkov-Baloš, M., Mihaljev, Ž., Mrđen, M., & Mašić, Z. (2001). Chemical, microbiological and mycotoxicological safety of animal diet mixes in the period from January 1999 to January 2000 (in Serbian with the summary in English). Veterinarski glasnik, 55 (1-2), 61-67.

Dimić, G., & Škrinjar, M. (1995). Toxigenic moulds and mycotoxins in mixture of species and corn black pepper used in meat industry (in Serbian with the summary in English). Tehnologija mesa, 5, XXXVI, 302-395.

Duletić, S., Muntañola-Cvetković, M. (1989). Investigations of trichothecenes in the world and in Yugoslavia with special review to T-2 toxin (in Serbian with the summary in English). ANU B&H, Special Publications, LXXXIX, Department of Medical Sciences, 14, 9-22.

Jajić, I., Terzić, V., Jurić, V., & Radanov Pelagić, V. (2001). Mycotoxins of maize, soya bean and sunflower meal 1996/2001 (in Serbian with the summary in English). A periodical of Scientific Research on Field and Vegetable Crops, Institute of field and vegetable Crops, Novi Sad, 194-200.

Jurić, V., & Pupavac, V. (1996). Mycotoxins as indicators of quality of the feedstuffs (in Serbian with the summary in English). Biotehnologija u stočarstvo,12, 161-168.

Jurić, V., Kovčin, S., Pupavac, V., & Jovanović, R. (1994). Maize grain contamination with mycotoxins in the region of Vojvodina (in Serbian with the summary in English). Contemporary Agriculture, 4, 87-91.

Kordić, B., Panin, M., Kandić, S., Lončarević, A., & Muntañola-Cvetković M. (1986). Results of a three year microbiological and mycotoxicological examination of feeds in SR of Serbia (in Serbian with the summary in English). ANU B&H, Special Publications, LXXX, Department of Medical Sciences, 12, 17-28.

Lazić-Savić, N., & Šutić, M. (1989). Sources of fruit product contamination with *Aspergillus flavus* and aflatoxin (in Serbian with the summary in English). ANU B&H, Special Publications, LXXXIX, Department of Medical Sciences, 14, 131-140.

Leslie, J.F. (1995). *Gibberella fujikuroi*: available populations and variable traits. Canadian Journal of Botany, 73 (Suppl. 1), S282-S291.

Lević, J. (2004). The Genus *Fusarium* - isolation and identification of species, genes and mycotoxins (monograph in preparation).
Lević, J., Ivanović, D., & Stanković, S. (2003). Seed-borne parasites of maize, sorghum and pearl millet and control measures (in Serbian with the summary in English). Biljni lekar, 6, 570-577.
Lević, J., Ivanović, D., Penčić, V., & Bača, F. (1993b). Effects of *Fusarium* spp. and *Ostrinia nubilalis* Hbn. association on maize hybrids. Programme and Abstracts of Maize and Sorghum Eucarpia - 16th Conference on Breeding and Molecular Biology: Accomplishments and Future Promises, Bergamo, Italy, VII-P17.
Lević, J., Penčić, V., Ivanović, D., & Bočarov-Stančić, A. (1993a). Maize ear and associated fungi in field and crib environment. Journal of Scientific Research in Biology, 45, 37-44.
Lević, J., Tamburić-Ilinčić, Lj., & Petrović, T. (1997). Maize kernel infection by *Fusaria* in the period 1994-1996. Cereal Research Communication, 25 (3/2), 773-775.
Ljubojević, A., & Šutić, M. (1989). Aflatoxins as contaminants of some vegetable products (in Serbian with the summary in English). ANU B&H, Special Publications, LXXXIX, Department of Medical Sciences, 14, 141-148.
Lozanić, S.M. (1904). Maize as Food (in Serbian), ed. Državna štamparija Kraljevine Srbije, Beograd, 23.
Marić, A. (1981). *Fusarium* diseases of wheat and corn in Eastern Europe and the Soviet Union. In: Nelson, P.E., Toussoun, T.A., & Cook, R.J. (Editors) *Fusarium*: Diseases, Biology, and Taxonomy. The Pennsylvania State University Press, University Park and London, 77-93.
Mašić, Z., Bočarov-Stančić, A., Pavkov, S., & Zurovac-Kuzman, O. (1997). Gas-chromatographic determination of type A trichothecene mycotoxins in extract of *Fusarium* spp. isolated from Yugoslav corn hybrid (in Serbian with the summary in English). Acta Veterinaria, 47 (1), 23-32.
Mašić, Z., Bočarov-Stančić, A., Sinovec, Z., Đilas, S., & Adamović, M. (2003). Mycotoxinis in feed for animals in the Republic of Serbia (in Serbian with the summary in English). Book of Proceedings of the 10th Symposium Food Technology forAnimal Safety and Quality, Vrnjačka Banja, Serbia and Montenegro, 290-298.
Mašić, Z., Jakić-Dimić. D., Stanaćev, V., & Sinovec, Z. (2002). Survey of quality of swine feed mixes (in Serbian with the summary in English). Veterinarski glasnik, 56, 41-52.
Mašić, Z., Pavkov, S., Orlić, D., Stankov, M., Simić, V., & Rackov O. (1994). Results of mycotoxicological examinations of poultry mixes (in Serbian with the summary in English). Veterinarski glasnik, 48, 531-534.
Mašić, Z., Rajić, I., Mrđen, M., Škrinjar, M., Pavkov, S., & Gagrčin (1989). Results of examination of meat, kidneys, liver and lungs in swine concerning the presence of ochratoxin A residue in an industrial slaughter-house in Vojvodina (in Serbian with the summary in English). ANU B&H, Special Publications, LXXXIX, Department of Medical Sciences, 14, 215-218.
Milić, D., Penčić, V., Kordić, B., Lončarević, A., Zaletel, I., & Filipović, V. (1969). Pathological status in pigs fed on mixtures containing maize infected with *Gibberella zeae* (in Serbian with the summary in English). Contemporary Agriculture, 5-6, 501-506.
Milošević, M., Konstantinović, B., Zlokolica, M., & Draganić, M. (1995). Mycoflora of wheat, barley and maize seeds (in Serbian with the summary in English). Plant Protection, 213, 221-227.
Mirocha, C.J., & Christensen, C.M. (1974). Oestrogenic Mycotoxins Synthesized by *Fusarium*. In: Purchase, I.F.H. (Editor) Mycotoxins, Elsevier Scientific Publishing Company, Amsterdam-Oxford-New York, 129-147.
Mrđen, M., Mašić, Z., Miloradović, S., & Rogožarski, D. (1990). Most common contaminants of animal food (in Serbian with the summary in English). Veterinarski glasnik, 44, 1095-1100.
Muntañola-Cvetković, M., Duletić, S., Bočarov-Stančić, A., & Franić-Mihajlović, D. (1991). Incidence and characteristics of *Fusarium* populations producing T-2 toxin in Serbia. Prehrambeno-tehnološka i biotehnološka revija, 29, 79-83.
Noory, M.J. (1983). Mycopopulations in corn grains and sunflower seeds, and toxicity of *Aspergillus* and *Fusarium* species isolated from these products. Ph.D. Thesis, University of Belgrade, Faculty of Sciences, Belgrade, 184.
Oberan, Lj. (1989). Zearalenone production by *Fusarium* isolates from agricultural cultures in Yugoslavia (in Serbian with the summary in English). ANU B&H, Special Publications, LXXXIX, Department of Medical Sciences, 14, 105-118.
Protić, N. (2000). Mycotoxin in grain of and wheat products as a result of infection with *Fusarium* spp. fungi (in Serbian with the summary in English). Ph.D. Thesis. University of Belgrade, Faculty of Agriculture, Belgrade-Zemun, 139.

Radić, Đ. (1872). All about Maize (in Serbian), ed. Društvo za poljsku privredu, Beograd, 87.

Radišić, M., Bočarov-Stančić, A., & Vukoje, M. (1994). Presence of mycotoxins in pure brewing barley varieties (in Serbian with the summary in English). PIVAAM, 27, 140-144.

Rajić, I. (1963). Alimentary intoxication of horses in the area of Sivac veterinary station (in Serbian). Veterinarski glasnik, 6, 554-557.

Ranojević, N. (1914). Dritter Beitrag zur Pilzflora Serbiens. Annales Mycologici (Band XII), 393-421.

Škrinjar, M. (2004). Presence of mycotoxins in wine, unpublished.

Škrinjar, M., & Dimić, G. (1992). Ochratoxigenicity of *Aspergillus ochraceus* group and *Penicillium verricosum* var. *media*. Acta Microbiologica Hungarica, 39, 257-261.

Škrinjar, M., & Horvat-Skenderović, T. (1989). Contamination of dry sausage with moulds, aflatoxins, ochratoxin and zearalenone (in Serbian with the summary in English). Tehnologija mesa, 2, 53-59.

Škrinjar, M., & Horvat-Skenderović, T. (1992). Ochratoxin A production in dry sausage by *Penicillium verrucosum* var. *cyclopium* strains. Book of Proceedings of the 38th International Congress of Meat Science and Technology, Clermont Ferrand, France, 4, 831-834.

Škrinjar, M., Danev, M., Dimić, G. (1995b). Investigation on the presence of toxigenic fungi and aflatoxins in raw milk. Acta Alimentaria, 24, 395-402.

Škrinjar, M., Danev, M., Petrevski, B., Došen-Bogićević, V., & Dimić, G. (1999a). Occurrence of ochratoxin A - producing moulds and ochratoxin A in some meat products. Book of Proceedings of the 45th International Congress Meat Science and Technology, Yokohama, Japan, 2, 610-611.

Škrinjar, M., Dimić, G., & Kolarević, I. (1999b). Contamination of raw materials for confectionary industry with toxigenic moulds and ochtatoxin A (in Serbian). Book of Abstracts of the Yugoslav Congress Food, Farmaceutical and Chemical Engineering, Novi Sad, Yugoslavia, 152.

Škrinjar, M., Došen- Bogićević, V., & Vujković, I. (2001). Mycopopulations, aflatoxins and ochratoxin A in some semi-dry sausages. Book of Proceedings of the 47th International Congress of Meat Science and Technology, Krakow, Poland, 2, 272-273.

Škrinjar, M., Injac, V., Sedej, I., & Kocić, S. (2003a). Mycological and mycotoxicological quality of some agricultural products taken from "healthy food" stores. Book of Abstracts of FEMS Congress of European Microbiologists, Ljubljana, Slovenia, P2-71, 160.

Škrinjar, M., Kolarević, I., Dimić, D., & Jovanović, O. (2000). Presence of moulds and some mycotoxins in raw materials used in confectionary industry (in Serbian with the summary in English). Žito i hleb, 27 (1), 9-14.

Škrinjar, M., Šarac, M., Dimić, G., & Filipović, N. (1997). Study on the presence of moulds and some mycotoxins in wheat flour. Romanian Biotechnological Letters, 2, 477-487.

Škrinjar, M., Stubblefield, R.D., & Vujičić, I.F. (1992b). Ochratoxigenic moulds and ochratoxin A in forages and grain field. Acta Veterinaria Hungarica, 40, 185-190.

Škrinjar, M., Stubblefield, R.D., Stojanović, E., & Dimić, G. (1995a). Occurrence of *Fusarium* species and zearalenone in dairy cattle feeds in Vojvodina. Acta Veterinaria Hungarica, 43, 259-267.

Škrinjar, M., Stubblefield, R.D., Vujičić, I.F., & Stojanović, E. (1992a). Distribution of aflatoxin-producing moulds and aflatoxins in dairy cattle feed and raw milk. Acta Microbiologica Hungarica, 39, 175-179.

Škrinjar, M., Stubblefield, R.D., Vujičić, I.F., Jurić, V., & Stojanović E. (1990). Contamination of cattle feed by aflatoxin B_1, ochratoxin A and zeaeralenone (in Serbian with the summary in English). Acta Veterinaria, 40, 105-112.

Škrinjar, M., Vujičić, I., Stubblefield, R., Jurić, V., & Mašić, Z. (1989a). The presence of aflatoxin B_1 in forage in Vojvodina (in Serbian with the summary in English). ANU B&H, Special Publications, LXXXIX, Department of Medical Sciences, 14, 171-180.

Škrinjar, M., Vujičić, I., Stubblefield, R., Jurić, V., Stojanović, E., & Mašić, Z. (1989b). Contamination of foodstuffs with ochratoxin A in Vojvodina (in Serbian with the summary in English). ANU B&H, Special Publications, LXXXIX, Department of Medical Sciences, 14, 171-180.

Škrinjar, M., Vukojević, J., Đorđević, M., & Krasić, V. (2003b). Isolation and identification of toxigenic moulds from some spices used in meat industry. Book of Abstracts of the 3rd Balkan Conference of Microbiology, Istanbul, Turkey, P-246, 442.

Stanković, S., Moretti, A., Logrieco, A., Lević, J., & Petrović, T. (2004). Biological and toxicological features of *Fusarium verticillioides*, *F. proliferatum* and *F. subglutinans* from Serbia (unpublished).

Stojanović, T. (1999). Wheat and flour quality from mycological and mycotoxicological aspect (in Serbian with the summary in English). Ph.D. Thesis. University of Novi Sad, Faculty of Technology, Novi Sad, 176.

Stojanović, Z., Popović, M., Kovčin, S., & Gagrčin, M. (1978). Effect of mouldy maize contaminated by *Fusarium* toxins on performance and health of pigs (in Serbian with the summary in English). Contemporary Agriculture, 7-8, 5-18.

Stojsin, V., Balaz, F., Lević, J., Petrović, T., Bagi, F., & Atlas, S. (2000). *Fusarium* root rot of sugar beet. Book of Abstracts of the 6th European *Fusarium* Seminar and Third COST 835 Workshop (Agriculturally Important Toxigenic Fungi), Berlin, Germany, 87.

Šutić, M., Svilar, N., Ivanović, D., Oljačić, E., & Tosman, N. (1989). Test results on mycotoxins in confectionary products and bread (in Serbian with the summary in English). ANU B&H, Special Publications, LXXXIX, Department of Medical Sciences, 14, 103-107.

Žakula, R., Todorović, M., Škrinjar, M., & Jerenić V. (1985). Moulds in raw and axillary materials in cherry cocktail juice in certain production stages (in Serbian with the summary in English). A Periodical of Scientific Research of Faculty of Technology, Novi Sad, 16, 215-222.

R. MATEO, Á. MEDINA, J. V. GIMENO-ADELANTADO &
M. JIMÉNEZ

AN OVERVIEW ON THE STATUS OF TOXIGENIC FUNGI AND MYCOTOXINS IN SPAIN

1. INTRODUCTION

Mycotoxins are a group of secondary metabolites which are produced by various filamentous fungi, and which can produce a toxic response in animals or humans after ingestion of contaminated food. The main toxigenic fungi isolated from foods and feeds belong to the genera *Penicillium*, *Aspergillus*, *Fusarium* and *Alternaria*. Spores of these moulds are almost universally present everywhere, they fall on plants and, under appropriate conditions, they can germinate and generate mycelia causing crop spoilage and/or production of mycotoxins.

In Spain, there are not many reports on the occurrence of toxigenic fungi in foods and feedstuffs. Some studies were directed to pre-harvest corn (Sanchis, Viñas, Jiménez, & Hernández, 1984; Jiménez, Sanchis, Santamarina & Hernández 1985; Jiménez, Sanchis, Mateo & Hernández, 1986). Others studies were addressed mainly to stored cereal grains (Sanchis, Viñas, Jiménez, Calvo & Hernández, 1982; Jiménez, Santamarina, Sanchis & Hernández, 1987), nuts and sunflower seeds (Jiménez, Mateo, Querol, Huerta & Hernández, 1991) or bananas ready for sale (Jiménez, Logrieco & Bottalico, 1993). Recently, the mycobiota occurring on Spanish grape fruits is being studied with interest (Sáez *et al.*, 2003).

The presence of toxigenic fungi in food does not imply that mycotoxins have been produced or will be produced as this event requires a special situation with regard to environmental conditions (temperature, water activity, substrate nature and time). There is not much information with regard to the occurrence of mycotoxins in food and feed marketed in Spain. The most relevant mycotoxins are aflatoxins, ochratoxins, patulin, fumonisins, trichothecenes and zearalenone. In the eighties, most research interest was focused mainly on aflatoxins (Sanchis *et al.*, 1984; Jiménez *et al.*, 1991). Later on, great interest was aimed towards possible contamination with fumonisins of maize products (Sanchis, Marin & Ramos, 2001) and beer (Torres, Sanchis and Ramos, 1998). Recently, several surveys have pointed out that ochratoxin A is present in Spanish products such as bread, beer, wine, must and grape juices and imported coffee (Burdaspal and Legarda, 1998; Burdaspal & Legarda, 1999; Legarda & Burdaspal, 1998; Legarda & Burdaspal, 1999; López, González-Peñas, Jiménez & Bello, 2002).

Spain is located South-western of Europe in the Iberian Peninsula (about 40 ºN, 4 ºW). The Balear and Canary Islands also belong to Spain. Average temperature vary depending on the particular climatic area but it is around 15-20ºC. Summers are

A. Logrieco and A. Visconti (eds.),
An Overview on Toxigenic Fungi and Mycotoxins in Europe, 219-235.

hot in interior (about 20-35ºC) and more moderate along the periphery. Winters are cold in interior (0-15ºC) and warmer along coasts. The North is humid while the Centre and the South are dry. The domestic agricultural and agricultural-based commodities, more susceptible to undergo contamination by toxigenic fungi and mycotoxin occurrence are cereals (wheat, maize, oats, barley, rice), nuts (almonds, hazelnuts), sunflower seeds, bananas, grape fruits and dry vine fruits, apple products, wine, must, grape juice and beer. Roasted and instant coffee, pistachio nuts, peanuts, wine, and beer are relevant imported commodities where mycotoxin occurrence has been reported.

The goal of this overview is to summarize the status of Spain with regard to contamination with toxigenic fungi and mycotoxins of food and feed produced or marketed in this country. Special emphasis has been put on data from the last years. Spanish regulations on the topic are also shown.

2. OCCURRENCE OF TOXIGENIC FUNGI

2.1. Cereals

Occurrence of species of the genera *Aspergillus*, *Penicillium* and *Fusarium* in samples of cereal grains (wheat, maize, oats, rice and barley), stored after crop and ready to be used as ingredients for foodstuffs or flour, was studied by Jiménez *et al* (1987) and Figure 1 shows the percentage of contaminated samples. It is remarkable that a high proportion of samples, mainly maize and wheat (100%) but also oats (80%), is infected with *A. flavus* or *A. parasiticus*, two aflatoxigenic species (Figure 1A). *Aspergillus* section Nigri was present in about 80% of maize samples but its incidence in the remaining samples was below 20%. *A. ochraceus* was isolated with lower frequency and mainly in barley (about 40% of samples). Contamination with *Penicillium* spp. was less relevant, being maize the most contaminated cereal, followed by barley. The most frequent species were *P. variabile*, *P. islandicum* and *P. purpurogenum* (Figure 1B). With regard to the genus *Fusarium,* maize kernel samples were contaminated with *F. verticillioides* (60%), *F. graminearum* (42%), *F. oxysporum* (20%) and *F. culmorum* (15%) (Figure 1C). *F. sporotrichioides* and *F. avenaceum* were found only in barley.

The mycobiota occurring in malt barley and malt taken from Spanish breweries was studied by Medina, Jiménez, Sáez, Hinojo, and Mateo (2002). Figure 2 shows the percentage of grains contaminated with different fungal species. *Penicillium* and *Aspergillus* species. were more abundant in malt grains than in malt barley. On the contrary, *Alternaria* spp. predominated in malt barley grains. The genus *Fusarium* was present in about 10 % of grains in both products. As shown in Table 1, the most relevant toxigenic *Aspergillus* species were *A. flavus* and *A.* section *Nigri. A. ochraceus* was found in malt barley but not in malt grains.

The main *Penicillium* species were *P. funiculosum, P. griseofulvum and P. purpurogenum*.

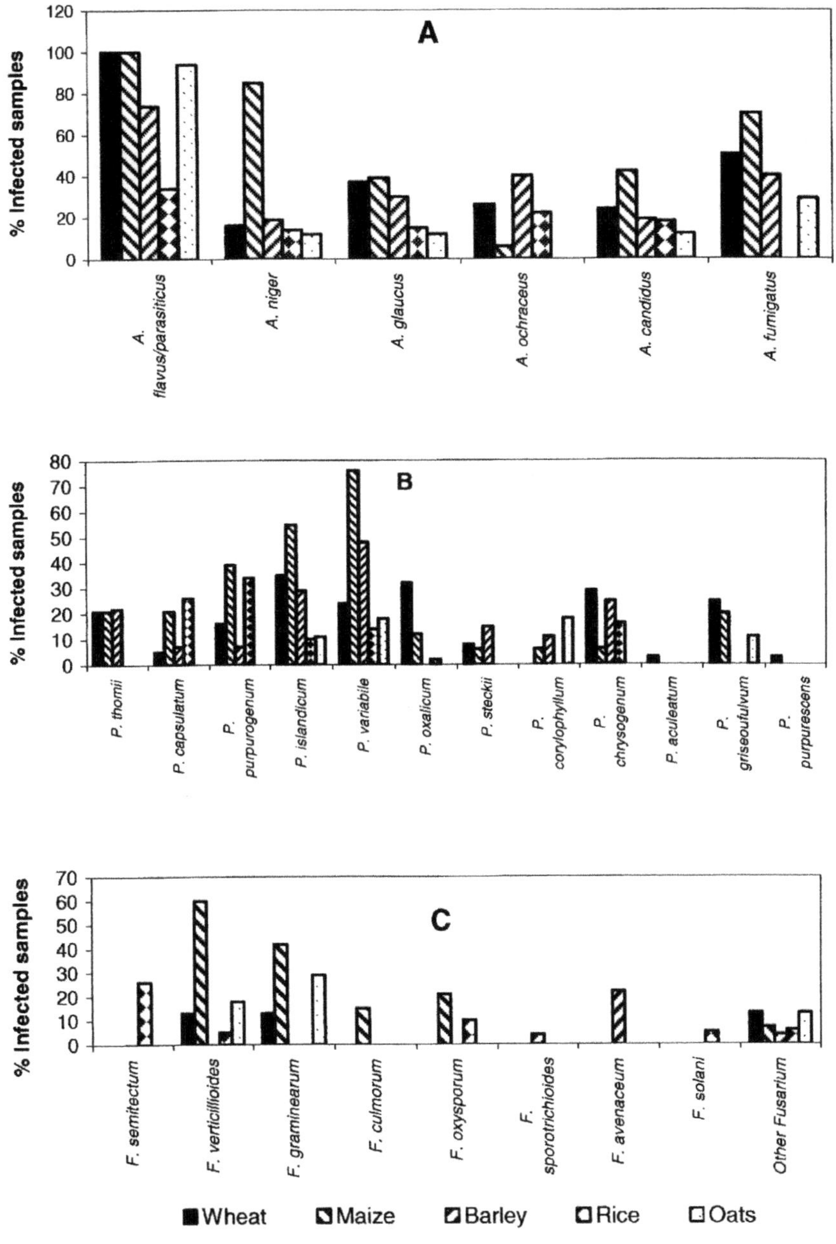

Figure 1. Occurrence of fungi in cereal grains in Spain. A: Aspergillus *spp; B:* Penicillium *spp; C:* Fusarium *spp.*

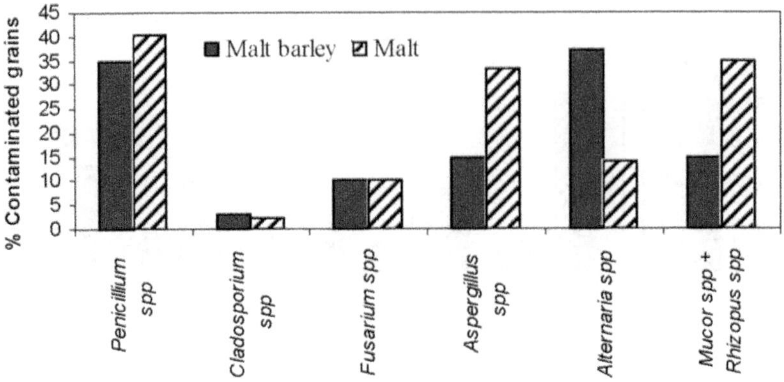

Figure 2. Occurrence of fungi in malt barley and malt from Spanish breweries

Table 1. Distribution of fungal species (%) among each genus in malt barley and malt grains

Species	Malt barley	Malt
Aspergillus flavus	53	83
Aspergillus section *Nigri*	6	6
A. ochraceus	7	—
Other *Aspergillus*	34	11
Penicillium funiculosum	32	26
P. griseofulvum	8	4
P. purpurogenum	22	14
Other *Penicillium*	38	56
Fusarium culmorum	12	3
F. graminearum	4	13
F. oxysporum	5	28
F. acuminatum	19	4
F. equiseti	7	6
Other *Fusarium*	53	46

The most interesting, from the toxicological viewpoint, *Fusarium* species isolated in these commodities were *F. culmorum, F. graminearum*, and *F. oxysporum.*

2.2. Nuts and sunflower seeds

Almonds, hazelnuts, pistachio nuts, peanuts and sunflower seeds marketed in Spain were studied with respect to occurring mycobiota (Jiménez *et al.*, 1991). Their results with regard to percentage of infected samples are summarized in Figure 3.

Figure 3A deals with occurrence of *Aspergillus* spp. The degree of infection by *A. flavus/parasiticus* in samples of pistachio nuts, almonds and hazelnuts and of infection by *Aspergillus* section Nigri in hazelnuts and almonds is remarkably high. The occurrence of *A. ochraceus* was quite lower, being almonds the most infected nuts.

Potential toxigenic *Penicillium* species were found in these nuts with low frequency, though contamination by *Penicillium* was important in pistachio nuts, sunflower seeds and peanuts (Figure 3B). Contamination with *Fusarium* spp. was not relevant. *Alternaria* spp. was isolated in about 40% of hazelnuts and 30% of pistachio nut samples. *Trichothecium roseum* was also isolated in less than 15 % of samples and was not detected in peanuts and sunflower seeds (Figure 3C).

2.3. Bananas

Bananas sold in Spain can be domestic products from the Canary Islands, which belong to Spain, or imported, mainly from South or Central America. Table 2 lists the percentage of samples from the Canary Islands infected with various *Fusarium* species. The number of screened samples was 36. Other genera like *Aspergillus, Penicillium* and *Alternaria* were also detected (Jiménez *et al.*, 1993). The incidence level of *F. verticillioides* was high. They produced, in general, undetectable or very low levels of fumonisins in cultures, although some of them produced moniliformin (Jiménez, Huerta & Mateo, 1997). *F. oxysporum, F. proliferatum* and *F. graminearum* were also isolated.

2.4. Grape fruits

The mycobiota of Spanish wine grapes is being currently studied due to the reported presence of ochratoxin A in wines and must (Burdaspal & Legarda, 1999). Some of the results obtained by Sáez *et al.* (2003) in 41 samples of grape fruits grown in Eastern Spain are shown in Table 3.

The most relevant potential toxigenic fungi were *Alternaria* spp. and *Aspergillus* section Nigri. The last group, encompassing mainly *A. niger* var *niger, A. carbonarius*, and other *Aspergillus* can be involved in production of ochratoxin A in grapes and related beverages. *A. ochraceus* and *P. verrucosum* were not detected. The relative proportion of these fungi varies depending on the grape varieties. *A.* section Nigri occurred more frequently in the Bobal, Monastrell and the white sweet Moscatel varieties, which is used to make dessert wines. This fungal group was not isolated in the Cencibel variety.

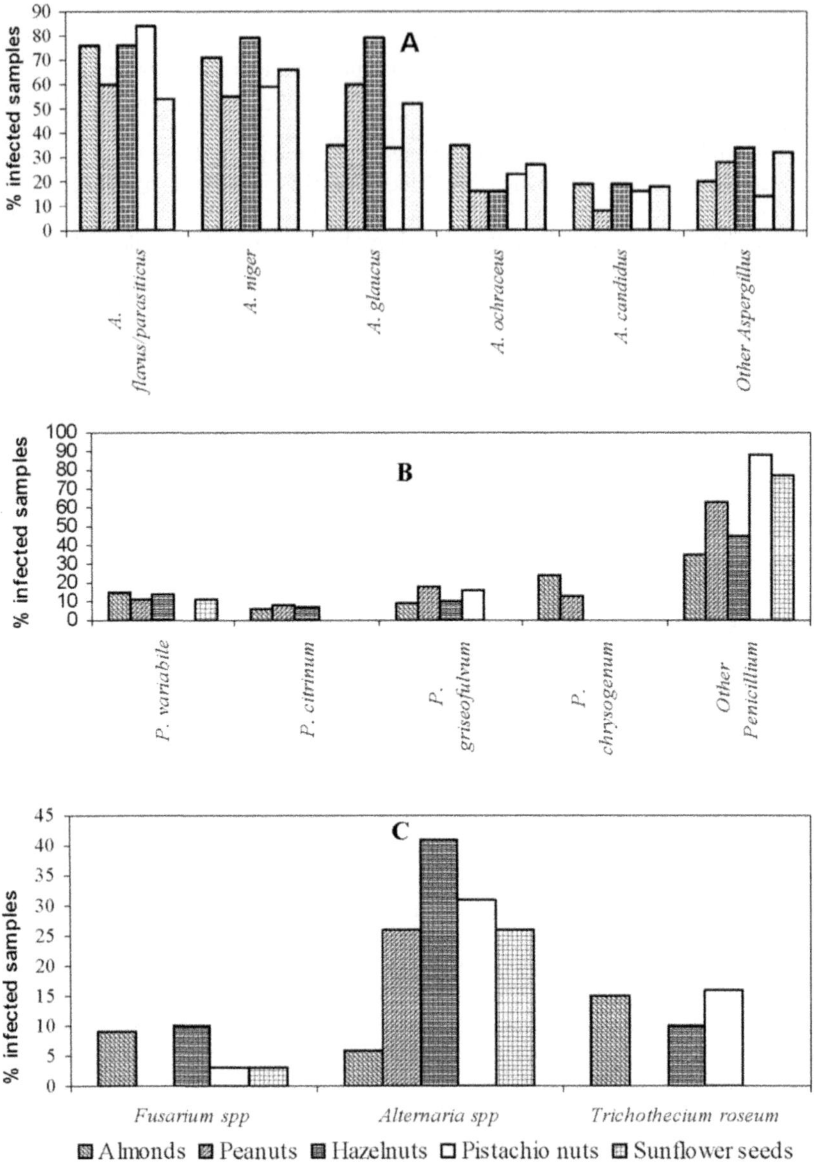

Figure 2. Occurrence of fungi in nuts and sunflower seeds. A: Aspergillus *spp; B:* Penicillium *spp; C: Other fungi.*

Table 2. Occurrence of Fusarium *spp in banana fruits from Canary Islands*

Fusarium *species*	*% infected samples*
F. semitectum var majus	*91.7*
F. verticillioides	*41.7*
F. solani	*19.4*
F. oxysporum	*27.8*
F. proliferatum	*16.7*
F. graminearum	*11.1*
F. camptoceras (white)	*13.9*
F. subglutinans	*5.5*
F. dimerum	*11.1*

Table 3. Percentage of infected samples in five grape varieties from Valencia (Eastern Spain)

	Grape variety					
Fungi	*Bobal*	*Cencibel*	*Garnacha*	*Monastrell*	*Moscatel*	*Mean*
Alternaria spp	27.6	13.8	6.8	18.5	12.96	11.38
Cladosporium spp	21.8	25.9	38.6	20.75	7.7	22.39
Aspergillus section *Nigri*	16.1	0	2.27	18.51	15.38	11.38
Phoma spp	0	3.45	0	0	0	0.59
Fusarium spp	1.15	1.72	0	1.85	4.35	1.83
Rhizopus spp	0	5.17	6.8	0	0	2.04
Penicillium spp	1.15	0	0	3.3	11.4	3.23
A. flavus	2.56	0	0	0	0	0.62
Acremonium spp	0	0	0	9.43	0	2.07

2.5. Dried vine fruits

Table 4 shows the results of a screening study carried out on 50 samples of raisins, currants and sultanas purchased in Barcelona (Spain) by Abarca, Accensi, Bragulat, Castellá and Cabañes (2003). Nearly all samples were contaminated with black aspergilli, 98% with *A. niger* var. *niger* and 58% with *A. carbonarius*. These species, and mainly the last one, were considered to be the origin of OTA in these products. *Mucorales* were common (76%). Other moulds isolated in less than 9 % of plated pieces were *Aspergillus flavus, A. versicolor, Alternaria alternata, Monilia* sp., *Eurotium* sp., *Trichoderma* sp., *Penicillium purpurogenum, P. digitatum and Penicillium* spp.

However, neither *P. verrucosum* nor *P. ochraceus* were isolated from these samples.

3. OCCURRENCE OF MYCOTOXINS

Data concerning the incidence of mycotoxins in food are very dependent on the sensitivities of the analytical methodologies used by analysts and researchers. As limits of detection have decreased with time due to the use of increasing performance in instrumentation and better extraction and cleanup procedures, the frequency of positive samples (those where mycotoxins have been detected) has also increased with time. Data available on mycotoxin occurrence in food and feed produced or marketed in Spain concern only a restricted but very important group of mycotoxins.

Table 4. Occurrence of aflatoxins in food and feed in Spain

Commodity	*Ref.*	*N/+*	*Aflatoxin B_1 (ng/g)*	*Aflatoxin B_2 (ng/g)*	*Aflatoxins B_1+B_2+G_1+G_2 (ng/g)*
Maize (fresh)	1	55/0	-	-	-
Maize (stored)	1	154/0	-	-	-
Maize-based food/feed	2	100/1(feed)	150	80	-
Maize (mouldy)	3	40/13	-	-	0.15-3520
Poultry feeds (mouldy)	3	6/4	-	-	0.38-112.5
Oilseed rape	2	20/1	0.25	-	-
Pistachio nuts	4,5	24/1	15	-	84
Groundnuts	4,5	44/3	>5	-	>10
Almonds	5	5/0	<0.6	-	-
Tiger nuts	5	5/0	<0.6	-	-
Wheat flour	5	36/0	<0.6	-	-
Rice	5	14/0	<0.6	-	-
Milk (sterilized and raw)	4	94/0	-	-	<0.05 (M_1)

N/+: number of assayed samples/number of positive samples. Refs.: 1, Muñoz, Cardelle, Pereiro and Riguera (1990); 2, Sanchis *et al.* (2001); 3, Jiménez and Mateo (2001); 4, Departamento de Sanidad del Gobierno Vasco (1995); 5, Laboratorio de Salud Pública de Valencia, 'Consellería de Sanitat, Generalitat Valenciana' (personal communication, July 10, 2003).

3.1. Aflatoxins

There are few reports with regard to occurrence of aflatoxins in food and feed in Spain. As listed in Table 4, levels of aflatoxins usually found in cereals, cereal-based feeds, nuts and milk were low or below the maximum allowable limits (MAL). One sample of maize-based feed showed high levels of aflatoxin B_1 and aflatoxin B_2. Very high levels for sum of aflatoxins were registered in mouldy maize and poultry feed samples (Jiménez & Mateo, 2001). Aflatoxin M_1 was always below the MAL in all screened milk samples. Based on these results, it may be deduced that high levels of aflatoxins are not likely to occur in Spanish foods.

3.2. Fumonisins

These mycotoxins are produced by some *Fusarium* species, mainly *F. verticillioides* and maize is the crop where they are likely to occur. Table 5 deals with occurrence of fumonisinsB_1 (FB_1) and B_2 (FB_2) in food in cereals (maize, barley, and wheat) and maize-based products. Beer as a cereal-based beverage has also been screened (Torres, Sanchis, & Ramos, 1998).

Table 5. Occurrence of fumonisin B_1 and fumonisin B_2 in food, feed and beverages

		Fumonisin B1			*Fumonisin B2*		
Food sample	*Ref.*	*N/+*	*% Positives*	*Range (μg/g)*	*N/+*	*% Positives*	*Range (μg/g)*
Maize/humans	1	55/48	87.3	0.2-19.2	55/22	40	0.2-5.9
Maize/animals	2	42/31	73.8	nd-8.6	17/5	29.4	nd-0.7
Maize(mouldy)	4	39/16	41.0	52.2-1678	39/16	41.0	nd-92
Corn grits	2	15/3	20	nd-0.09	-	-	-
Snacks	2	32/17	53.1	nd-0.2	-	-	-
Corn flour	2	12/5	41.7	nd-0.938	-	-	-
Corn flakes	2	42/6	14.2	nd-0.10	-	-	-
Toasted corn	2	30/2	6.7	nd-0.027	-	-	-
Pop corn grain	2	6/0	0	nd	6/0	0	nd
Corn screenings	2	27/26	96.3	nd-5.46	15/7	46.7	nd-0.5
Barley	1	29/21	72.4	0.2-11.6	1	3.5	0.5
Wheat	1	17/8	47.1	0.2-8.8	1	5.9	0.2
Sorghum	4	1/1	100	182	1/1	100	53
Swine feed	1	47/42	89.4	0.4-11.6	0	0	nd
Horse feed	1	20/14	70	0.4-23.6	1	5	0.3
Poultry feed	1	2/2	100	1.1-1.4	2/0	0	nd
Beer	3	96/28	29.2	nd-0.086*	-	-	-

N/+: number of analyzed samples/number of positive samples; nd: not detected; *: ng/ml. Refs.: 1, Castellá *et al.* (1999) ; 2, Sanchis *et al.* (2001); 3, Torres *et al.* (1998); 4, Jiménez and Mateo (2001).

Some mouldy samples, which were suspected to be related to farm animal disorders, exhibited high fumonisin levels. At present, there is no regulation in the E.U. and Spain on these mycotoxins but maximum limit has been proposed at 1000 ng/g for FB_1 plus FB_2 in Switzerland (Food and Agriculture Organisation, 1997). Taking into account this reference, the problem of high FB_1 levels focuses on unprocessed maize and other cereals and feeds.

3.3. Trichothecenes and zearalenone

There is very little information about the natural contamination by trichothecenes and zearalenone in food products in Spain. Table 6 shows data concerning the occurrence of deoxynivalenol (DON) and zearalenone (ZEA) in maize and feed. Mouldy samples submitted to the laboratory because of suspicion of being associated to animal health problems were more severely contaminated (Jiménez & Mateo, 2001) than randomly screened samples taken from typical field open storehouses built in Galicia (North-western Spain) named 'horreos', where DON and ZEA were detected in 5.8 % and 11.7 % of stored samples, respectively (Muñoz *et al.*, 1990).

Table 6. Incidence of deoxynivalenol (DON) and zearalenone (ZEA) in maize

Sample	*Mycotoxin*	*N/+*	*% Positive*	*Range (mg/kg)*	*Ref.*
Mouldy maize	DON	20/7	35	0.05-7.2	1
Store maize	DON	154/9	5.8	0.004-0.315	2
Mouldy maize	ZEA	20/7	35	15.6-95.5	1
Store maize	ZEA	154/18	11.7	0.7-9.9	2
Mouldy poultry feeds	ZEA	3/2	66.7	20-225	1

N/+: number of analyzed samples/number of positive samples: Refs: 1, Jiménez and Mateo (2001); 2, Muñoz *et al.* (1990).

3.4. Ochratoxin A (OTA)

This mycotoxin has received particular attention all over the world in the recent years due to its nephrotoxicity, teratogenicity, possible carcinogenicity (International Agency for Research on Cancer, 1993) and widespread occurrence in a variety of commodities. Studies for OTA occurrence have been performed in some food products from the Spanish market. The technique usually applied for its analysis is liquid chromatography with fluorescence detection with or without enhanced sensitivity by post-column addition of ammonia.

A screening survey was carried out in 93 bread samples taken from Spanish shops (Legarda & Burdaspal, 1989). With the exception of four samples all of them had been baked with wheat flour. Concentration values were given on dry matter basis. All samples were contaminated with OTA but levels were usually low. Only 4 samples surpassed 1.0 ng/g. The mean and the highest level were 0.45 and 7.371 ng/g, respectively. The last level exceeds the MAL for this toxin in the Spanish and EU regulation (3 ng/g). Bread samples from twelve other countries were included for comparison purposes and OTA was detected as well in all of them, indicating the widespread incidence of the toxin. The authors assessed that bread consumption by Spanish consumers accounts for about 1.8% of the Provisional Tolerable Weekly Intake (PTWI) set at 100 ng/kg body weight by the FAO/WHO Expert Committee on Food Additives (Joint Expert Committee of Food Additives, 1995).

Table 7 shows the results of a survey achieved on coffee marketed in Spain by Burdapal and Legarda (1998). OTA was found in all 29 samples of roasted coffee (both ground and grain) and 8 samples of instant coffee. Five samples were decaffeinated roasted coffee and three were decaffeinated instant coffee. The claimed limit of detection of the method was 0.11 ppb. The overall mean was 1.0 and the range was 0.19-5.64 ng/g. Means and ranges were 0.88 ng/g (0.22-5.64 ng/g) for roasted coffee, and 0.50 ng/g (0.19-1.08 ng/g) for instant coffee. Coffee consumption in Spain was supposed to contribute to 1.42 (roasted) and 0.17% (instant) of the PTWI (Joint Expert Committee of Food Additives, 1995).

.Table 7. Occurrence of OTA in roasted and instant coffee marketed in Spain[a]

	Samples		*OTA level (ng/g)*		
Type of coffee	*Analysed*	*Positives*	*Mean*	*Minimum*	*Maximum*
Roasted	29	29	0.88	0.22	5.64
Instant	8	8	0.50	0.19	1.08
Total	38	38	1.01	0.19	5.64
Decaffeinated [b]	8	8	0.55	0.22	1.29

[a] : Burdapal and Legarda (1998); [b]: Five roasted and three instant coffee samples.

Table 8 shows data about the occurrence of OTA in Spanish wines. OTA was often found in 46.4 % of white wines and 58.3% of red ones from Navarra (a region located in North Spain) (López *et al.*, 2002). Another screening study carried out on national and imported wines, marketed in Spain by Burdaspal and Legarda (1999), provided very high occurrence levels due to the lower detection limit of the LC-FLD method (0.003 ng/ml). The ratio of positive samples was 93.8% and 92.3% in dessert and red wines, respectively, but was also high in the remaining wine types. Dessert wine showed both the highest level and mean (2.54 ng/ml and 1.048 ng/ml, respectively). White wines showed the lowest OTA levels. Grape juices and grape musts were also contaminated with OTA. The ranges were 0.015 to 0.102 ng/ml in grape juices and undetectable (<0.003 ng/ml) to 0.176 ng/ml in grape musts. According to Burdaspal and Legarda (1999) wine consumption by a Spanish regular wine drinker of 60 kg weight may account for about 1% of the PTWI set by the Joint Expert Committee of Food Additives (1995).

OTA has been found in a great proportion of beer samples taken in Spanish shops and supermarkets. Legarda and Burdaspal (1998) carried out a survey on beers marketed in Spain including beer brewed in Spain and abroad. Their results are summarised in Table 9. OTA was detected in all except one of Spanish beer samples and in all imported samples.

Table 8. Occurrence of OTA in wines, musts and grape juices marketed in Spain

		Table Samples		OTA level (ng/ml)		
Sample	Ref.	N/+	% Positives	Mean	Minimum	Maximum
Type of wine						
Red [a]	1	28/13	46.4	0.189	<0.050	0.326
White [a]	1	12/7	58.3	0.108	<0.050	0.208
Red [b]	2	91/84	92.3	0.054	<0.003	0.603
White [b]	2	69/45	65.2	0.020	<0.003	0.257
Rosé [b]	2	32/29	90.6	0.031	<0.003	0.161
Aperitif [b]	2	47/35	74.5	0.040	<0.003	0.254
Sparkling [b]	2	12/10	83.3	0.012	<0.003	0.037
Dessert [b]	2	16/15	93.8	1.048	<0.003	2.540
Grape juice	2	10/10	100	0.044	0.015	0.102
Grape must	2	8/		0.046	<0.003	0.176

N/+: number of analyzed samples/number of positive samples; [a] : samples were from Navarra, a region located in North Spain.[b] : Most samples were marketed in Spain but part of them had been manufactured in other countries. Refs: 1, López *et al.* (2002); 2, Burdaspal and Legarda (1999).

The overall highest level was found in a Dutch sample (0.121 ng/ml). The highest level in Spanish samples was 0.075 ng/ml. In a second study by Medina *et al.* (2003) (Table 9) a high percentage of positive samples (above 80%) was also detected in Spanish and foreign (imported) beer, confirming the preceding data. This time, the overall highest level (0.2 ng/ml) was found in a German beer sample and the highest level found in Spanish beer was about 0.15 ng/ml.

4. REGULATIONS

As in other European Union (E.U.) countries and since Spain joined this organisation, the Spanish Government has assumed the E.U. directives and regulations with regard to food contaminants (including mycotoxins) as national regulations so that they are common to other E.U. countries. However, some specific national regulations have been published in the Spanish 'Boletín Oficial del Estado' and will be mentioned.

Table 10 shows the maximum tolerable levels for aflatoxins in some products. Royal Decree 475/1988 (Boletín Oficial del Estado, 1988) laid down the values of 5 ng/g for aflatoxin B1 and 10 ng/g for the sum of aflatoxins B1+B2+G1+G2 in food for human consumption, but these limits are not longer in force. At present, regulations have set different maximum permitted levels, as shown in Table 10, which vary depending on the products: for aflatoxin B1 the range is 2 to 8 ng/g. For the sum of aflatoxins B1, B2, G1 and G2, maximum permitted levels range from 4 to 15 ng/g. For aflatoxin M1 in milk, the MAL is 0.05 ng/ml.

Table 9. Occurrence of OTA in beer marketed in Spain

		Samples		OTA level (ng/ml)		
Ref.	*Country of origin*	*N/+*	*%*	*Mean*	*Minimum*	*Maximum*
1	Spain	38/37	97.4	0.024	>0.005	0.075
	Germany	6/6	100	0.015	0.008	0.047
	Denmark	2/2	100	0.010	0.008	0.013
	The Netherlands	12/12	100	0.032	0.009	0.121
	Belgium	5/5	100	0.018	0.011	0.028
	United Kingdom	10/10	100	0.028	0.007	0.096
	France	4/4	100	0.016	0.012	0.022
	Ireland	1/1	100	0.013	—	—
	Czech Rep.	1/1	100	0.046	—	—
	Mexico	1/1	100	0.010	—	—
2	Spain	31/26	83.9	0.036	nd	0.147
	Germany	15/10	66.7	0.045	nd	0.204
	Denmark	8/6	75	0.037	nd	0.096
	The Netherlands	12/12	100	0.025	0.011	0.132
	Belgium	13/11	84.6	0.059	nd	0.120
	Scotland	1/1	100	0.201	—	—
	Undefined (E.U.)	1/1	100	0.055	—	—
	U.S.A.	4/3	75	0.057	nd	0.115
	Mexico	2/2	100	0.067	0.010	0.032

N/+: Analysed samples/positive samples; — : not applicable; nd: not detected. Ref.: 1, Burdaspal and Legarda (1999); 2, Medina *et al.* (2003).

The updated E.U. Regulations affecting nuts and cereals appeared in 2001 in Commission Regulation (C.R.) 466/2001, and in 2002 in C.R. 257/2002 (Commission Regulation 2001, 2002a). Maximum levels for aflatoxins in spices (*Capsicum* spp, *Piper* spp., *Myristica fragrans*, *Zingiber officinale* and *Curcuma longa*) were established by C.R. 472/2002 (Commission Regulation, 2002b) and they are shown in Table 10.

Sampling protocols and characteristics of analytical methods for official control in groundnuts, pistachio nuts, Brazil nuts, dry fruits, nuts, cereals, milk and derivatives were regulated by Royal Decree 90/2001 (Boletín Oficial del Estado, 2001) based on Directive 1998/53/EC (Commission Directive, 1998). Order SCO/338/2003 published in the Boletín Oficial del Estado (2003a) provides a modification of annexes 1 and 2 of the Royal Decree 90/2001 concerning rules for sampling and analytical protocols for spices. With regard to OTA, Spain has adopted the sampling procedures for cereals, cereal-based food and dried vine fruits, rules for reference analytical methods and MAL established by the E.U. authorities in the Commission Directive 2002/26/EC (Commission Directive, 2002) and C.R. 472/2002 (Commission Regulation, 2002b) (Table 11). The pertinent Spanish regulation is Royal Decree 294/2003, published in Boletín Oficial del Estado (2003b). Commission Directive 2003/78/EC (Commission Directive, 2003) has regulated sampling and analytical methods and C.R. 1425/2003/EC (Commission Regulation, 2003) has set up MALs for patulin in some products, mainly apple and its derivatives (Table 12). Limits range from 10 to 50 ng/g, depending on the products. At this time no specific Spanish regulation has appeared concerning patulin, but the E.U. regulation is applicable.

Table 10. Maximum allowable levels (MAL) of aflatoxins in groundnuts, pistachio nuts, Brazil nuts, other shell and dried fruits, cereals, milk and their derivatives, and spices laid down by regulations.

Commodity	*Aflatoxin*	*MAL (ng/g)*	*Regulations*
Nuts ready for human consumption	B_1 $B_1+B_2+G_1+G_2$	2 4	C.R. 466/2001 [a]; C.R. 257/2002 [b]
Groundnuts before treatment	B_1 $B_1+B_2+G_1+G_2$	8 15	
Other nuts/dry fruits before treatment	B_1 $B_1+B_2+G_1+G_2$	5 10	
Cereals and cereal-based products destined to or ready for direct human consumption	B_1 $B_1+B_2+G_1+G_2$	2 4	
Milk	M_1	0.05 [c]	
Spices	B_1 $B_1+B_2+G_1+G_2$	5 10	Order SCO/338/2003 [d] C.R. 472/2002

[a] : Commission Regulation (2001); [b] : Commission Regulation (2002a); [c] : ng/ml. [d]: Boletín Oficial del Estado (2003a).

Table 11. Maximum allowable limits for OTA in some commodities.

Commodity	*MAL (ng/g)*	*Regulations*
Raw cereal grains	5	
All products derived from cereals	3	Royal Decree 294/2003 [a]; C.R. 472/2002 [b]
Dried vine fruits (currants, raisins and sultanas)	10	

[a]: Boletín Oficial del Estado (2003b); [b] : Commission Regulation (2002b).

Table 12. E.U. regulations for patulin

Food Products	*MAL (ng/g)*
Fruit juices and fruit nectar, in particular apple juice, and fruit juice ingredients in other beverages/concentrated fruit juice after reconstitution	50
Spirit drinks, cider and other fermented drinks derived from apple or containing apple juice	50
Solid apple products, including apple compote, apple pure for direct consumption	25
Apple juice and solid apple products for infants and young children/other baby food	10

6. ACKNOWLEDGEMENTS

The authors wish to thank to Dr. A. Logrieco (CNR, Bari, Italy) for the organization of the Workshop 'Mycotoxins in Food Safety: an Overview on Toxigenic Fungi and Mycotoxins in Europe' held in Martina Franca (Italy) and kind attention during the event, and also to the European Commission (COST Action 835) for providing support to attend the Workshop (R. Mateo). Project AGL 2001-2974-C05-01 from the Spanish 'Ministerio de Ciencia y Tecnología' is also highly appreciated.

7. AFFILIATIONS

Rufino Mateo and José V. Gimeno-Adelantado: Departamento de Química Analítica, Facultad de Química, Universitat de Valencia. Dr. Moliner 50, E-46100 Burjassot, Valencia, Spain

Angel Medina and Misericordia Jiménez: Departamento de Microbiología y Ecología, Facultad de Biología, Universitat de Valencia. Dr. Moliner 50, E-46100 Burjassot, Valencia, Spain

8. REFERENCES

Abarca, M. L., Accensi, F., Bragulat, M. R., Castellá. G., & Cabañes, F.J. (2003). *Aspergillus carbonarius* as the main source of ochratoxin A contamination in dried vine fruits from the Spanish market. Journal of Food Protection, 66, 504-506.

Boletín Oficial del Estado. (1988). Real Decreto (RD) 475/1988 de 13 de mayo por el que se establecen los límites máximos permitidos de las aflatoxinas B_1, B_2, G_1 y G_2 en alimentos de consumo humano, BOE 121, May 20, 1988, p. 15329. Madrid: Boletín Oficial del Estado.

Boletín Oficial del Estado. (2001). Real Decreto (RD) 90/2001 de 2 de febrero por el que se establecen los métodos de toma de muestras y de análisis para el control oficial del contenido máximo de aflatoxinas en cacahuetes, frutos de cáscara, frutos desecados, cereales, leche y los productos derivados de su transformación. BOE 47, February 23, 2001 (pp. 6996-7000). Madrid: Boletín Oficial del Estado.

Boletín Oficial del Estado. (2003a). Orden SCO/388/2003 de 25 de febrero por la que se modifican los anexos I y II del Real Decreto 90/2001, de 2 de febrero. BOE 50, February 27, 2003 (pp. 7839-7840). Madrid: Boletín Oficial del Estado.

Boletín Oficial del Estado. (2003b). Real Decreto (RD) 294/2003 de 7 de marzo por el que se establecen los métodos de toma de muestras y de análisis para el control oficial del contenido de ocratoxina A en cereales y uvas pasas. BOE 60, March 11, 2003 (pp. 9482-9485). Madrid: Boletín Oficial del Estado.

Burdaspal, P. A., & Legarda, T. M. (1988). Ocratoxina A en muestras de café comercializado en España. Alimentaria, 296 (October), 31-35.

Burdaspal, P. A., & Legarda, T. M. (1999). Ocratoxina A en vinos, mostos y zumos de uva elaborados en España y en otros países europeos. Alimentaria, 299 (January-February), 107-113.

Castellá, G., Bragulat, M. R., & Cabañes, F.J. (1999). Surveillance of fumonisins in maize-based feeds and cereals from Spain. Journal of Agricultural Food Chemistry, 47, 4707-4710.

Commission Directive (1998). No 1998/53/EC of 16 July 1998 laying down the sampling methods and the methods of analysis for the official control of the levels for certain contaminants in foodstuffs. Official Journal of the European Communities, No L 201, 17/07/1998 (pp. 93–101).

Commission Directive (2002). No 26/2002/EC of 13 March 2002 laying down the sampling methods and the methods of analysis for the official control of the levels of ochratoxin A in foodstuffs. Official Journal of the European Communities, No L 75, 16/03/2002 (pp. 38 – 43).

Commission Directive (2003). No 78/2003/EC of 11 August 2003 laying down the sampling methods and the methods of analysis for the official control of the levels of patulin in foodstuffs. Official Journal of the European Union, No L 203, 12/08/2003 (pp. 40–44).

Commission Regulation (2002a). No 257/2002 of 12 February 2002 amending Regulation (EC) No 194/97 setting maximum levels for certain contaminants in foodstuffs and Regulation (EC) No 466/2001 setting maximum levels for certain contaminants in foodstuffs. Official Journal of the European Communities, No L 41, 13/02/2002 (pp. 12–15).

Commission Regulation (2002b). No. 472/2002 of 12 March 2002 amending Regulation (EC) No 466/2001 setting maximum levels for certain contaminants in foodstuffs. Official Journal of the European Communities, No L 75, 16/03/2002 (pp. 18–20).

Commission Regulation (2003) No 1425/2003 of 11 August 2003 amending Regulation (EC) No 466/2001 as regards patulin. Official Journal of the European Union, No L 203, 12/08/2003 (pp. 1–3).

Departamento de Sanidad del Gobierno Vasco (1995). Vigilancia de la contaminación química de alimentos en la Comunidad Autónoma del País Vasco: 1990-1995: Micotoxinas: aflatoxinas. Retrieved July 20, 2003 from http://www.euskadi.net/sanidad/publicaciones/datos/vigila9511.pdf.

Food and Agricultural Organisation. (1997). World-wide regulations for mycotoxins 1995. A compendium. Food and Nutrition Paper. Vol. 64, Rome: Food and Agricultural Organisation (FAO).

International Agency for Research on Cancer. (1993). Monographs on the evaluation of carcinogenic risks to humans, some naturally occurring substances: food items and constituents, heterocyclic aromatic amines and mycotoxins. Vol. 56 (pp. 489-521). Lyon: International Agency for Research on Cancer (IARC).

Jiménez, M., Logrieco, A., & Bottalico, A. (1993). Occurrence and pathogenicity of *Fusarium* species from banana fruits. Journal of Phytopathology, 137, 214-220.

Jiménez, M., Huerta, T., & Mateo, R. (1997). Mycotoxin production by *Fusarium* isolated from bananas. Applied and Environmental Microbiology, 63, 364-369.

Jiménez, M., Mateo, R., Querol, A., Huerta, T., & Hernández, E. (1991). Mycotoxins and mycotoxigenic moulds in nuts and sunflower seeds for human consumption. Mycopathologia, 115, 121-127.

Jiménez, M., & Mateo, R. (2001). Occurrence of toxigenic fungi and mycotoxins in agricultural commodities in Spain. In A. Logrieco (Ed.), Occurrence of toxigenic fungi and mycotoxins in plants, food and feed in Europe (pp.173-190). Brussels: European Commission, Directorate-General for Research, COST Action 835.

Jiménez, M., Santamarina, M. P., Sanchis, V., & Hernández, E. (1987). Investigation of mycoflora and mycotoxins in stored cereals. Microbiologie-Aliments-Nutrition 5,105-109.

Joint Expert Committee of Food Additives. (1995). Joint Expert Committee of Food Additives (JECFA) 44th Meeting, Technical Report Series, 859, Geneve: Food and Agriculture Organization/World Health Organization.

Legarda, T. M., & Burdaspal, P. A. (1998) Ocratoxina A en cervezas elaboradas en España y en otros países europeos. Alimentaria, 291 (April), 115-122.

Legarda, T. M., & Burdaspal, P. A. (1999). Presencia de ocratoxina A en muestras de pan comercializado en España y en muestras procedentes de doce países extranjeros. Alimentaria, 301 (April), 89-96.

López de Cerain, A., González-Peñas, E., Jiménez, A. M., & Bello, J. (2002). Contribution to the study of ochratoxin A in Spanish wines. Food Additives and Contaminants, 19, 1058-1064.

Medina, A., Jiménez, M., Sáez, J. M., Hinojo, M. J., & Mateo, R. (2002). Evaluation of the occurrence of toxigenic fungi in barley and malt in Spain. Abstracts from: The Seventh International Mycological Congress (IMC7) (2002, August 11-17) (p. 257). Oslo: The Seventh International Mycological Congress.

Medina, A., Sáez, J. M., Vargas-Pérez, R., Gimeno-Adelantado, J. V., Mateo, R., & Jiménez, M. (2003) Assessment of ochratoxin A in beers from the Spanish market [CD ROM]. Abstracts from: Cuarto Congreso Latinoamericano de Micotoxicología (September 23-25). La Habana: Centro Nacional de Sanidad Agropecuaria (CENSA).

Muñoz, L., Cardelle, M., Pereiro, M., & Riguera, R. (1990). Occurrence of corn mycotoxins in Galicia (Northwest Spain). Journal of Agricultural and Food Chemistry, 38, 1004-1006.

Sáez, J. M., Medina, A., González, G., Llorens, A., Mateo, R., & Jiménez, M. (2003). Evaluation of the mycobiota involved in the presence of ochratoxin A in wines [CD ROM]. Abstracts from: Cuarto Congreso Latinoamericano de Micotoxicología (September 23-25). La Habana: Centro Nacional de Sanidad Agropecuaria (CENSA).

Sanchis, V., Viñas, I., Jiménez, M., Calvo, M. A., & Hernández, E. (1982). Mycotoxin-producing fungi isolated from bin-stored corn. Mycopathologia, 80, 89-93.

Sanchis, V., Viñas, I., Jiménez, M., & Hernández, E. (1984). Aflatoxin and A. flavus in preharvest corn in Valencia (Spain). Microbiologie-Aliments-Nutrition, 2, 193-197.

Sanchis, V., Marín, S., & Ramos, A. J. (2001). Occurrence of toxigenic fungi and related mycotoxins in Spain, In A. Logrieco, (Ed.). Occurrence of toxigenic fungi and mycotoxins in plants, food and feed in Europe. (pp.191-199). Brussels: European Commission, Directorate-General for Research, COST Action 835.

Torres, M. R, Sanchis, V., & Ramos, A. J. (1998). Occurrence of fumonisins in Spanish beers analyzed by an enzyme-linked immunosorbent assay method. International Journal of Food Microbiology, 39, 139-143.

N. GURSOY & M. BICICI

A REVIEW ON CURRENT SITUATION OF TOXIGENIC FUNGI AND MYCOTOXINS FORMATION IN TURKEY

1. INTRODUCTION

Mycotoxins have received worldwide attention because of the significant losses associated with their influence on animal and human health, agriculture and consequent national and international economic implications. Mycotoxins are toxic secondary metabolites of fungal origin which have respectively small molecules (molecular weight under 700 Da) with a highly diverse biologic activity and chemical structure. Mycotoxins tend to be genotypically specific for groups of species belonging to the same genus. Moreover, the same toxin compound can be produced by fungi from different genera. Mycotoxins are mainly produced by species of genera *Aspergillus*, *Fusarium* and *Penicillium* (Moss, 1991; Abarca, Bragulat, Castellá, Accensi & Cabañes, 2000; Brera, Caputi, Miraglia, Iavicoli, Salerno & Carelli, 2002; Council for Agricultural Science and Technology [CAST], 2003).

Fungi, which produce toxins in agricultural crops and foodstuffs, are classified within field fungi and storage fungi depending on their ecological requirements for growth (Bankole, 1994). Field fungi require grain moisture, which is above 20% in cereals and generally cause ear rot diseases and toxin production occurs when the crop are still in the field. The important field fungi genera are *Fusarium*, *Alternaria* and *Cladosporium*. The storage fungi generally grow in grains that are preserved in a moisture content equilibrium of 70-90% relative humidity and the major genera are *Aspergillus* and *Penicillium*. They are associated with crops in the field, but also with plant surfaces, plant debris and atmosphere and the other substrates where the water activity is relatively low (Edmond, 2002; Bankole & Adebanjo, 2003).

The Aflatoxins and OTA are among the mycotoxins most important for agriculture and they are naturally occurring mycotoxins produced by fungi during production and storage (Bankole & Adebanjo, 2003).

Aflatoxins are often found in tropical and subtropical climates where typical environmental conditions exist for the growth of mycotoxigenic fungi and subsequent production of toxins at high levels. The formation of aflatoxin contamination is sporadic and depends on climatic conditions. Aflatoxigenic fungi exist in soil as hyphae, conidia and sclerotia, which act as primary inoculum for infecting crops. Aflatoxins occur particularly in cereals, groundnuts, pistachios, and dried figs (Strange, 1991; CAST, 2003). Aflatoxins are undoubtedly the most important mycotoxins and are produced mainly by three species of section Flavi: *A.*

A. Logrieco and A. Visconti (eds.),
An Overview on Toxigenic Fungi and Mycotoxins in Europe, 237-246.

flavus, *A. parasiticus* and *A. nomius* (Deacon, 1997; Shenasi, Candlish & Aidoo, 2002). In addition, Goto, Wicklow &Ito (1996), described for the first time production of Aflatoxin B1 and B2 and cyclopiazonic acid in a strain of *A. tamarii* and later, they detected four strains capable of synthesizing aflatoxin B1 and B2. There are four major aflatoxins referred to as B1, B2, G1, and G2. Aflatoxin M1 and M2 are hydroxylated metabolites of aflatoxin B1 and B2, found in milk dairy cows exposed to aflatoxin contaminated feed. Aflatoxins are known to be hepatotoxic, carcinogenic, teratogenic and mutagenic. Naturally occurring mixtures of aflatoxins have been classified as class 1 human carcinogens (Abarca *et al.*, 2000; International Agency for Research on Cancer [IARC], 1993).

Ochratoxins are receiving special attention throughout the world due to their nephrotoxic nature. The OTA is more toxic and more frequently detected than Ochratoxin B and Ochratoxin C (Biro, Solti, Barna-Vetro, Bago, Glavits, Szabo & Fink-Gremmels, 2002). OTA is produced by different species of *Aspergillus* and *Penicillium*, although it was originally isolated from *A. ochraceus* (Van der Merwe, Steyn, Fourie, Scoot & Thero, 1965). Production of OTA has been associated within the genus *Aspergillus* to seven species of the section *Circumdati* and other sections *Wentii*, *Usti* and *Terrei* (Varga, Kevei, Rinyu, Teren & Kozakiewicz, 1996). Some species of *Penicillium* have been implicated in the production of OTA such as *P. viridicatum*, *P. brevicompactum*, *P. chrysogenum*, *P. aurantiogriseum*, *P. verrucosum* and *P. expansum* that they belong to the genus *Eupenicillium* (Pitt, 1987; Kozakiewicz, Paterson & Bridge, 1993; Leistner & Pitt, 1977; Krivobok Seigle-Murandi, Steiman & Marzin, 1987; Mills, Abramson, Frohlich & Marquardt, 1989). OTA is a natural contaminant in many crops and foodstuffs such as cereals and cereal products, dried fruits (raisin and figs), wine, coffee and cocoa (Romani, Sacchetti, Lopez, Pinnavaia & Rosa, 2000; Joosten Goetz, Pittet, Schellenberg & Bucheli, 2001; Pietri, Bertuzzi, Pallaroni & Piva 2001). According to Sedmikova, Reisnerora, Dufkova, Burta, and Jilek (2001), OTA can cause an increase of the mutagenic ability of aflatoxin B1 when occurring simultaneously in the same crop. OTA is immunosuppressive, teratogenic, genotoxic and mutagenic and possibly carcinogenic to animal and human (IARC, 1993; Abarca *et al.*, 2000; Bankole & Adebanjo, 2003).

For food quality and safety, the identification of mycoflora is essential. Detection to fungal species level provides important data for the production of their mycotoxins. The present study is aimed to review mycoflora and naturally occurrence of aflatoxins and ochratoxin A contamination on some important crops and to give an information on the general mycotoxigenic situation in Turkey.

2. TOXIGENIC FUNGI AND MYCOTOXINS IN TURKEY

Turkey is one of the most important food producing countries of the World and toxigenic fungi growth and sequent mycotoxins formation is a common problem on susceptible food commodities as well as other countries that have favorable conditions. Mycotoxins are found in a wide range of food commodities around the World especially in countries with climates of high temperature and humidity. In addition, poor harvesting, processing and storage conditions encourage fungal growth and mycotoxin formation. The process of fungal growth and mycotoxin formation on agricultural crops and foodstuffs may change depending on contaminated fungi species, variety of crops, ecologic conditions, harvesting, drying, processing, transportation, and storage conditions of crops. Today many countries are conducting studies to determine mycologic problems and provide solutions regarding their own mycoflora and related mycotoxins formation.

Some regions of Turkey have subtropical climate with high ambient temperature and relative humidity, providing optimal conditions for the growth of toxigenic fungi and consequently mycotoxin formation. Mycoflora and mycotoxin formation on agricultural crops have currently been intensively investigated in Turkey (Alperden, Aran, Topal, Eke, Kara & Karaali, 1990; Topal, 2003; Topal, Aran & Pembeci, 1999).

According to recent published records, totally 251 species belonging to *Aspergillus* and *Penicillium* species are reported to occur in different regions of Turkey, among which *A. niger* is the most common species. This is followed by *A. flavus, A. fumigatus, A. versicolor, A. ochraceus, A. terreus, A. wentii, P. chrysogenum* and *P. freguentas* (Asan, 2000). In another study by Topal (2003), the distribution and incidence of mycoflora and mycotoxin formation were determined on 73 different kinds of crops and foodstuffs, covering all agricultural regions of Turkey. The most prevalent fungi species isolated of some important Turkish food commodities belonged to the genera *Penicillium* 65%, *Aspergillus* 19%, *Fusarium* 2%, *Deuteromycetes* 11%, *Zygomycetes* 2% and *Basidiomycetes* 1%. Mycotoxins determined by TLC methods were mainly roquefortine C, sterigmatocystine, cyclopiazonic acid, penitrem A, mycophenolic acid, penicillic acid, patulin, HT-2 toxin, T-2 toxin, aflatoxin (B_1, B_2), citrinin and fumigillin.

Exposure to aflatoxins and OTA is mainly on maize, pistachios, hazelnuts, groundnuts, raisin and dried figs in Turkey (Bicici, 1980; Alperden *et al.*, 1990; Ozay, Aran & Pala, 1995; Ozilgen & Ozdemir, 2001; Gursoy & Bicici, 2003; Aksoy, Sabır, Eltem, Kıraç, Sarıgul, Meyvacı, Ateş & Cakır, 2003; Eltem, Aksoy, Altındisli, Sarıgul, Taşkın, Askun *et al*, 2003).

2.1.1. Wheat

Wheat (*Triticum aestium* L.) represents the biggest part of cereal crops production in Turkey. Turkey has a significant place among the World's biggest cereal producing countries with an average annual production of 20.000.000 tons in the World (State Instıtute of Statistics Prime Ministry Republic [DIE], 2001; Food and Agriculture Organization [FAO], 2002).

Sometimes, environmental conditions such as relative humidity, warm temperatures and rainfall, especially during the ear growth period of wheat, cause infection by *Fusarium*, *Aspergillus*, *Penicillium* and *Alternaria* species in Turkey. Additionally, *Fusarium* species can induce head blight disease in the flowering period of wheat under such environmental conditions.

A recent study determined the most common fungi flora on wheat kernels used as animal feed, in which were *Alternaria*, *Fusarium*, *Aspergillus*, *Rhisophus* and *Mucor* spp. (Karabulut & Değirmencioğlu, 2002). Cukurova is one of the important wheat producers regions in Turkey. In a recent study, fungal infections and aflatoxins were determined by CD-ELISA and HPLC method and aflatoxins level was between 0.18-3.5 ppb in 7 samples of 43 freshly harvested wheat kernels in this region. These values are found below the legal practice limit in Turkey. Fungi species are isolated as 75.8% *Fusarium*, 15.6% *Aspergillus* and 8.5% *Penicillium* species by using special medium, species keys and MIS (Microbial Identification System). The most frequent mycoflora are defined as *F. graminearum*, *F. oxysporum*, *F. moniliforme*, *F. solani*, *F. culmorum*, *A. flavus*, *A. niger*, *A. terreus*, *A. nidulans* and *P. chrysogenum* (Gursoy & Bicici, 2003).

2.1.2. Maize

Maize (*Zea mays* L.) is one of the three biggest produced cereals of Turkey. Maize is an excellent substrate for fungi growth and mycotoxin contamination and considered to be risky crop for aflatoxins and ochratoxins formation (DIE, 2001; Alperden *et al.*, 1990). The production of corn poses special problems such as many strains of mycotoxigenic *Fusarium*, *Aspergillus* and *Penicillium*, which grow at higher rates in tropical and subtropical climates. The fungi invade plants in the field with the help of insects and wind. Inoculation has been shown to take place through silk infection and insect damage, followed by the glumes and kernel surface (Marsh & Payne, 1984).

Maize is commonly cultivated in the Cukurova region and 15.8% of the total maize production of Turkey originates from this region (DIE, 2001). According to Gursoy & Bicici (2003), the most common fungi, such as 61.5% *Fusarium*, 19.3% *Aspergillus* and 19.1% *Penicillium,* were identified on 73 freshly harvested maize kernels tested. Aflatoxins were detected among 0.7-50 ppb in 17 samples of examined maize kernel samples by using CD-ELISA and HPLC method in the Cukurova region. In another study, a total of 167 maize samples collected from various regions were surveyed and fungi contamination level was identified as *A. flavus*, *A. niger*, *F. oxysporum*, *P. variable* and *Rhisopus spp.* on all samples. Aflatoxins levels were found among 3-70 ppb in 46% of the examined samples. OTA was detected in trace amount (Alperden *et al.,* 1990).

2.1.3. Hazelnuts

Hazelnuts (*Corylus avellana* L.) are one of the important exportation crops of Turkey. Turkey ranks the first producer country in the World. The shelled hazelnut production between 1997-2001 is about 600.000 tons, which amounts to nearly 80% production of the World (DIE, 2001; Union of Agricultural Cooperatives [Fiskobirlik], 2002). *C. colurna* var. *glandulifera* (Turkish hazelnut), *C. avellana* var. *pontica* (common hazelnut) and *C. maxima* (the giant hazelnut) are economically important Turkish hazelnut varieties (Ozdemir, 1997; Fiskobirlik, 2002).

Various fungi species affect fruit growth, drying, harvest and storage period in hazelnuts. In some studies, *T. roseum, A. flavus, A. parasiticus, A. niger, A. tamarii, P. commune, P. brevicompactum, P. aurantiogriseum, P. janczewiskii, Eurotium*, and *Cladosporium* species are reported as the most prevalent mycoflora on hazelnut particularly during the drying period in the Black Sea region of Turkey. *Rhizopus, Acremonium, Paecilomyces, Mucor, Alternaria, Verticillium* and *Pestalotiopsis* are less common fungi species on hazelnuts in Turkey (Alperden *et al*., 1990; Ozilgen & Ozdemir, 2001; Heperkan, 2003).

Hazelnuts are traditionally sun-dried and this may cause fungal growth and subsequent mycotoxin formation due to a prolonged drying period under humid and rainy weather conditions in Turkey (Ozilgen & Ozdemir, 2001). *A. flavus* and *A. parasiticus* development on the surface of hazelnut shells starts on the shell surface while the crops are still on tree. Then fungal contamination increases as a result of the contact of crops with soil in the drying period. Storage conditions and especially storing in heaps encourage contamination impairing the quality of crops.

In a study, 130 hazelnut samples collected from trees, ground, harvest and storage were examined for fungi contamination and aflatoxin formation. Aflatoxin B1 was found on harvested and stored samples with amount of 1.8-2.4 ppb. According to results of this study, the highest aflatoxin concentration was detected on samples which were stored for 6 months (Alperden *et al*., 1990).

2.1.4. Pistachios

Pistachios are one of the most important products of Turkey and commonly grow in the South Eastern Anatolia region. Turkey ranks the fourth with 40.000 tons pistachio production in the World. The most common problem of pistachios exportation is aflatoxin contamination.

In a recent study, aflatoxin B1 contamination is detected by ELISA and HPLC method in the South Eastern Anatolia region. 38 pistachio kernel samples were examined and aflatoxin B1 was detected at levels between 2.1-5.8 ppb in 16 kernel samples. According to this study the most prevalent fungi species identified were belonging to *Aspergillus, Penicillium, Fusarium, Paecilomyces, Stemphylium, Cladosporium* and *Thielaviopssis* species (Calıskan, 2003). In another study, *A. flavus* contamination and mycoflora were examined in 143 pistachios samples from the South Eastern Anatolia region. According to results, 38% *A. flavus* development was determined and 33% A. flavus isolates showed ability to produce aflatoxins in storage condition. Dominant mycoflora was reported among *Aspergillus*, *Penicillium, Cladosporium* and *Ulacladium* species (Alperden *et al.*, 1990).

2.1.5. Figs

Most of the figs (*Ficus carica* L.) producing countries are situated in areas with Mediterranean ecological conditions because of the climatic requirements of fig cultivation. Turkey is one of the largest fig producing countries of the World (DIE, 2001; FAO, 2002). Economically significant Turkish dried fig cultivars are Sarılop, Bursa Siyahı (Black Bursa) and Yediveren.

Aflatoxin contamination has been a problem in dried figs since the 1980's. Aflatoxin formation in figs may start in the orchard and sequent sun-drying provides favorable conditions for aflatoxin production in the field (Alperden *et al.,* 1990; Aksoy *et al*, 2003).

Another mycotoxigenic problem is the formation of Ocratoxin A (OTA) on dried figs in Turkey. OTA is commonly produced in available substrates by *Aspergillus* and *Penicillium* species. *Aspergillus* species which are isolated from dried figs of Aegean region are *A. foetidus* var. *pallidus*, *A. niger*, *A. amawori*, *A. aculeatus*, *A. carbonarius*, *A. ficuum*, *A. ochraceus*, and *A. tubingensis* (Alperden *et al.,* 1990; Aksoy *et al.*, 2003). The Commission of European Communities [EC], (2002b), in accordance with regulation put into effect on 5, April 2002, defined a maximal OTA level of 10 ppb for dried grape. Moisture content and water activity (Aw), the important factors in mycotoxin formation, was investigated in 103 fig samples. OTA level and incidence of OTA were reported as 5.20-8.30 μg/kg and 3% in samples (Ozay & Alperden, 1991).

2.1.6. Grapes

Grapes (*Vitis vinifera* L.) are important exportation crops of Turkey, which rank first with 360.000 tons production of raisin (Sultani) in the World (DIE, 2001; FAO, 2002). The Aegean region is in the first place regarding 28% of the processing and covering 41% of the grape production in Turkey. Sultani and Round Seedless grapes cover 50-55% vineyard area of the Aegean region and consist of 90% Round Seedless and 10% Sultani Turkish grape varieties (Eltem *et al.*, 2003).

Turkey was faced with problems of OTA contamination in raisin for the first time in 1996-97 to. For this reason fungi development was investigated in fresh and raisin samples were taken from selected vineyards in 1998-2001. OTA was detected at 0.24-1.5 ppb toxin level in 15 samples (28.8%) and toxin was absent in 37 samples (71.2%) when 52 freshly harvested grapes were investigated totally. Out of 61 raisin samples OTA was detected at 0.14-25.5 ppb toxin level in 20 samples (32.8%) and in 41 samples (67.2%) toxin was absent . Nearly 1500 fungal strains were identified and determined among 105 species and 3 varieties fungi imperfect. The most frequently isolated species produce black spores, which have the same frequency in veraison, harvest and after drying periods as *A. foetidus* var. *pallidus*, *A. aculeatus*, *A. tubingensis* and *A. awamori*. Although there is no information about toxin production of *A. foetidus* var. *pallidus* in the literature, high toxin production has been determined in *A. foetidus var. pallidus* strains, most commonly isolated from grapes in the Aegean region (Eltem *et al.,* 2003).

Table 1. The maximum microbial toxin level of significant contaminant of Turkish Food Codex.

Food stuffs	*Maximum level (ppb)*				
	Aflatoxin			*Ochratoxin A*	*Patulin*
	B_1	$B_1+B_2+G_1+G_2$	M_1		
Hazelnuts, groundnuts, oil seeds, figs, grapes, dried-fresh fruits and their products.	5	10	-	-	-
Cereals and cereals products .	2	4	-	-	-
Milk	-	-	0.05	-	-
Milk powder	-	-	0.5	-	-
Cheese	-	-	0.25	-	-
Baby's food and milk products	-	-	0.05	-	-
Baby's food and foodstuffs	1	2	-	-	-
Spices	5	10	-	-	-
Cereal grains	-	-	-	5	
All cereal borne products	-	-	-	3	-
Dried grape (Raisin)	-	-	-	10	-
Apple juice and it's products	-	-	-	-	50

3. NATIONAL GUIDELINES IN TURKEY

The problems due to mycotoxins contamination of food commodities are a worldwide issue. Mycotoxins can occur in all steps of the food chain and are difficult to remove once entered into the food chain. The control of mycotoxins is a complex process and an integrated management is needed, since toxigenic fungi can infect crops before harvest or contaminate during the post-harvest periods. Governments have implemented different types of control procedures including the setting of limits to protect public health from mycotoxins. Current inability in accomplishing pre-harvest control of mycotoxins has made post-harvest control mandatory.

Some measures are used for reduction or prevention of mycotoxins, such as Hazard Analysis and Critical Point (HACCP) and Hurdle Technology System. Application of HACCP, in accordance with the law regulation put into effect on 10 June 1995and based on "Food production, consumption and inspection", became obligatory for all products in Turkey as well as in USA, Canada, Australia and Japan. Legal practices, needed for foodstuffs and agricultural crops, are organized by the Ministry of Agriculture Rural Affairs, Ministry of Health, the Turkish Standarts Institution and their substructures.

Turkish food codex standards were determined and applied only for aflatoxin, OTA and patulin according to this legal practices. The allowed maximum microbial toxin levels of significant contaminants of Turkish Food Codex are presented in Table 1.

4. ACKNOWLEDGEMENTS

The authors are grateful to the COST ACTION 835 which was held in 24-25 October 2003, Martina Franca-Italy. Their thanks also go to Dr. Antonio Logrieco for his invitation and excellent organization.

5. REFERENCES

Abarca, M.L., Bragulat, M.R., Castellá, G., Accensi, F. & Cabañes, F.J. (2000). Emerging mycotoxin-producing fungi. Rev. Iberoam. Micol. Vol. 17, pp. 63-68.

Aksoy, U., Sabır, E., Eltem, R., Kıraç, S., Sarıgul, N., Betül, K.B., Ates, M. & Cakır, M. (2003). Researches on potential ochratoxin A contamination in dried figs. I. National Mycotoxin Symposium, Istanbul-Turkey, pp.41-46.

Alperden, I., Aran, N., Topal, S., Eke, D., Kara, M. & Karaali, A. (1990). Systematic Analysis of Mycoflora Turkish Foodstuffs. NATO-TU Mycotoxins Project. Joint-Final Report of Mycology and System Analysis Subprojects. Gebze-Kocaeli, Turkey: TUBITAK MRI., Printing Office.

Asan, A. (2000). Check list of *Aspergillus* and *Penicillium* species reported from Turkey. Turk Journal Bot. TUBITAK, Vol 24, pp.151-167.

Bankole, S.A. & Adebanjo, A. (2003). Mycotoxins in food in West Africa: current situation and possibilities of controlling it. African Journal of Biotechnology. Vol. 2(9), pp. 254-263.

Bankole, S.A. (1994). Changes in moisture content, fungal infection and kernel germinability of maize in storage. Int. J. Trop. Plant Disease. Vol. 12, pp. 213-218.

Bicici, M. (1980). Investigations on diseases and aflatoxin caused by *Aspergillus flavus* Link and diseases by *Aspergillus niger* Van Tieghem in the field harvesting and drying periods of groundnuts (Doctoral dissertation, Cukurova University, 1980).

Biro, K., Solti, L., Barna-Vetro, I., Bago, G., Glavits, R., Szabo, E. *et al.* (2002). Tissue distribution of ochratoxin A as determined by HPLC and ELISA and histopathological effects in chickens. Avian Pathology. Vol 31, pp.141-148.

Brera, C., Caputi, R., Miraglia, M., Iavicoli, A., Salerno, A. & Carelli, G. (2002). Exposure assessment to mycotoxins in workplaces: aflatoxins and ochratoxin A occurence in airborne dusts and human sera. Vol. 73(1), pp. 167-173.

Calıskan, I. (2003). Application of the ELISA and HPLC tests for detection of aflatoxin in pistachio (Master dissertation, Gaziantep University, 2003).

CAST. (2003). Mycotoxins: Risks in Plant, Animal, and Human Systems. Council for Agricultural Science and Technology, Ames, Iowa, USA. Printed in the United States of America. Task Force Report No. 139.

Deacon, J.W. (1997). Modern Mycology (3rd edition), Blackwell Science, Oxfort.

DIE. (2001). Agricultural Structure (Production, Price, Value). State Instıtute of Statistics Prime Ministry Republic of Turkey, Ankara.

EC. 2002b. Commission Regulation (EC) Amending Regulation (EC), No. 466/201 Setting maximum levels for certain contaminants in foodstuffs, Commision of European Communities, Brussels, SANCO/146/2002-rev 1.

Edmond, E. C. (2002). Update of surveys, regulation and toxic effects of mycotoxins in Europe, Toxicology Lertters. 127, pp. 19-28.

Eltem, R., Aksoy, U., Altındisli, A., Sarıgul, N., Taşkın, E., Askun, T. Ates, Meyvacı, B., Arasıler, Z., Turgut, H & Kartal, N. (2003). Determination of ochratoxin A in Sultanas of the Aegean region. I. National Mycotoxin Symposium, Istanbul-Turkey, pp. 54-59.

FAO. (2002). Nutrition Country Profiles-Turkey. Food and Agriculture Organization, USA. Retrived Decembre 23, 2003, from http://www.fao.org.

Fiskobirlik. (2002). Union of Agricultural Cooperatives for the Sale of Hazelnut and Records of Research Center, Giresun-Turkey.

Goto, T., Wicklow, D.T. & Ito, Y. (1996). Aflatoxin and cyclopiazonic acid production by a sclerotium producing *Aspergillus tamarii* strain. Appl. Environ. Micb., Vol. 62, pp. 4036- 4038.

Gursoy (Posluoglu), N. & Bicici, M. (2003). Determination of fungal infections on wheat and maize grains and some of their mycotoxins in Cukurova region. I. National Mycotoxin Symposium, Istanbul-Turkey, pp. 68-74.

Heperkan, D. (2003). Mycotoxins in fodds and their importance for Turkey. I. National Mycotoxin Symposium, Istanbul-Turkey, pp. 1-7.

IARC. (1993). Some naturally occuring substances: Food items and constituents, heterocyclic amines and mycotoxins. IARC monographs on evolution of carcinogenic risk to humans, Lyon, France, International Agency for Research on Cancer 56.

Joosten, H.M.L., Goetz, J., Pittet, A.,Schellenberg, M. & Bucheli, P. (2001). Production of ochratoxin A by *Aspergillus carbonarius* on coffee cherries. International Journal of Food Microbiology. Vol. 65, pp. 39-44.

Karabulut, O.A. & Degirmencioglu, T. (2002). Inhibition of toxin producing fungi on wheat grain used as animal feed. Review of Uludag Univ. Agricultural Faculty. Vol. 16, pp.129-138.

Kozakiewicz, Z., Paterson, R.R.M. & Bridge, P.D. (1993). Novel approaches to the identification of mycotoxin producing Penicillium species. In: Scudamore (Ed.). Occurrence and significance of mycotoxins (pp. 64-75), Central Science Laboratory, London.

Krivobok, S., Seigle-Murandi, F., Steiman, R. & Marzin, D. (1987). Screening methods to detect toxigenic fungi in liquid medium. Jof Microbiol Methods. Vol. 7, pp. 29-36.

Leistner, L. & Pitt, J.I. (1977). Miscellaneous Penicilliumtoxins. En: Rodricks JV, Hesseltine CW, Mehlmann MA (Eds.). Mycotoxins in human and animal health. Park Forest South IL, Pathotox Publishers, pp. 639-653.

Marsh, S.F. & Payne, G.A. (1984). Preharvest infection of corn silk and kernels by *Aspergillus flavus*. Phytopatholog , Vol.74, pp. 1284-1289.

Mills, J.T., Abramson, D., Frohlich, A.A. & Marquardt, R.R. (1989). Citrinin and ochratoxin A production by Penicillium spp. from stored durum wheat. Can.J. of Pl. Pat., Vol. 11, pp. 357-360.

Moss, M.O. (1991). The environmental factors controlling mycotoxin formation. In: Smith JE, Henderson RS (Eds.) Mycotoxins and animal foods (pp.37-56).Boca Raton, CRC Press Inc.

Ozay, G. & Alperden, I. (1991). Aflatoxin and Ochratoxin A contamination of dried figs (Ficus carica L.) from the 1988 crop. TUBITAK, Marmara Research Center, Dept. of Food and Refrigeration Technology, Gebze-Turkey.

Ozay, G., Aran, N. & Pala, W. (1995). Influence of harvesting and drying techniques on microflora and mycotoxin contamination of figs. Nahrung, 39(2), pp.156-165.

Ozdemir, M. (1997). Critical evaluvation of properties of Turkish hazelnuts. Food Technology, Vol. 2(10), pp. 46-52.

Ozilgen, M. & Ozdemir, M. (2001). A review on Grain and Nut Deterioration and Design of the Dryers for Safe Storage with Special Reference to Turkish Hazelnuts. Critical Reviews in Food Science and Nutrition, Vol. 41(2), pp. 95-132.

Pietri, A., Bertuzzi, T., Pallaroni, L. & Piva, G. (2001). Occurence of ochratoxin A in Italian wines. Food Additives and Contaminants, Vol. 18, pp. 647-654.

Pitt, J.I. (1987). *Penicillium viridicatum, Penicillium verrucosum* and production of ochratoxin A. Appl Environ Microbiol ,Vol. 53:, pp.266- 269.

Romani, S., Sacchetti, G., Lopez, C.C., Pinnavaia, G.G. & Rosa, M.D. (2000). Screening of occurence of ochratoxin A in green coffee beans of different origins and types. Journal of Agr. and Food Chemstry, Vol. 18, pp. 3616-3619.

Sedmikova, M., Reisnerora, H., Dufkova ,Z., Burta, I. & Jilek, F.(2001). Potential hazard of simultaneous occurence of aflatoxin B1 and ochratoxin A. Vet. Med. Vol. 46, pp. 169-174.

Shenasi, M., Candlish, A.A.G. & Aidoo, K.E. (2002) The production of aflatoxins in fresh date fruits and under simulated storage conditions. Journal of Sci. Food Agric.,Vol. 82, pp. 848-853.

Strange, R.N. (1991). Naturally occurence of mycotoxins in groundnuts, cottonseed, soya and cassava. In: Smith, J.E. and Henderson, R.S (Eds). Mycotoxins and Animal Foods (pp. 341-362) . Boca Raton; CRC press Inc.

Topal, R.S. (2003). Dominant mycoflora distrubution and the mycotoxin profiles of Turkey's agricultural regions and products. I. National Mycotoxin Symp., Istanbul-Turkey, pp.117-125.

Topal, R.S., Aran, N. & Pembeci, C. (1999). Mycotoxin profiles of Turkish agricultural mycoflora. Gıda, Vol. 24(2), pp. 129-137.

Van der Merwe, K.J., Steyn, P.S., Fourie, L., Scoot, D.B. & Thero, J.J. (1965). Ochratoxin A, a toxigenic metabolite produced by *Aspergillus ochraceus* Wilh., Nature 205, pp. 1112-1113.

Varga, J., Kevei,.E., Rinyu, E., Teren, J. & Kozakiewicz, Z. (1996). Ochratoxin production by *Aspergillus* species. Appl. Environ. Microbiol., Vol. 62, pp. 4461-4464.

K. A. SCUDAMORE

OCCURRENCE OF MYCOTOXINS IN UK FOOD COMMODITIES

1. INTRODUCTION

Mycotoxins found in the UK arise from home grown commodities and imports. Those arising from imports will be typical of mycotoxins experienced by other importing countries and will vary depending on source or because of seasonal variation. An independent assessment of UK Government funded Research for the period 1993-1996 (Scudamore, 1999) includes the results from surveys for aflatoxins, ochratoxin A, patulin, fumonisins, sterigmatocystin, mycotoxins in ethnic foods (Patel *et al.*, 1996) and multi-occurrence of mycotoxins. That report addresses both UK-grown and imported foods.

The UK Food Standards Agency (FSA) was set up in April 2000 to protect public health and to restore public confidence in the way food safety decisions are made. Its job is to protect consumers from risks, and otherwise to look after their interests in all matters related to foods. Within this role the FSA commission both surveys and research in the field of mycotoxins. The results of recent surveys are published as soon as available on their website at http://www.food.gov.uk/.

2. MYCOTOXIN OCCURRENCE

The principal mycotoxins of concern arising from food crops grown within the UK are deoxynivalenol, nivalenol, HT2-toxin, T2-toxin and zearalenone in cereals at harvest, ochratoxin A in stored commodities and patulin in apple juice. Moniliformin has also been reported in cereals. The occurrence of aflatoxins in UK cereals or other crops is extremely unusual because the climate does not encourage either the growth of *Aspergillus flavus* or the development of the toxins. However, aflatoxins are encountered in imported nuts, spices, figs, maize and other food commodities. Concentrations of aflatoxin M_1 occurring in milk and dairy products have been reduced to very low concentrations and frequency by the control of aflatoxins in imported feedstuffs (Scudamore 1999). Poorly stored animal feed components will be at risk from a much wider range of mycotoxins but there is little evidence of any significant transfer or these or of derived metabolites to human food.

The occurrence of ochratoxin A in UK stored grain occurs because it is not always possible to dry all grain at harvest rapidly enough to appropriate storage water activity. Because the climate at harvest is unpredictable, it is uneconomic to provide drying facilities that can cope with all eventualities so that sometimes intake

A. Logrieco and A. Visconti (eds.),
An Overview on Toxigenic Fungi and Mycotoxins in Europe, 247-252.

exceeds drying capacity. Another possibility is that grain moistures may rise during storage through inadequate management practice. Surveys have shown that contamination of wheat and barley by ochratoxin A in the UK is generally at a low level, although approximately 2-3% of samples exceed the EC limit for whole cereals (Scudamore *et al.*, 1999; MacDonald *et al.* 2004). Stored wheat in excess of about 18% moisture content is nearly always invaded by *Penicillium verrucosum* and other moulds although only *P. verrucosum* has been shown to produce ochratoxin A. Studies show that some reduction can occur during processing such as milling, bread baking (Osborne *et al.*, 1996, Scudamore *et al.*, 2004) and extrusion procedures (Scudamore *et al.*, in press).

10 trichothecenes mycotoxins and zearalenone were determined in 283 samples of wheat sampled at harvest in 2001 from fields of known agronomy within the UK. Results showed that the incidence of mycotoxins in 2001 were generally low with only deoxynivalenol, nivalenol, HT-2 toxin and zearalenone detected in more than 5% of the samples tested. Analysis of samples by region indicated a highly significant difference with trichothecene levels being lowest in Scotland and the North of England and highest in the South and East of England (Edwards, 2003).

There can be a significant loss of some mycotoxins during processing so that it is equally important to survey products that reach the consumer. A preliminary report of a survey for trichothecene mycotoxins and zearalenone (ZEA) in retail cereal products has been published (Patel, 2003). A total of 377 samples comprising a wide range of retail cereal products were obtained from retail outlets. Trichothecenes were quantified in 298 of these and ZEA in 53. A total of eight trichothecene mycotoxins were detected with deoxynivenol (DON) and nivalenol (NIV) being the most common. Two samples of breakfast cereals, 1 sample of flour and 10 samples of cereal-based snacks contained DON above the proposed EC action level with one sample of breakfast cereal containing 15-AcDON above the action level for DON. Zearalenone was quantified in 53 of the 377 samples at levels ranging from 8.0 to 231.8μg/kg and these results are shown in table 1. The highest levels were detected in samples of breakfast cereals and cereal-based snacks. All samples that contained ZEA also contained DON. No other trichothecenes were detected when DON was absent, except one oat-based breakfast cereal that contained HT2.

A recent survey has reported levels of trichothecenes in oat-based foods. In a preliminary study funded by the UK Food Standards Agency, an HPLC method was developed for determining the fumonisin content of human dietary samples (Chan *et al.*, 2003). The method comprised methanol/water extraction, immunoaffinity column clean-up, separation by reverse phase HPLC and detection by fluorescence after pre-column derivatisation. The limit of determination for fumonisin B_1 was 1 $\mu g\ kg^{-1}$. Composite dietary samples were prepared from the duplicate diets of 20 human subjects over the course of a month. Samples were selected from individuals with high cereal intake to increase the likelihood of detecting fumonisins in actual diet samples. Preliminary results showed levels of fumonisin B_1 ranged from <1.0 to 16.0 $\mu g\ kg^{-1}$ in the 20 diet samples. This data corresponds to an estimated daily intake of <0.03 to 0.5 $\mu g\ kg^{-1}$ bw per day. We believe this type of data to be the first of its kind to be reported in the UK, since the estimates here are calculated from actual human diet samples and not estimated statistically.

Table 1: Occurrence of zearalenone in survey of retail cereal products in the UK

Commodity/ Toxin	*Number of samples in which zearalenone was detected (<8µg/kg)*	*Range (µg/kg)*
Breakfast Cereals (n=60)	11	10.0 – 231.8
Biscuits (n=60)	2	10.7 – 11.8
Cakes (n=40)	59	15.8
Flour (n=40)	3	8.7 – 23.4
Polenta (n=8)	25	19.5 – 99.0
Savoury Snacks (n=84)	35	8.0 – 55.8
Baby Cereal (n=17)	16	11.6

Fumonisins enter the UK diet through imported maize from South America and Europe. The presence of fumonisins, zearalenone, aflatoxins and ochratoxin A in whole maize showed that levels of aflatoxins and ochratoxin A were low, but fumonisins and zearalenone, sometimes in high concentrations, occurred frequently (Scudamore and Patel, 2000). Although the number of samples examined was relatively small, concentrations of both fumonisins and aflatoxins were highest in Spain and Southern France but only occurred in small amounts in Northern France. The occurrence of zearalenone showed no regional difference. Patel *et al.*, 1997 determined the levels of fumonisins in UK maize-based foods and other cereals.

In a 1992 UK survey of patulin in apple juice, made from concentrate and fresh-pressed apple juice, over 73% of the samples of fresh-pressed apple juice tested contained detectable levels of patulin with nearly 27% containing over 50 $\mu g\ kg^{-1}$ with a maximum concentration of 434 $\mu g\ kg^{-1}$ (Ministry of Agriculture, Fisheries and Food, 1993). In response to concerns expressed by the COT the UK Food Advisory Committee recommended an 'action level' of 50 $\mu g\ kg^{-1}$ of patulin in apple juice and the industry produced a Code of Practice for the Production of Apple Juice (British Soft Drinks Association, 2002). Subsequent surveys indicated a significant decrease in the number of samples contaminated above 50 $\mu g\ kg^{-1}$ (Figure 1).

In the most recent UK survey of patulin in apple juice and apple products approximately 2% of samples of fresh-pressed apple juice contained over 50 $\mu g\ kg^{-1}$ with a maximum concentration of 78.9 $\mu g\ kg^{-1}$ (Jones and Toal, 2003).

Mycotoxins are known to concentrate in the liver and kidneys of various animals, and as these can be eaten by humans, it is important to be able to quantify the level of contamination.

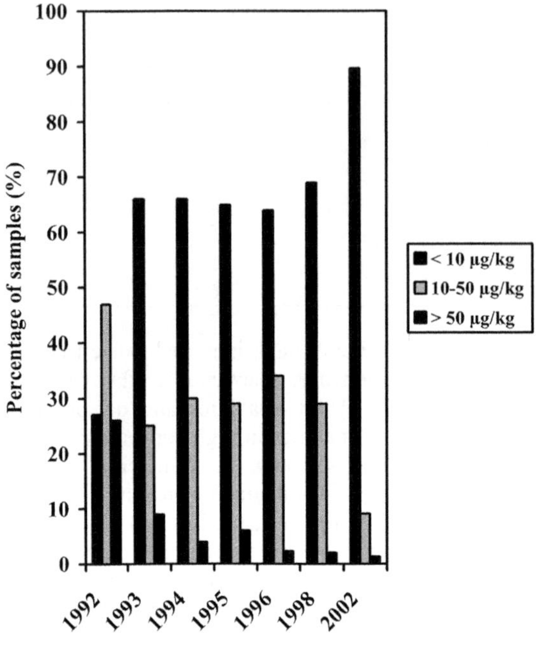

Fig. 1. Summary of UK surveillance results for patulin in apple juice.

A preliminary report of a new method has been described for the analysis of aflatoxin B_1, ochratoxin A, zearalenone and its metabolites α-zearalenol and β-zearalenol, nivalenol, deoxynivalenol and ochratoxin α in a range of offal-based food products (Driffield *et al.*, 2003). This involves a simple solvent extraction and clean-up step before analysis by HPLC-MS/MS to identify and quantify the mycotoxins present. After refining and optimising for the recovery of the specific mycotoxins the method will be used in a UK Food Standards Agency-funded survey for a range of mycotoxins in animal offal and animal offal-based foodstuffs.

During cereal milling mycotoxins are often concentrated into specific milling streams, commonly the bran or outer parts of the grains. These fractions are often diverted to animal feed for direct feeding or may be used with other imported commodities to produce compound feeds for livestock. A number of surveys during the past year have shown that many samples contain multiple mycotoxin contamination (e.g. Scudamore *et al.* 1997a {mycotoxins in raw ingredients used for animal feeding stuffs}, Nawaz *et al.* 1997,{Determination of *Alternaria* mycotoxins in oilseed rape meal and sunflower seed meal}, Scudamore *et al.*, 1998a {Determination of mycotoxins in maize and maize products}, Scudamore *et al.*

1998b {Determination of mycotoxins in rice bran) and Scudamore *et al.* 1997b {mycotoxins in pet feeds}).

3. CONCLUSIONS

Considerable effort has been put into reducing levels of mycotoxins occurring in UK produced food commodities. Collaboration between Government officials and industry has significantly reduced levels of patulin in apple juice since 1990, while enforcement of the aflatoxin regulations for animal feed has indirectly reduced levels of aflatoxin M_1 in milk. Considerable research has indicated many of the factors that lead to the formation of ochratoxin A during storage but recommendations have still to be fully taken up. The minimising of pre-harvest mycotoxin formation by *Fusarium* moulds probably represents the biggest current challenge for UK industry.

EDITORS NOTE: The Editors are indebted to the Author for providing this chapter at short notice and acknowledge consequently that it is not a complete review of recent UK studies and an assessment of the related fungi has not been possible.

5. REFERENCES

British Soft Drinks Association. 2002. The code of practice for the production of apple juice. 2nd Edition, 39 pp.

Chan, D., MacDonald, S.J. and Brereton, P., 2003, Assessment of fumonisins in human diets. Aspects of Applied Biology, 68, (Wellesbourne: The Association of Applied Biologists), pp.181-186.

Driffield, M., Hird, S.j. and MacDonald, S.J., 2003, The occurrence of a range of offal food products by HPLC-MS/MS. Aspects of Applied Biology, 68, (Wellesbourne: The Association of Applied Biologists), pp.205-210.

Edwards, S. G., 2003, *Fusarium* mycotoxins in UK wheat. Aspects of Applied Biology, 68, (Wellesbourne: The Association of Applied Biologists), pp.35-42.

Jones, B. D. and Toal, M. E., 2003, Mycotoxins in food: a UK reulatory perspective. Aspects of Applied Biology, 68, (Wellesbourne: The Association of Applied Biologists), pp.1-10.

MacDonald, S., Prickett, T. J., Wildey, K.B. and Chan, D., 2004, Survey of ochratoxin A and deoxynivalenol in stored grains from the 1999 harvest in the UK, Food Additives and Contaminants, 21, 172-181

Ministry of Agriculture, Fisheries and Food. 1993. Mycotoxins: Third Report. Food Surveillance Paper No. 18. London: HMSO. 46-50.

Nawaz, S., Scudamore, K. A., and Rainbird, S. C., 1997, Mycotoxins in Ingredients of Animal Feeding Stuffs: I. Determination of *Alternaria* mycotoxins in oilseed rape meal and sunflower seed meal. Food Additives and Contaminants, 14 249-262.

Osborne, B. G., Ibe, F. I., Brown, G. L., Patagine, F., Scudamore, K. A., Banks, J. N., and Hetmanski, M. T., 1996, The Effects of Milling and Processing on Wheat Contaminated with Ochratoxin A. Food Additives and Contaminants, 13, 141-153.

Patel, S., Hazel, C. M., Winterton, A. G. M., and Gleadle, A. E., 1997, Surveillance of fumonisins in UK maize-based foods and other cereals. Food Additives and Contaminants, 14, 187-191.

Patel, S., Hazel, C. M., Winterton, A. G. M., and Mortby, E., 1996, Survey of ethnic foods for mycotoxins. Food Additives and Contaminants, 13, 833-841.

Patel, S., 2003, Mycotoxins in UK food products. Aspects of Applied Biology, 68, (Wellesbourne: The Association of Applied Biologists), pp.133-138.

Scudamore K A, Banks J and Guy R (2004 in press), The fate of ochratoxin A in the processing of whole wheat grain during extrusion. Food Additives and Contaminants, 00, 000-000.

Scudamore K A, Banks J and MacDonald (2003), The fate of ochratoxin A in the processing of whole wheat grains during milling and bread production. Food Additives and Contaminants, 20, 1153-1163.

Scudamore, K. A., 1999, 'Mycotoxins' A Report of MAFF Funded Applied Research and Surveillance, 1993-1996 (London: MAFF), 167 pages

Scudamore, K. A., and Patel, S., 2000, Survey for aflatoxins, ochratoxin A, zearalenone and fumonisins in maize. Food Additives and Contaminants, 17, 407-416.

Scudamore, K. A., Hetmanski, M. T., Chan, H. K., and Collins, S., 1997a, Occurrence of mycotoxins in raw ingredients used for animal feeding stuffs in the United Kingdom in 1992. Food Additives and Contaminants, 14, 157-173.

Scudamore, K. A., Hetmanski, M. T., Nawaz, S., Naylor, J., and Rainbird, S., 1997b, Determination of mycotoxins in pet foods sold for domestic pets and wild birds using linked-column immunoassay clean-up and HPLC. Food Additives and Contaminants, 14, 175-186.

Scudamore, K. A., Nawaz, S., and Hetmanski, M. T., 1998a, Mycotoxins in Ingredients of Animal Feeding Stuffs: II. Determination of mycotoxins in maize and maize products. Food Additives and Contaminants, 15 30-55.

Scudamore, K. A., Nawaz, S., Hetmanski, M. T., and Rainbird, S.C., 1998b, Mycotoxins in Ingredients of Animal Feeding Stuffs: III. Determination of mycotoxins in rice bran. Food Additives and Contaminants, 15, 185-194.

Scudamore, K. A., Patel, S., and Breeze, V., 1999, Surveillance of stored grain from the 1997 United Kingdom harvest for ochratoxin A. Food Additives and Contaminants, 16, 281-290.

GPSR Compliance

The European Union's (EU) General Product Safety Regulation (GPSR) is a set of rules that requires consumer products to be safe and our obligations to ensure this.

If you have any concerns about our products, you can contact us on ProductSafety@springernature.com

In case Publisher is established outside the EU, the EU authorized representative is:

Springer Nature Customer Service Center GmbH
Europaplatz 3
69115 Heidelberg, Germany

Batch number: 10755292

Printed by Printforce, the Netherlands